THE EVOLUTION
OF THE IGNEOUS ROCKS

By N. L. BOWEN

WITH A NEW INTRODUCTION BY J. F. SCHAIRER
CARNEGIE INSTITUTION OF WASHINGTON

DOVER PUBLICATIONS, INC.
NEW YORK

This Dover edition, first published in 1956, is an
unabridged and unaltered republication of the first
edition with a new introduction by J. F. Schairer
and a complete bibliography of the writings of
N. L. Bowen.

It is published through special arrangement with
Princeton University Press.

International Standard Book Number: 0-486-60311-3
Library of Congress Catalog Card Number: 56-14011

Manufactured in the United States of America
Dover Publications, Inc.
180 Varick Street
New York, N.Y. 10014

INTRODUCTION

This volume has had a profound influence on the younger generation of geologists because it has emphasized the importance of the point that sound principles of physical chemistry underlie geological processes. It showed how a knowledge of the equilibrium relations in silicate systems, in conjunction with field observations on the rocks and studies of the rock-forming minerals, helped to elucidate the nature and mechanism of the processes involved in rock origins. This volume first appeared in 1928 and was based on a course of lectures given to advanced students in the Department of Geology at Princeton University in the spring of 1927. Although much additional information has been acquired since that time, both in the laboratory and in the field, Bowen's sound application of the principles of physics and chemistry is as cogent today as it was then.

The student of the earth wants to know not only the nature of the rock-forming materials but is even more concerned with the processes by which minerals and rocks form and are modified by subsequent changes. Nearly all rock-forming minerals except quartz are not of simple fixed composition, but are complex solid solutions. This makes studies of their composition and stability relations more difficult, but this very complexity and the resulting abrupt or progressive changes in mineral composition in response to a changing environment may provide many clues to the nature and mechanism of the processes involved. The answers are in the rocks themselves, but a knowledge of the processes may provide the key that unlocks the secrets.

The desirability of experimental studies in the laboratory as an important adjunct to geological field observations has long been urged by Bowen. Experiment is a necessary check on inference from observations on the natural materials and in turn provides a chemical basis for hypotheses on origins, which may be tested in the field and modified to give a nearer approach to the mechanism of rock genesis. Some have doubted the value of laboratory phase-equilibrium studies in geology because under natural conditions equilibrium is not always attained. The only practical method of studying the physical chemistry of geological processes is to determine equilibrium relations first and then to evaluate the factors that lead to failure of equilibrium under natural conditions, together with the magnitude and direction of their effects.

In the year 1910, Norman L. Bowen came to the Geophysical Laboratory of the Carnegie Institution of Washington as a young student to use the facilities of the laboratory in making a phase-equilibrium study of a silicate system. He was permitted to use these results for a thesis for the degree of Doctor of Philosophy at the Massachusetts Institute of Technology. For forty years he has pursued his laboratory studies and has carried his results to the field to check them with the rocks themselves. By 1915, when he published a paper on *The Later Stages of the Evolution of Igneous Rocks*, his reputation among petrologists was established. Besides his many papers reporting the results of phase-equilibrium studies on specific systems of rock-forming minerals, his papers *The Problem of the Anorthosites* (1917), *Crystallization Differentiation in Igneous Magmas* (1919), *Diffusion in Silicate Melts* (1921), *Genetic Features of Alnoitic Rocks* (1922), *The Behavior of Inclusions in Igneous Magmas* (1922), and *The Origin of Ultrabasic and Related Rocks* (1927) were outstanding contributions to the literature of petrology, which preceded the publication of his monumental book *The Evolution of the Igneous Rocks* in 1928.

The clarity of his presentation of the problems and the bearing of the phase-equilibrium data on their solution have done much to emphasize the importance of close cooperation between the physical chemist in the laboratory and the geologist in the field. Since 1928, Bowen has continued his studies in the laboratory and in the field. This skilled and resourceful experimenter, in collaboration with his colleagues, has added much to our knowledge of the silicates of ferrous iron and the relations between the rock-forming olivines and pyroxenes, the relations between early-crystallizing minerals and the late-crystallizing minerals of rocks and the system $NaAlSiO_4$—$KAlSiO_4$—SiO_2, which he has called petrogeny's "residua" system, the phase relations in systems of alkali aluminosilicates, the relations in the quaternary system Na_2O—CaO—Al_2O_3—SiO_2, the alkali feldspar system and this system with silica in the presence of water, and the bearing of these studies on the origin of granites. He has also extended the application of laboratory data to the chemistry and mineralogy of progressive metamorphism. From the complete bibliography of N. L. Bowen given at the end of this Introduction to this reprint edition of his book, the reader may observe in detail his many contributions to geological science.

Even with these many advances since 1928, *The Evolution of the*

Igneous Rocks is still an invaluable starting point for serious students of rock origins. It is a MUST reference work for all geologists and a serious but delightful introduction to chemical geology for all advanced students in geology at our universities. For many years it has been out of print. I am delighted that this classic is now available again and consider it a privilege to write this introduction to the reprint edition.

J. F. Schairer

Geophysical Laboratory
Carnegie Institution of Washington

BIBLIOGRAPHY OF NORMAN L. BOWEN

Diabase and aplite of the cobalt-silver area. *Jour. Canadian Mining Inst.*, 95–106. 1909.

Diabase and granophyre of the Gowganda Lake District, Ontario. *Jour. Geol.*, **18**, 658–674. 1910.

Silver in Thunder Bay District. *20th Rept. Bureau of Mines, Ontario*, 119–132. 1911.

Notes on the salt industry of Ontario. *20th Rept. Bureau of Mines, Ontario*, 247–258. 1911.

The binary system $Na_2Al_2Si_2O_8$ (nephelite, carnegieite)—$CaAl_2Si_2O_8$ (anorthite). Abstract of thesis submitted to the faculty of Mass. Inst. of Technol. in partial fulfillment of the requirements for the degree of Doctor of Philosophy. 14 pp. 1912.

The composition of nephelite. *Amer. Jour. Sci.*, **33**, 49–54. 1912.

The order of crystallization in igneous rocks. *Jour. Geol.*, **20**, 457–468. 1912.

The binary system: $Na_2Al_2Si_2O_8$ (nephelite, carnegieite)—$CaAl_2Si_2O_8$ (anorthite). *Amer. Jour. Sci.*, **33**, 551–573. 1912.

The melting phenomena of the plagioclase feldspars. *Amer. Jour. Sci.*, **35**, 577–599. 1913.

Die Schmelzerscheinungen bei den Plagioklas-Feldspaten. *Z. anorg. Chem.*, **82**, 283–307. 1913.

A geological reconnaissance of the Fraser River valley from Lytton to Vancouver, British Columbia. *Geol. Surv., Canada, Summary Rept.*, 108–114. 1913.

The order of crystallization in igneous rocks. *Jour. Geol.*, **21**, 399–401. 1913.

(with Olaf Andersen) The binary system MgO—SiO_2. *Amer. Jour. Sci.*, **37**, 487–500. 1914.

(with Olaf Andersen) Das binäre System Magnesiumoxyd—Silicium—2-oxyd. *Z. anorg. Chem.*, **87**, 283–299. 1914.

The ternary system: diopside—forsterite—silica. *Amer. Jour. Sci.*, **38**, 207–264. 1914.

Das ternäre System: Diopsid—Forsterit—Silicium—2-oxyd. *Z. anorg. Chem.*, **90**, 1–66. 1914.

Crystallization-differentiation in silicate liquids. *Amer. Jour. Sci.*, **39**, 175–191. 1915.

The crystallization of haplobasaltic, haplodioritic, and related magmas. *Amer. Jour. Sci.*, **40**, 161–185. 1915.

Das ternäre System: Diopsid—Anorthit—Albit. *Z. anorg. Chem.*, **94**, 23–50. 1916.

The later stages of the evolution of the igneous rocks. *Jour. Geol.*, **23**, Suppl., 1–89. 1915.

The sodium-potassium nephelites. *Amer. Jour. Sci.*, **43**, 115–132. 1917.

The problem of the anorthosites. *Jour. Geol.*, **25**, 209–243. 1917.

Adirondack intrusives. *Jour. Geol.*, **25**, 509–512. 1917.

The significance of glass-making processes to the petrologist. *Jour. Wash. Acad. Sci.*, **8**, 88–93. 1918.

Crystals of barium disilicate in optical glass. *Jour. Wash. Acad. Sci.*, **8**, 265–268. 1918.

The identification of "stones" in glass. *Jour. Amer. Ceram. Soc.*, **1**, 594–605. 1918.

Devitrification of glass. *Jour. Amer. Ceram. Soc.*, **2**, 261–278. 1919.

Optical properties of anthophyllite. *Jour. Wash. Acad. Sci.*, **10**, 411–414. 1920.

Tridymite crystals in glass. *Amer. Mineral.*, **4**, 65–66. 1919.

Abnormal birefringence of torbernite. *Amer. Jour. Sci.*, **48**, 195–198. 1919.

Cacoclasite from Wakefield, Quebec. *Amer. Jour. Sci.*, **48**, 440–442. 1919.

Crystallization-differentiation in igneous magmas. *Jour. Geol.*, **27**, 393–430. 1919.

Echellite, a new mineral. *Amer. Mineral.*, **5**, 1–2. 1920.

Differentiation by deformation. *Proc. Nat. Acad. Sci.*, **6**, 159–162. 1920.

Diffusion in silicate melts. *Jour. Geol.*, **29**, 295–317. 1921.

Preliminary note on monticellite alnoite from Isle Cadieux, Quebec. *Jour. Wash. Acad. Sci.*, **11**, 278–281. 1921.

Genetic features of alnoitic rocks from Isle Cadieux, Quebec. *Amer. Jour. Sci.*, **3**, 1–34. 1922.

Two corrections to mineral data. *Amer. Mineral.*, **7**, 64–66. 1922.

The reaction principle in petrogenesis. *Jour. Geol.*, **30**, 177–198. 1922.

(with G. W. Morey) The melting of potash feldspar. *Amer. Jour. Sci.*, **4**, 1–21. 1922.

The behavior of inclusions in igneous magmas. *Jour. Geol.*, **30**, 513–570. 1922.

The genesis of melilite. *Jour. Wash. Acad. Sci.*, **13**, 1–4. 1923.

(with M. Aurousseau) Fusion of sedimentary rocks in drill-holes. *Bull. Geol. Soc. Amer.*, **34**, 431–448. 1923.

(with J. W. Greig) The system: Al_2O_3—SiO_2. *Jour. Amer. Ceram. Soc.*, **7**, 238–254. 1924.

(with J. W. Greig and E. G. Zies) Mullite, a silicate of alumina. *Jour. Wash. Acad. Sci.*, **14**, 183–191. 1924.

The Fen area in Telemark, Norway. *Amer. Jour. Sci.*, **8**, 1–11, pls. I–III. 1924.

(with G. W. Morey) The binary system sodium metasilicate—silica. *Jour. Phys. Chem.*, **28**, 1167–1179. 1924.

The mineralogical phase rule. *Jour. Wash. Acad. Sci.*, **15**, 280–284. 1925.

(with J. W. Greig) The crystalline modifications of NaAlSiO₄. *Amer. Jour. Sci.*, **10**, 204–212. 1925.

The amount of assimilation by the Sudbury norite sheet. *Jour. Geol.*, **33**, 825–829. 1925.

(with G. W. Morey) The ternary system sodium metasilicate—calcium metasilicate—silica. *Jour. Soc. Glass Tech.*, **9**, 226–264. 1925.

(with J. W. Greig) Discussion on "An X-ray study of natural and artificial sillimanite." *Bull. Amer. Ceram. Soc.*, **4**, 374–376. 1925.

Concerning "Evidence of liquid immiscibility in a silicate magma, Agate Point, Ontario." *Jour. Geol.*, **34**, 71–73. 1926.

Properties of ammonium nitrate: I. A metastable inversion in ammonium nitrate; II. The system: ammonium nitrate—ammonium chloride; III. A note on the system: ammonium nitrate—ammonium sulphate. *Jour. Phys. Chem.*, **30**, 721–737. 1926.

(with R. W. G. Wyckoff) A petrographic and X-ray study of the thermal dissociation of dumortierite. *Jour. Wash. Acad. Sci.*, **16**, 178–189. 1926.

(with R. W. G. Wyckoff and J. W. Greig) The X-ray diffraction patterns of millite and of sillimanite. *Amer. Jour. Sci.*, **11**, 459–472. 1926.

The carbonate rocks of the Fen area in Norway. *Amer. Jour. Sci.*, **12**, 499–502. 1926.

Die Carbonatgesteine des Fengebietes in Norwegen. *Centrbl. Min. Geol.*, Abt. A, 241–245. 1926.

Review of "Über die Synthese der Feldspatvertreter," by W. Eitel. Leipzig, 258 pp., IV pls., 1925. *Amer. Jour. Sci.*, **11**, 280. 1926.

An analcite-rich rock from the Deccan traps of India. *Jour. Wash. Acad. Sci.*, **17**, 57–59. 1927.

(with G. W. Morey) The decomposition of glass by water at high temperatures and pressures. *Jour. Soc. Glass Tech.*, **11** (Trans.), 97–106. 1927.

The origin of ultrabasic and related rocks. *Amer. Jour. Sci.*, **14**, 89–108. 1927.

The evolution of the igneous rocks. Princeton University Press, Princeton, New Jersey, x+334 pp. 1928.

Geologic thermometry. In "The laboratory investigation of ores," edited by E. E. Fairbanks. McGraw-Hill, New York, Chapter 10, pp. 172–199. 1928.

(with J. F. Schairer) The system: leucite—diopside. *Amer. Jour. Sci.*, **18**, 301–312. 1929.

(with J. F. Schairer) The fusion relations of acmite. *Amer. Jour. Sci.*, **18**, 365–374. 1929.

(with F. C. Kracek and G. W. Morey) The system potassium metasilicate—silica. *Jour. Phys. Chem.*, **33**, 1857–1879. 1929.

Central African volcanoes in 1929. *Trans. Amer. Geophys. Union*, 10th and 11th Annual Meetings, pp. 301–307. Nat. Res. Council, Washington, D.C. 1930.

(with G. W. Morey and F. C. Kracek) The ternary system K₂O—CaO—SiO₂. (With correction.) *Jour. Soc. Glass Tech.*, **14**, 149–187. 1930.

(with J. F. Schairer and H. W. V. Willems) The ternary system: Na_2SiO_3 Fe_2O_3—SiO_2. *Amer. Jour. Sci.*, **20**, 405–455. 1930.

(with E. Posnjak) Magnesian amphibole from the dry melt: A correction. *Amer. Jour. Sci.*, **22**, 193–202. 1931.

(with E. Posnjak) The role of water in tremolite. *Amer. Jour. Sci.*, **22**, 203–214. 1931.

(with G. W. Morey) "Devitrite." Letter to Editor, *Glass Industry*, June, 1931.

(with J. F. Schairer) The system FeO—SiO_2. *Amer. Jour. Sci.*, **24**, 177–213. 1932.

Crystals of iron-rich pyroxene from a slag. *Jour. Wash. Acad. Sci.*, **23**, 83–87. 1933.

Vogtite, isomorphous with wollastonite. *Jour. Wash. Acad. Sci.*, **23**, 87–94. 1933.

(with J. F. Schairer and E. Posnjak) The system, Ca_2SiO_4—Fe_2SiO_4. *Amer. Jour. Sci.*, **25**, 273–297. 1933.

The broader story of magmatic differentiation, briefly told. In "Ore deposits of the Western States," Amer. Inst. Min. Met. Eng., New York, Chapter III, Pt. II, pp. 106–128. 1933.

(with J. F. Schairer and E. Posnjak) The system, CaO—FeO—SiO_2. *Amer. Jour. Sci.*, **26**, 193–284. 1933.

Note: Non-existence of echellite. *Amer. Mineral.*, **18**, 31. 1933.

Viscosity data for silicate melts. *Trans. Amer. Geophys. Union*, 15th Annual Meeting, pp. 249–255. Nat. Res. Council, Washington, D.C. 1934.

(with J. F. Schairer) The system, MgO—FeO—SiO_2. *Amer. Jour. Sci.*, **29**, 151–217. 1935.

The igneous rocks in the light of high-temperature research. *Sci. Monthly*, **40**, 487–503. 1935.

(with J. F. Schairer) Preliminary report on equilibrium-relations between feldspathoids, alkali-feldspars, and silica. *Trans. Amer. Geophys. Union*, 16th Annual Meeting, pp. 325–328. Nat. Res. Council, Washington, D.C. 1935.

(with J. F. Schairer) Grünerite from Rockport, Massachusetts, and a series of synthetic fluor-amphiboles. *Amer. Mineral.*, **20**, 543–551. 1935.

(with J. F. Schairer) The problem of the intrusion of dunite in the light of the olivine diagram. *Rept. XVI International Geol. Congress*, 1933, pp. 391–396. Washington, D.C. 1936.

"Ferrosilite" as a natural mineral. *Amer. Jour. Sci.*, **30**, 481–494. 1935.

(with R. B. Ellestad) Nepheline contrasts. *Amer. Mineral.*, **21**, 363–368. 1936.

(with J. F. Schairer) The system, albite—fayalite. *Proc. Nat. Acad. Sci.*, **22**, 345–350. 1936.

Review of "Interpretative petrology of the igneous rocks," by Harold Lattimore Alling. McGraw-Hill, New York, xv+353 pp., 48 figs., 11 pls., 1936. *Amer. Mineral.*, **21**, 813–814. 1936.

Recent high-temperature research on silicates and its significance in igneous geology. *Amer. Jour. Sci.*, **33**, 1–21. 1937.

A note on aenigmatite. *Amer. Mineral.*, **22**, 139–140. 1937.

(with R. B. Ellestad) Leucite and pseudoleucite. *Amer. Mineral.*, **22**, 409–415. 1937.

(with F. C. Kracek and G. W. Morey) Equilibrium relations and factors influencing their determination in the system K_2SiO_3—SiO_2. *Jour. Phys. Chem.*, **41**, 1183–1193. 1937.

(with J. F. Schairer) Crystallization equilibrium in nepheline—albite—silica mixtures with fayalite. *Jour. Geol.*, **46**, 397–411. 1938.

Lavas of the African Rift Valleys and their tectonic setting. *Amer. Jour. Sci.*, **35–A**, 19–33. 1938.

(with J. F. Schairer) The system, leucite—diopside—silica. *Amer. Jour. Sci.*, **35–A**, 289–309. 1938.

Appendix V. Rept. of the Committee on Research in the Earth Sciences, Div. of Geol. and Geogr., Nat. Res. Council. 3 pp. 1938.

Mente et malleo atque catino. Presidential address, 18th Annual Meeting, Mineralogical Society of America. *Amer. Mineral.*, **23**, 123–130. 1938.

(with N. M. Fenneman, T. W. Vaughan, and A. L. Day) A possible program of research in geology. *Proc. Geol. Soc. Amer.*, 143–155. 1938.

Geology and chemistry. *Science*, **89**, 135–139. 1939.

Progressive metamorphism of siliceous limestone and dolomite. *Jour. Geol.*, **48**, 225–274. 1940.

Geologic temperature recorders. *Sci. Monthly*, **51**, 5–14. 1940.

Certain singular points on crystallization curves of solid solutions. *Proc. Nat. Acad. Sci.*, **27**, 301–309. 1941.

Physical controls in adjustments of the earth's crust. In "Shiftings of the sea floors and coast lines," Univ. Penn. Bicentennial Conference. Univ. Penn. Press, Phila., pp. 1–6. 1941.

Presentation of the Penrose Medal to Norman Levi Bowen. *Proc. Geol. Soc. Amer.*, 79–87. 1942.

(with J. F. Schairer) The binary system $CaSiO_3$—diopside and the relations between $CaSiO_3$ and akermanite. *Amer. Jour. Sci.*, **240**, 725–742. 1942.

Petrology and silicate technology. *Jour. Amer. Ceram. Soc.*, **26**, 285–301. 1943.

Phase equilibria bearing on the origin and differentiation of alkaline rocks. *Amer. Jour. Sci.*, **243–A**, 75–89. 1945.

Magmas. *Bull. Geol. Soc. Amer.*, **58**, 263–280. 1947.

(with J. F. Schairer) Melting relations in the systems Na_2O—Al_2O_3—SiO_2 and K_2O—Al_2O_3—SiO_2. *Amer. Jour. Sci.*, **245**, 193–204. 1947.

(with J. F. Schairer) The system anorthite—leucite—silica. *Bull. Soc. Geol. Finlande*, **20**, 67–87. 1947.

The granite problem and the method of multiple prejudices. *Geol. Soc. Amer. Mem.* **28**, 79–90. 1948.

Phase equilibria in silicate melts including those containing volatile constituents. Committee on Geophys. Sci., Res. and Devel. Board, Panel on Geology. 6 pp. 1948.

(with O. F. Tuttle) The system MgO—SiO_2—H_2O. *Bull. Geol. Soc. Amer.*, **60**, 439–460. 1949.

Memorial to Rollin Thomas Chamberlin. *Proc. Geol. Soc. Amer.*, 135–144. 1949.

(with O. F. Tuttle) The system $NaAlSi_3O_8$—$KAlSi_3O_8$—H_2O. *Jour. Geol.*, **58**, 489–511. 1950.

(with O. F. Tuttle) High-temperature albite and contiguous feldspars. *Jour. Geol.*, **58**, 572–583. 1950.

The making of a magmatist. *Amer. Mineral.*, **35,** 651–658. 1950.

Presentation of the Roebling Medal of the Mineralogical Society of America to Herbert E. Merwin. *Amer. Mineral.*, **35,** 255–257. 1950.

Presentation of the Wollaston Medal to N. L. Bowen by C. E. Tilley. Acceptance by N. L. Bowen. *Abstr. Proc. Geol. Soc., London*, No. 1463, 103–105. 1950.

Obituary notice. Charles Whitman Cross. *Quart. Jour. Geol. Soc., London*, **105,** lv–lvi. 1950.

Presentation of the Roebling Medal of the Mineralogical Society of America to Norman L. Bowen by A. F. Buddington. Acceptance by Norman L. Bowen. *Amer. Mineral.*, **36,** 291–296. 1951.

Review of "Silicate melt equilibria," by W. Eitel. Rutgers Univ. Press, New Brunswick, New Jersey, x + 159 pp., 200 figs., 1951. *Amer. Mineral.*, **36,** 785–787. 1951.

Presentation of the Mineralogical Society of America Award to Orville Frank Tuttle. *Amer. Mineral.*, **37,** 250–253. 1952.

Review of "Theoretical petrology: A textbook on the origin and evolution of rocks," by Tom F. W. Barth. Wiley, New York; Chapman & Hall, London; 387 pp., 1952. *Science*, **115,** 443. 1952.

Review of "Principles of geochemistry," by Brian Mason. Wiley, New York; Chapman & Hall, London; 276 pp., 1952. *Science*, **116,** 209. 1952.

Review of "Igneous and metamorphic petrology," by F. J. Turner and J. Verhoogen. McGraw-Hill, New York, 1st ed., ix + 602 pp., 92 figs., 1951. *Jour. Geol.*, **60,** 1952.

Review of "The origin of metamorphic and metasomatic rocks," by Hans Ramberg. Univ. Chicago Press, Chicago, Ill., xvii + 317 pp., 1952. *Chem. and Eng. News*, **31,** 3679. 1953.

Experiment as an aid to the understanding of the natural world. *Proc. Acad. Nat. Sci., Phila.*, **106,** 1–12. 1954.

PREFACE

THIS *volume is based on a brief course of lectures given to advanced students in the Department of Geology at Princeton in the spring of* 1927. *Although considerably expanded over the substance of the lectures themselves the same general restriction of subject matter is observed as was observed in this special course. A knowledge of the accumulated facts of petrology, such as would be obtained in a general course, is assumed. There is nothing of the description and classification of rocks, nothing of the subdivision of intrusive bodies according to their outward form, nothing of many subjects that occupy much space in standard texts. The reason for this lies largely in the special purpose for which most of the material here presented was first brought together but partly also in my conviction that those sections of any new text which deal with the subjects mentioned are for the most part a profitless repetition of the similar sections of older texts.*

Through avoidance of the descriptive and classificatory side of the science the subject matter has become largely interpretative. It is an attempt to interpret the outstanding facts of igneous-rock series as the result of fractional crystallization. The use of the term "evolution" in the title is intended to designate only a process of derivation of rocks from a common source and not to imply that detailed knowledge of the process which the term connotes when applied to organic development. While rocks themselves remain the best aid to the discussion of their origin by fractional crystallization, much light is thrown upon the problem by laboratory investigations of silicate melts. In this study I have tried to give the bearing of the pertinent facts from both sources. It was my hope that, before anything of the kind here offered was written, all of the diagrams it would be necessary to use would be determined diagrams. Yet I offer no apology for the use of deduced diagrams where this is still necessary. Vogt's pioneer work with such diagrams has more than justified their use. Attack with their aid may be regarded as a skirmishing which feels out the strength and the weakness of our adversaries, the rocks, and thus lays a necessary foundation for a more serious campaign of experimental attack, concentrated upon those points where progress is most likely to be made.

The book is divided into two parts. In Part I are given those aspects of fractional crystallization of magmas where facts determined in the laboratory are susceptible of fairly direct application to the natural problems. In Part II various problems are discussed in which the amount of extrapolation from ascertained fact is relatively great or where the diagrams used are mainly deduced. The conclusions reached in such matters are thus to be regarded as resting on a less certain foundation. Again, some subjects have been relegated to Part II because they are considered to be of relatively minor importance in the problem as a whole. Discussion of the effects of volatile components will be found there for that reason.

Upon the question of the relative importance of fractional crystallization, as compared with other processes, in the derivation of igneous rocks I can lay no claim to an open mind. Anatole France has said that there may be times when an open mind is itself a prejudice. I believe that that time has come in petrology as far as the question of fractional crystallization is concerned. But upon the relative importance of the various factors that may induce crystal fractionation there is much room for an open mind. There is a common impression that I am a proponent of crystal settling as opposed to other methods of crystal fractionation but the impression has never had any justification. In my earliest writings on the subject I set down side by side the various methods of crystal fractionation that had been proposed and reached no decision as to their relative importance in the general problem though their relative importance in a few specific occurrences may be plain enough. I still set them down side by side in discussing the general problem.

In treating some of the relations involved in fractional crystallization I have adopted the method of taking the statements of various objectors and discussing in considerable detail the questions which they raise. This lends to some of the subject matter an air of controversy that may, in some respects, seem undesirable. Yet in other respects it may be a matter of satisfaction that the hypothesis of fractional crystallization is susceptible of such detailed discussion. To attempt a discussion of some hypotheses of igneous-rock derivation is to tilt with windmills.

The extent to which I am indebted to the writings of Harker, Lacroix, Niggli, Goldschmidt, Daly and many others will be plain to any reader. My thanks are due to some of my colleagues and especially to Washington, Morey and Greig for helpful discussion of many problems. To the members of the staff and the students in Geology at Princeton, to whom the lectures were given, I am indebted for many suggestions.

CONTENTS

PART ONE

CONTENTS

CONTENTS

CONTENTS

PART ONE

CHAPTER I

THE PROBLEM OF THE DIVERSITY
OF IGNEOUS ROCKS

THE accumulation of detailed knowledge of the mineralogical and chemical characters of igneous rocks has led to a generalization which has now been accepted by petrologists for some four decades. It is that the rocks of a given region, that have been intruded at a definite period, tend to exhibit certain similarities of mineral or chemical composition which persist even in the presence of diversity and which mark them off more or less distinctly from the rocks of another region or from rocks of the same region intruded at another period. Thus in New England and in adjacent portions of Canada there occur isolated stocks and plugs of Palaeozoic igneous rocks showing a wide range of composition but with a distinct general tendency to be rich in Na_2O. Petrologists have conveniently designated such a regional grouping of related igneous rocks as a "petrographic province."[1] The rocks of the region just mentioned are strongly contrasted with, say, the Coast Range intrusives of Western Canada and Alaska which show general tendencies of a different character and thus constitute a distinct petrographic province. As more and more examples have accumulated of rock associations of the kind which led to the concept of petrographic provinces petrologists have come to realize that the similarity of characters exhibited in any given association must be connected with a community of origin. That the rocks have been derived from a single original magma, responding to the influence of external conditions, is the assumption commonly made as to the nature of this community of origin. The supposed derivation of different rocks from a single magma has been called differentiation, and the processes whereby the different rocks have arisen have been called the processes of differentiation.

The concept of differentiation is thus an hypothesis proposed to explain various rock associations. The only rival hypothesis ever proposed was the doctrine of the mixing of two fundamental magmas

1 Judd, *Quart. Jour. Geol. Soc.*, 42, 1886, p. 54. The term "province" is not without objectionable features because it emphasizes place too much whereas time is of equal importance. There may be in a single geographic area several petrographic provinces of different ages. We shall use the term rock association to designate a group of rocks associated in the field and of the same age.

(basaltic and rhyolitic) but this has been found to fail so completely that the concept of differentiation has come to be regarded as a fact as well established as the observed rock associations themselves. Only the processes which bring about differentiation are ordinarily regarded as of hypothetical character.

In the earlier days of speculation as to the factors which brought about a diversity of associated rocks, a splitting of the magma into complementary fractions with possible further splitting of the fractions was the explanation appealed to. At first this was apparently not correlated with the definite physical process of liquid immiscibility but was rather a vague notion based on a dualistic philosophy. The actual nature of the variation in any rock association is not the sharp partitioning that such a "splitting" would lead to. It is rather a continuous variation. To be sure, in any given association perfectly continuous variation ordinarily fails, but by piecing together the facts of related associations one finds convincing evidence for continuous variation. The members of rock associations are thus related to each other as members of a series[1] and the division of the series into members is purely arbitrary. Igneous rocks are not, however, to be referred to a single series; indeed, it is the existence of different series that marks off petrographic provinces and emphasizes the fact of differentiation. The real problem of differentiation is thus the explanation of these natural series, which represent continuous variation. Not only is there a continuous variation within a series, but viewed as a whole, rocks appear to show no sharp demarcation of one series from another, yet the concept of a series is none the less useful, just as the concept of a rock type is no less useful because it is an arbitrary subdivision of a continuous series. In a general way it appears that in any given province some determining factor has brought it about that a simple serial relation is comparatively evident and it is only when a number of provinces are compared that the transition from series to series becomes evident. The problem is thus the explanation of this polyvariant condition with a local tendency towards relatively simple series.

Among the series into which rocks may be divided we may mention

> gabbro, diorite, quartz diorite, granodiorite, granite
> gabbro, diorite, monzonite, syenite
> basalt, nephelite-basalt, melilite basalt, phonolite

There are many others but these three may give a concrete idea of the kind of natural grouping that constitutes a rock series.

There was formerly a tendency to believe that each of these series had its own distinctive parental magma which was usually assumed to be of approximately the average composition of the assemblage. Many who

1 Brögger, *Die Eruptivgesteine des Kristianiagebietes*, 1, 1894, pp. 169 ff.

were unwilling to carry the subdivision to such extreme lengths still adhered to the view that there were two great branches, the alkaline and the subalkaline, with distinct parental magmas and with no association of types of such a nature as to indicate any genetic connection between the branches. The adherents of such views are no longer so numerous, because detailed studies have brought out the intimate association of types belonging to the supposedly antagonistic branches. For purposes of discussion division into these two great branches is often useful.

To Daly, in particular, we owe the demonstration, apparently satisfactory, that basaltic magma is a constant member of all these associations and that there is no essential difference in the basaltic magma of the various associations. Partly for this reason and partly on geologic grounds he considers that basaltic magma is the parental magma of all igneous-rock series, except certain pre-Cambrian rocks. The facts are not such as to enforce belief in the parental nature of basaltic magma but they are sufficiently definite that many petrologists now entertain the belief favorably and include it in their general scheme of rock derivation. In the present discussion the parental nature of basaltic magma is taken as a fundamental thesis and other rock-types are developed principally by fractional crystallization. Nevertheless this assumption is not fundamental in the sense that the whole system of the derivation of rock types by fractional crystallization would fall to the ground were the parental nature of basaltic magma disproved. Fractional crystallization would still remain the best explanation of the kind of relation shown between the various members of rock series. The reasons for preferring a thoroughly basic, presumably basaltic, parental magma are, however, strong and will become apparent as the discussion proceeds.

The possible factors that may have led to the formation of different rocks from a single magma have been listed by many petrologists. Apart from vague suggestions of a "splitting" which is not referred to any known process, the factors appealed to are definite physico-chemical processes that are known to occur in various complex mixtures under appropriate conditions. A gradation of composition, in a completely liquid mass, resulting from a gradation of temperature in the mass (Soret effect) is among the possibilities considered, but there is every reason to believe that the greatest theoretical magnitude of this effect would be very small and that even this small effect would never be attained. The production of an appreciable effect would require a considerable temperature gradient, which condition carries with it the necessity of the rapid loss of heat and the onset of crystallization before diffusion has the opportunity to establish even the small effects that are possible in unlimited time. With the onset of crystallization, phase equilibrium controls the composition of the liquid.

Gradients of composition in a liquid, produced by the force of gravity, must likewise be of very small magnitude and there is the same barrier to their establishment in any intrusive mass in the time available before crystallization. If there are any such masses as large permanent reservoirs of liquid in depth it is reasonable to suppose that they may normally exhibit the composition gradient demanded by gravity but the actual magnitude of the possible composition differences is not such as to account for the differences observed in rock series.

Variation of composition in the liquid magma has been supposed to originate as a result of a pressure gradient, in so far as this may affect the concentration of volatile components. This question is considered in a subsequent chapter on volatile components.

In addition to the processes involving gradients of composition in a single phase there are the processes involving the separation of distinct phases. These may be gaseous, liquid or solid and the processes involved are respectively gaseous transfer, liquid immiscibility and crystallization. The importance of liquid immiscibility is discussed in the next chapter. Gaseous transfer is discussed in a subsequent chapter which treats of the importance of volatile constituents. The rest of the volume is taken up with a discussion of crystallization in silicate systems, including natural magmas, and the correlation of the observed facts of rock series with the results of fractional crystallization.

In addition to the effects that may be produced as a result of the inherent properties of the magma there are the effects of the contamination of the magma with foreign material. This can be appropriately discussed only in connection with fractional crystallization. The question of the importance of this action, assimilation, is treated at some length after the principles and the main results of fractional crystallization have been set down.

CHAPTER II

LIQUID IMMISCIBILITY IN SILICATE MAGMAS

IT IS a well-known fact that many substances which are capable of mixing as liquids in all proportions at high temperatures may separate into two liquids upon cooling. It is natural that, in seeking an explanation of associated magmas, petrologists should early have turned to this process, but it is remarkable that the concept should still enjoy considerable popularity even after the accumulation of many facts regarding the detailed relations of rocks and of theoretical studies of the manner in which this process should go forward.

In no case has any petrologist advocating this process been able to point out exactly how it is to be applied to any particular series of rocks. It is usually merely stated that the original magma split up into *this magma* and *that magma*. Apparently the authors of such statements do not realize that they have not in any way described or discussed a process but have merely restated, with a maximum of indirection, the observational fact that *this rock* and *that rock* are associated in the described field.

The extreme of advocacy of immiscibility is found in the maintenance of the origin of monomineralic rocks such, for example, as a pure olivine rock, through the separation of a pure olivine liquid from a basaltic liquid. The most elementary considerations of phase equilibrium show that, when such complete immiscibility occurs, there can be no mutual lowering of melting points between the phases concerned, and yet serious proposals have been made of the unmixing of a pure olivine liquid from solution in a complex liquid at temperatures hundreds of degrees below the melting point of olivine, temperatures at which olivine liquid is, indeed, incapable of existence. Appeal to the possible effect of volatiles in lowering the freezing point of the olivine liquid helps the matter little, for it involves the assumption of a partition of the volatiles between the two liquids such that the olivine liquid acquires a concentration of volatiles many times that obtaining in the basaltic liquid. This assumption must be regarded as quite unwarranted by such knowledge as we have of the properties of these liquids and as altogether unsupported by the evidence of the quantities of volatiles associated with basaltic and dunitic rocks.

A feature of igneous rocks that has led some investigators to favor immiscibility is the fact that two adjacent rocks, that are evidently closely related, frequently show a very abrupt transition from the one to the other. Yet a brief consideration of liquid immiscibility should show that it is not as likely to give discontinuous variation as is crystallization. It is true that if two liquids that are only partially miscible are shaken together in a flask, two different liquids are formed, and if the flask be set aside they will become two separate layers with a definite bounding surface. If the temperature is kept constant these two distinct and sharply bounded layers will persist. However, if the immiscibility is the result of cooling a homogeneous solution, the behavior is not so simple. In this case a certain amount of immiscible globules should form in the liquid when a certain temperature is reached, and, even if time were allowed then for the collection of the globules as a separate layer, more immiscible globules would form *in each layer* as soon as cooling was resumed. And when cooling had proceeded to the point where crystallization ensued, a marked increase in the separation of immiscible globules would occur in association with, and as a necessary consequence of, the separation of crystals. We thus see that immiscibility is not a process taking place at an early stage of cooling, as a result of which a sudden separation of a liquid into two liquid layers occurs. The separation is rather a formation of small globules that grow slowly by diffusion and can collect as a separate layer only by comparatively slow movement in response to gravity. Neither is immiscibility a process that is completed at a very early stage in the cooling history, and of which all evidence is destroyed. It is a process that may begin very early but must continue until the later stages of crystallization, and the evidence of it would be as obvious and unfailing as the evidence of crystallization itself. The complete collection of *all* the immiscible liquid as a separate and distinct layer is as unlikely as the complete collection of a kind of crystals whose separation continues until a late stage.

We may illustrate these facts regarding immiscibility by discussing the simplest possible binary example. Fig. 1 presents the temperature-composition relations. When a liquid of composition x is cooled to the temperature FK, liquid of composition K, that is, a liquid rich in B, begins to separate from it, and as cooling proceeds the composition of the one liquid changes along FE and of the other along KD. The liquid represented by points on FE decreases in amount, and that represented by points on KD increases in amount. The first separation of liquid must be represented by the formation of minute nuclei that grow to larger and larger globules as the cooling proceeds, and as a result of the slow diffusion of material to these globules. There is no reason why this process should be accomplished any more rapidly for separated liquid

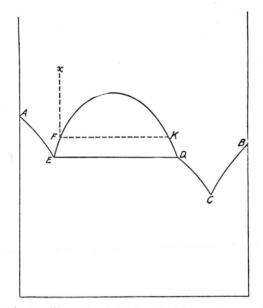

Fig. 1.   Diagram illustrating behavior of a binary mixture with partial miscibility.

than for separated crystals. If the separated globules were heavier than the general mass of liquid they would sink, and here enters the possibility of the growth of these globules to much larger dimensions than crystals, because two globules encountering each other may coalesce. The formation of very large globules in this manner would result in their more rapid accumulation as a separate layer. It should be noted, however, that this rapidity of accumulation could never result in the complete accumulation of all the globules as a separate layer. If, for example, cooling were interrupted at some temperature between FK and ED, and time allowed for the accumulation of all the globules as a separate layer, as soon as cooling was resumed new globules would form *in each layer*, and their accumulation by the slow process of gravitative adjustment would begin again. It is plain then that, whatever complications are assumed, the magma must arrive at the temperature ED in a blotchy condition, many of the blotches being of rather large dimensions as a result of the coalescence of globules. By large dimensions is meant a diameter several times, perhaps very many times, the diameter of the crystals in the average plutonic rock. At the temperature ED, when the liquids in equilibrium have the composition E and D, crystallization begins, crystals of A separating. It is important to note the nature of the first crystals separating, for it will be recalled that the liquid separating was rich in B. Those who advocate the separation of olivine, pyroxene, plagioclase, etc., as immiscible liquid tacitly assume a correspondence between the kind of material that would separate early as a liquid and the kind of material that we know from experimental and petrographic experience separates early as crystals. As a matter of fact there is no necessary relation, and the fact that correspondence must be assumed in each individual case is sufficient in itself to throw doubt on a process requiring such an assumption.

Continuing the consideration of the cooling of the mixture, which had been carried to the stage of the beginning of crystallization, at the temperature ED, we find that crystals of A would separate, and that as a necessary consequence more liquid of composition D would be formed and some liquid of composition E would be used up. These reactions would continue at constant temperature with the amount of crystals A and the amount of liquid D increasing at the expense of liquid E until finally all of liquid E would disappear, when the whole mass would be made up of about 80 per cent of crystals A and 20 per cent of liquid D. We thus see that up to a time when the mass is largely crystalline two liquids are present, and the crystalline product can not fail to show the blotchy condition that this predicates. The evidence of immiscibility would not be confined to rapidly chilled flow and dike rocks alone, though it would presumably be especially clear in them. Further cooling would result simply in the separation of more crystals of A with a consequent change in the composition of the liquid from D to C, where eutectic crystallization of both A and B would occur. It will be noted too that the liquid C, which is the *last* material to crystallize, is closely related in composition to the liquids K-D, the *first* material to separate as a liquid. This is important in connection with the well-recognized parallelism between "Differentiationsfolge" and "Kristallizationsfolge." If liquid immiscibility were a prominent factor, or even a subsidiary factor, in the differentiation of igneous rocks no such parallelism would exist.

THE SIGNIFICANCE OF GREIG'S WORK ON ACTUAL EXAMPLES OF UNMIXING IN SILICATES

One of the principal lines of evidence bearing upon the question of immiscibility in silicates is, as Vogt early emphasized, the results of experimental and industrial work with silicate melts. Until recently no example of unmixing in such melts had been encountered but the work of Greig has now revealed a number of examples of this phenomenon in a variety of mixtures. Without exception the mixtures are of what might be termed highly specialized composition and even in these the liquid fractions can exist only at excessively high temperatures. At lower temperatures two liquid fractions can no longer coexist, but this fact is not due to their becoming miscible but to the intervention of crystallization.

It is a matter of satisfaction to find immiscibility so well displayed and so thoroughly amenable to quantitative study, for it shows that when immiscibility occurs the resources of the modern laboratory are adequate to detect it. The mixtures in which it has been found are particularly intractable, yet the only reason for failure to discover the facts long since is that the temperature region concerned is relatively inaccessible from the point of view of carefully controlled experimentation.

In strong contrast with these restricted compositions showing immiscibility, stands the great range of silicate compositions that have been completely studied throughout the conditions where they show inhomogeneity of any kind, and this inhomogeneity always consists in the separation of crystals. No suggestion of the separation of a liquid phase has ever been observed. Considering the ready separation occurring in the less favorable, high-silica mixtures, its failure in all other mixtures must be regarded as decisive. But in addition to this evidence, which might be regarded as of the negative variety, there is the direct, positive evidence of the forms of the equilibrium diagrams which are such as to preclude the possibility of unmixing. No suggestion of a discontinuity in the crystallization surfaces is to be found and such discontinuities would be inevitable were immiscibility a fact, even though actual unmixing were prevented by, let us say, excessive sluggishness. Indeed it was the indications, obtained in earlier work, of discontinuities on the crystallization surface at high silica concentrations that led to the further investigations of Greig. These revealed that the discontinuities were due to immiscibility, as had been suspected.

Referring to the actual results obtained by Greig,[1] we find that, of important rock-forming oxides, CaO, MgO, FeO and Fe_2O_3 show immiscibility with SiO_2 but only at high silica concentrations, whereas Na_2O, K_2O and Al_2O_3 show complete miscibility with SiO_2 in all concentrations. No rocks are known which have the compositions required for the appearance of immiscibility on the basis of these data. Rocks of high silica content never have CaO, MgO and FeO as the principal additional constituents: on the contrary they are always rich in Al_2O_3, Na_2O, and

1 J. W. Greig, *Amer. Jour. Sci.*, 13, 1927, pp. 1-44, 133-54.

K₂O. Greig's work shows, moreover, that it requires but a small content of the latter group to completely neutralize the unmixing tendency of the former group. This relation is well summarized by him in a generalized ternary diagram which is here reproduced as Fig. 2. The complete

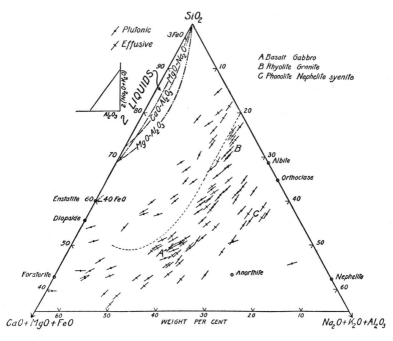

FIG. 2. Diagram (after Greig) to show the relation of the compositions of mixtures which show immiscibility to the compositions of igneous rocks. The extent of immiscibility in three binary and three ternary systems is indicated. The compositions of igneous rocks have been plotted by reducing the analyses to a water-free basis and adding the constituents not indicated on the diagram to CaO + MgO + FeO. The points lying below the dotted curve are Daly's averages; those above are extreme cases from Washington's Tables. The ratio of alkali to alumina is shown by the slope of the line passing through the point as indicated by the small diagram, upper left.

avoidance of the much restricted two-liquid area by igneous magmas is obvious.

In spite of the examples of immiscibility in silicates that this new work demonstrates, we must still adhere to the older conclusions regarding the bearing of experimental work on immiscibility in rock magmas.

The evidence from experiment, as far as it goes, is decidedly against limited miscibility in natural magmas.

SUPPOSED EXAMPLES OF IMMISCIBILITY IN NATURAL MAGMAS

However, it is to natural rocks, after all, that appeal must be made for a decision on this question. Certainly those with an abiding faith in unmixing of rock magmas are not likely to accept the evidence of experiment until a much closer approach is attained to the conditions represented by a natural magma and its surroundings. Therefore we must turn to the evidence of rocks themselves. It is natural to expect that our first task would be to consider the merits of those features of rocks that have been suggested as evidence of immiscibility, but we find that no one has offered any detailed statement of this evidence. Perhaps those features of rocks that have done most to suggest immiscibility are the intimate association and frequent alternation of lavas of different composition and the abrupt changes of composition often noted in deeper-seated rocks. It is not necessary to point out that these features could come about in a number of different ways and that some very good reason for preferring the immiscible relation should be offered.

It is apparently not realized by many petrologists that unmixing is not a mysterious process whereby one liquid separates instantaneously into two masses of liquid of different composition and with but one common boundary surface. Unmixing is, on the contrary, a manifestation of phase equilibrium just as is crystallization. As already pointed out, the new phase must begin to appear at dispersed nuclei, the individual units must grow as the result of slow accretion by diffusion and any collection into a distinct mass must take place slowly with the aid of gravity as it must in the case of crystals. In spite of these facts we can find, even at this late date, the serious statement that many magmas are too viscous for the influence of gravity on crystallized phases to be appreciable, and that therefore the differentiation of these magmas must have been brought about by unmixing.[1] If great viscosity prevents a gravitative effect, crystallization differentiation still has the effects of deformation to fall back on, but for unmixing liquids there is no alternative to gravity as a means of separation of phases.

A long period of suspension of one liquid in another is a necessary consequence of unmixing in any silicate system. No convincing reason is forthcoming as to why this period is so effectively concealed from observation. Having regard for the fact that unmixing, like crystallization, should be the result of cooling, we should find it manifested in varying degrees in batholith, dike, and flow, for they represent various rates of cooling. In lavas, particularly, we are entitled to expect that

1 B. Asklund, "Granites and Associated Basic Rocks of the Stavsjö Area," *Sveriges Geol. Undersök. Årsbok*, 17 (1923), No. 6, p. 106.

quick chilling would put an end to the previous slow growth and accumulation of separate phases under deep-seated conditions and that, since
this interruption may take place at any stage of the cooling process,
every step in the formation of separate phases would be revealed. When
we turn to lavas we find a wealth of information regarding steps in the
separation of crystals but none as to the separation of liquid phases.
Lavas would reveal interrupted unmixing, were it a general phenomenon,
just as plainly and as freely as they reveal interrupted crystallization.

The best type of evidence of the existence of one liquid as an immiscible separation from another, and one that would involve the element
of interpretation in minimum degree, would be the occurrence of glassy
globules in a glassy rock of different composition. Even here there would
be the possibility of interpreting the globules as inclusions of an adjacent
glassy rock that had become rounded, but if collateral evidence permitted the elimination of this possibility the indications would be very
strong that one liquid occurred as immiscible globules in another. If
immiscibility were a common phenomenon such globules would be common in glassy lavas. The facts are that they are utterly lacking.

Tanton recently described an example of a globule-bearing quartz
porphyry which, according to his claims, presented the type of relation
just described.[1] Bain has interpreted the globules as foreign inclusions,[2]
and yet another interpretation has been offered by Bowen.[3] The latter was
based on a single altered specimen, submitted by Tanton, which contained no discernible glass in either globule or matrix and the conclusion
then reached was that the rounded masses (globules) were residuals of
alteration. Greig has since made a detailed study of the occurrence in the
field and of numerous specimens there obtained and has reached the
conclusion that the globules are the result of spherulitic crystallization.[4]
It is, in fact, as typical an example of this as one could well find. He has
shown that there are no glassy globules but only globules whose shape
and general character are determined by spherulitic crystallization.
There are no bands that can be interpreted as the result of drawing out
of these globules when they were liquid, for the globules may occur in
typical rounded form where the banding is most marked and a single
globule may traverse several bands with complete indifference. Indeed,
there is no evidence that the globules existed when the mass was entirely
liquid. They came into existence as the result of partial crystallization
and are simply spherulites. In some altered facies of the rock the spherulites occur as residuals of alteration so that this interpretation is not
incorrect and is a quite reasonable one if only the altered facies is

1 T. L. Tanton, *Jour. Geol.*, 33, 1925, p. 629.
2 G. W. Bain, *Amer. Jour. Sci.*, 11, 1926, pp. 74-88.
3 N. L. Bowen, *Jour. Geol.*, 34, 1926, pp. 71-3.
4 J. W. Greig, *Amer. Jour. Sci.*, 15, 1928, pp. 375-402.

available for observation. It is, however, only half the story, Greig's observations showing why those residuals have the form of globules. They are spherulites which, in virtue of a difference of type or of degree of crystallization, have resisted alteration, at least of a certain kind, better than the surrounding matrix. This supposed example of immiscible globules in lava does not, therefore, withstand the test of critical examination.

Though evidence of unmixing would be commonest in lavas it would nevertheless be common enough in other rocks. There is, as a matter of fact, but one excessively rare type of structure in plutonic rocks that bears a resemblance to the product of crystallization of a liquid suspension and has been so interpreted. This is the orbicular structure. Bäckstrom, Daly, and others have expressed adherence to the view of this structure noted above and recently Asklund has appealed to immiscibility as the cause of variation in a series of rocks, the rare local occurrence of orbicular structure being the only tangible evidence bearing on the question that he is able to produce. He does, however, give a rather detailed discussion of the processes that he believes to have occurred, which is more than any other adherent of these views has ever done, and he therefore presents something definite for discussion. He pictures a process that involves the breaking-up of dioritic magma into two magmas, the one producing a hornblende granite and the other a norite. The norite and quartz-plagioclase nodules of the diorites represent this process interrupted by crystallization. The granitic magma is believed to have further unmixed, and various other complications are assumed which need not concern us here. The first process mentioned is essentially that advocated by several petrologists, that is, the splitting-up of dioritic magma into granitic and gabbroic magma. It is, for example, not infrequently invoked to explain the association of dolerite (or gabbro) and granophyre. The proponent of unmixing has always had a very marked advantage over him who would explain the diversity of igneous rocks in terms of crystallization. Since nothing is known of the successive stages of the supposed unmixing in such liquids, either from the evidence of natural magmas or from experimental evidence, one who claims unmixing can not be required to show the detailed correspondence of the result with the expectations of the process. He may merely make the bald statement that unmixing occurred and his task is finished, or, at least, so it has seemed hitherto.

As a matter of fact there is a very simple test, arising from theoretical considerations, that can be applied to any pair of liquids for which an immiscible relation may be proposed. It is to be remembered that unmixing is a manifestation of phase equilibrium and two liquids which constitute an immiscible pair are in equilibrium with each other. Not only are they in equilibrium with each other but both must be in equili-

brium with any additional phase that may be formed. Thus in the association, gabbro-granophyre, if it is assumed that their liquids constitute
an immiscible pair and if we imagine that they are cooled until crystallization begins in one of them, then the crystals formed in it should be
in equilibrium with the other liquid as well. At a stage when the gabbroid
liquid is in equilibrium with plagioclase of composition $Ab_{31}An_{69}$ (and
there is such a stage), the associated granophyre liquid should also be in
equilibrium, not merely with some plagioclase, but with the precise
plagioclase $Ab_{31} An_{69}$ and any crystals that might migrate across the
border into the granophyre liquid would be entirely at home there. We
know perfectly well, however, that there is no stage at which granophyre
liquid is at equilibrium with so basic a plagioclase. Any petrologist who is
willing to take the trouble of applying the same considerations to any
of the other crystalline phases of either gabbro or granophyre can readily
convince himself that there is no stage of crystallization at which the
crystals in equilibrium with either one would also be in equilibrium with
the other.

We thus find that the concept of immiscibility is a complete failure
as an explanation of the association, gabbro-granophyre, rhyolite-basalt
and similar associations for which it has been invoked. Indeed, no one has
ever advanced any good reason for advocating it. In the former association the granophyre is so plainly a crystallization residuum of the
gabbro, occupying crystallization interstices, that it is surprising to find
any other relation suggested. In the case of rhyolite and basalt we have
an association which, while frequent, is none the less a random one in
that they are far removed from the place and conditions of their origin,
and when they have accidentally come together while still at a high
temperature they show a disposition to mix which is wholly at variance
with an immiscible relation.

If we return now to the specific assumptions made by Asklund we
find that he gives a diagram indicating the relations between the various
immiscible liquids with respect to their feldspar content, which diagram
is reproduced here as Fig. 3. Asklund says, "In this diagram P_1 represents
the supposed parental magma. It was broken up into A_1 and G_1, each of
which broke up into separate magmas. A_1 was differentiated into B
(gabbro-magma) and granites richer in plagioclase than the particular
magmas of G_1. In deeper situations the sinking magma B was dissolved
in P_1, which consequently grew more basic, P_2. When intruded this complex magma broke up into A_2 and G_2 and later on A_2 broke up into N
(noritic gabbro) and hornblende-bearing granite, G_2 formed quartz
granites and coarse two-mica granite." The first portions of this we can
not test by any reference to the rocks because there are no corresponding
rocks, and only when noritic gabbro and hornblende-bearing granite are
pictured as an immiscible pair do we reach something tangible for dis-

cussion. We must regard the two liquids as existing in equilibrium and apply the test of joint equilibrium with a crystalline phase and we find that the concept breaks down completely under this test. On the other hand if we compare the diagram (Fig. 3) with Asklund's own diagram

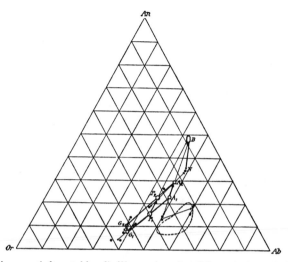

FIG. 3. Diagram (after Asklund) illustrating the differentiation of the Stavsjö rocks with respect to the feldspar relations. The dots indicate the analyzed rocks.

of a "Survey of the crystallization of the feldspars in the noritic gabbro," here reproduced as Fig. 4, we note a most remarkable correspondence between the directions of differentiation in this assumed unmixing and the direction actually taken by the liquid as it changes composition through crystallization. It is to be remembered that this is the actual course of crystallization as deduced by Asklund himself from the study of thin sections. It is plain enough from this diagram alone (Fig. 4) that fractional crystallization of the basic magma could give, if not the whole series of rocks, at least the principal types.

There is one aspect of the question of liquid immiscibility that apparently requires to be pointed out, though it should not. If one assumes that the association basalt-rhyolite, without intermediate types, demonstrates the immiscibility of these two liquids, the acme of absurdity is reached if one then assumes that the basaltic liquid resulting from this unmixing can dissolve solid granite. The two assumptions are utterly incompatible. It is because liquid silicates and liquid platinum are immiscible that one can make silicate melts in a platinum crucible without danger of attack on the crucible. It is because one silicate liquid is normally miscible with any other silicate liquid that one can not make a silicate melt in a fire-clay crucible without attack on the crucible. And so if basalt and rhyolite liquids are immiscible then basaltic liquid can not dissolve granite or, *per contra*, if basaltic liquid does dissolve granite,

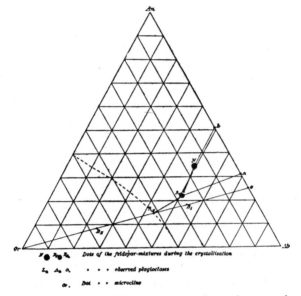

Fig. 4.   Diagram (after Asklund) illustrating the observed course of crystallization of the feldspars in the noritic gabbro.

then basalt and rhyolite can not constitute a pair of immiscible liquids. An avenue of escape from these conclusions may seem to exist in the possibility that basalt and rhyolite are miscible when very hot and become immiscible only on cooling. Very hot basaltic liquid could, on this assumption, dissolve granite but colder basalt could not. But the facts indicate that basaltic liquid within its crystallization range is capable of attacking (dissolving) granitic and related material. The immiscible relation must, therefore, exist, on this assumption, only at still lower temperatures or well within the crystallization range of basalt, and at this stage the test of joint equilibrium with a crystalline phase of definite composition is particularly applicable, which condition we have found to fail.

A final loophole remains. Basalt and rhyolite may be assumed to show the immiscible relation at very high temperatures but to be miscible at lower temperatures. Such a relation between a pair of liquids is not unknown and if any petrologist should choose to assume that this condition obtains for basalt and rhyolite it would be very difficult to prove him wrong, but the assumption would have nothing else to recommend it.

On the basis of immiscibility of any kind it is impossible to build up an adequate explanation of the associated members of rock series which is the fundamental problem of petrology.

CHAPTER III

FRACTIONAL CRYSTALLIZATION

GENERAL CONSIDERATIONS

THE evidences of fractional crystallization or crystallization-differentiation in magmas are not, as many seem to believe, the fact that at times it can be demonstrated that certain minerals are concentrated in the lower layers of an igneous mass. Such observations are serviceable principally in throwing light on one of the methods whereby fractionation can originate. The real evidences are universally present in igneous rocks, whether they reveal the method or not, and consist in the *mineral associations and antipathies* which they characteristically display.

Considering antipathies first we find that there are certain actual incompatibilities such as that between quartz and nephelite which have a purely chemical basis and need not be further considered here. In addition to these there are incompatibilities of a different kind, based wholly on petrogenic factors. We know, for example, that a "rock" made up of quartz (or free SiO_2) and very basic plagioclase can be and has been made in the laboratory but is unknown among igneous rocks in nature. There is no *a priori* reason why this should be so, nor is the fact rationally accounted for by any of the various theories of petrogenesis that have been suggested except the theory of fractional crystallization. According to this theory they belong too far apart in a crystallization sequence to be associated in such quantities as to make up the entire rock. For a similar reason igneous rocks consisting of muscovite and very basic plagioclase or of pyroxene and quartz are unknown and many other petrologic incompatibilities might be mentioned.

Besides these real incompatibilities there are antipathies which, while not strong enough to make themselves universally felt, are obvious enough in a general survey of igneous rocks. Rocks rich in basic plagioclase tend to be poor in orthoclase and vice versa. Those rich in orthoclase never have olivine in important amounts. Abundance of biotite means poverty in basic plagioclase.

Turning to the association tendencies of igneous-rock minerals we find again the control of crystallization-differentiation. Rocks rich in quartz tend to be rich in orthoclase or sodic plagioclase or both. If a rock has hornblende as its most important ferromagnesian constituent it is likely to have intermediate plagioclase as its principal feldspathic component. If the feldspathic mineral is basic plagioclase, pyroxene with or without olivine tends to be the principal femic constituent. These, the ordinary, everyday facts about rocks, are the principal evidence of crystallization-differentiation. To be sure there are many mineral association-tendencies in the quantitatively less important rocks that can not at present be so readily traced to crystallization-differentiation, but those of the dominant igneous series give indications that are plain enough. The detailed significance of these, which is merely suggested in the above, can be brought out only in the fuller discussion of crystallization on later pages. Suffice it to say here that those minerals that belong to the same general period of crystallization tend to be associated and those belonging to remote periods ordinarily fail of association. The controlling factors are thus analogous to those which determine that little girls ordinarily play "London Bridge" with other little girls, occasionally with their mothers, seldom with their grandmothers and never with their great-grandmothers. Not only is it true that these association and antipathetic tendencies point to the control of crystallization but they make it clear that igneous activity can not have its origin in remeltings of any random crustal material but is in agreement with geological evidence pointing to basic material as parental magmatic substance.

The full evidence that basic magma undergoing fractional crystallization can give rise to the mineral associations of igneous rocks can be developed only in the discussion on later pages.

FACTORS BRINGING ABOUT FRACTIONATION
DURING CRYSTALLIZATION

Fractionation or differentiation in a crystallizing mass may be brought about in two ways: through the localization of the crystallization of a certain phase or phases and through the relative movement of crystals and liquid.

The outer parts of a body of magma are ordinarily cooler than the interior and there should therefore be a period during which crystallization is taking place only near the border. The separation of crystals from the peripheral parts of the mass necessarily brings about an impoverishment of the liquid in the constituents separating. A composition gradient is thus set up in the liquid and a diffusion of substance into the peripheral liquid will take place, with a tendency to bring it to a composition uniform with that of the main mass of liquid in which crystallization has not yet begun. But a liquid of such composition must precipitate some of its substance under the temperature conditions prevailing at the border so that the net result of this diffusion process would be the continued growth of the crystals of early separation at the border and continued impoverishment of the whole mass of liquid in the material of these crystals. Real though this tendency must be, it can readily be shown that diffusion of substance, as compared with diffusion of temperature, is altogether too slow to permit the accomplishment of significant results in the time available before complete solidification of the mass has occurred. The constant of diffusivity of temperature for rocks is of the order of magnitude 0.01 cm^2 per second. The highest measured coefficient of diffusivity of mass in silicate melts is of the order of magnitude 0.25 cm^2 per day, and for some silicates it is much less.[1] The rate of diffusion of temperature is thus at least 4000 times as great as the rate of diffusion of substance. With such a contrast of rates it can readily be shown that only effects of most insignificant magnitude can be produced in a cooling mass by diffusion toward the chilled margin.[2]

There is another possible process of feeding new liquid into the cool border that is not necessarily so slow and in a general way might be presumed to give a similar result. This is convective circulation. Some circulation of this kind is a practical necessity in the great majority of cooling masses of magma and it may on occasion bring about the result suggested for it. It should be noted, however, that such molar flow, in contrast with the molecular flow of diffusion, is in no wise selective as to the materials which participate in it. When cool liquid at the border moves to give place to warm liquid from below it must carry along its suspended crystals. It would seem that only crystals actually attached to the frozen border could receive a contribution to their growth from the incoming fresh liquid. The laboratory analogy frequently cited is indubitably of this nature. In a laboratory vessel containing a saturated solution, crystals may deposit on the walls and circulation of the liquid may cause growth of these crystals until a considerable deposit of the salt occurs on the walls. Border facies of certain restricted igneous masses, say of a pegmatitic nature, may possibly be formed by such a process but the great majority of large-scale examples do not suggest this origin. They may, indeed, be enriched in early-formed crystals but these occur typically as discrete grains embedded in the general mass of later crystallization. They could have been nothing but suspended crystals at early stages and must have participated in any liquid circulation that existed. They could not at that time, then, have had attachment of any kind, so that it would seem that the case can not be regarded as analogous to the deposit on the laboratory vessel cited. However, since the

1 N. L. Bowen, "Diffusion in Silicate Melts," *Jour. Geol.*, 29, 1921, pp. 295-317.
2 *op. cit.*, pp. 312-16.

evidence of the rocks usually points merely to the fact of fractional crystallization rather than to the method whereby it is accomplished, it is essential to have in mind all processes of fractionation by crystallization.

The factors which are adequate to produce important results in the way of fractionation of a crystallizing magma would appear to be thus reduced to those involving the relative movement of crystals and liquid. The movement may take place under the influence of gravity or under the influence of deformative forces. The former process must have its principal importance during the comparatively early stages of crystallization. The rôle of the liquid is the dominantly passive one of permitting motion of the crystals, though of course a downward movement of any crystal connotes the upward movement of its volume equivalent of liquid.

The second process, relative movement of crystals and liquid through deformative agencies, must reach the height of its effectiveness only in the middle and late stages of crystallization. The liquid now assumes the active rôle and the crystalline mesh may merely permit liquid to pass through it or it may be broken down and its constituent grains more closely packed by rotational movements with consequent expulsion of the residual liquid.

In addition to these two methods of separating crystals and liquid there is a localized separation taking place about individual growing crystals during their zoning whereby inner zones are effectively deterred from participation in the equilibria involved in the later stages of crystallization.

Of the reality of the two processes of gravitational and deformational differentiation during crystallization and perhaps also of the zonal effect no question is likely to be raised. Of their general adequacy to explain the diversity of igneous rocks much doubt is frequently expressed. Many investigators make the mistake of regarding the question of crystallization-differentiation and of the sinking of crystals as identical, when for proper discussion the deformational, the gravitational and the zonal effects should be considered. Certainly the assemblages of minerals that we know as rocks are such as to indicate that the forces in control caused a grouping of minerals of the same general period of crystallization and a separation of those of remote periods, and this is the result toward which the three processes mentioned would tend.

The principal objections raised against crystallization-differentiation have not been concerned with its adequacy to produce these results. They have turned rather upon the fact that many igneous masses of great dimensions fail to show differences of composition between one part and another. In the case of deformational differentiation possible reasons for its absence are so obvious that examples of lack of differentiation have not been urged against it. In the case of the gravitative effect, however, absence of differentiation in certain masses has been regarded as proof that gravity can never give the results referred to it. Such examples are, however, equally cogent evidence against any other process of differentiation, indeed against the fact of differentiation itself.

At first sight it might appear that, since gravity never takes a holiday, it is somewhat more reasonable to assume that a process of, let us say, gas transportation, was inoperative in individual instances than that gravity was inoperative. The appearance is, however, deceptive for it is not a question of lack of gravity but a lack of the gravitative effect, and to account for this it is only necessary to assume that the factors opposing the gravitative effect were dominant. Nor can the necessity of such an assumption be regarded as a weakness peculiar to the hypothesis of the general prevalence of gravitative control, for the same assumption of incompetence has to be made for any other process as applied to those particular instances. Indeed it is not merely this general assumption that will of necessity be the same for various processes but the specific assumptions as to the factors opposing differentiation will likewise be similar. Thus, in the case of gravity, it may be assumed that motion of the liquid kept it stirred or that the viscosity was too great to permit significant effects in the time available, and precisely these assumptions must be made to explain lack of results on the part of practically any other process.

The great majority of exposures of igneous rocks do not, indeed, furnish evidence of the manner of their genesis, but considered as members of the series of rocks with which they are associated they furnish evidence of crystallization-differentiation usually without indicating which method of fractionation has dominated. The evidences of fractional crystallization must consist in a detailed comparison of the results to be expected from the process with the features of rocks as we find them. It is the purpose of this work to attempt this comparison. When a similar comparison is made for any other hypothesis of the factors controlling igneous differentiation and a similar degree of correspondence is found, that hypothesis may then be regarded as having an equal basis of probability and only then.

CHAPTER IV

CRYSTALLIZATION IN SILICATE SYSTEMS

THE general problem of the crystallization of igneous magmas is the problem of crystallization in polycomponent systems. It has frequently been stated in the past, and is still occasionally stated, that crystallization in polycomponent systems is a simple matter, that it begins with the separation of one of the components, which is later joined by another, the pair by a third, and so on until final crystallization of the polycomponent eutectic occurs. This is said to be analogous to determined facts in alloy systems and aqueous salt solutions. As a matter of fact a glance at a collection of alloy diagrams and a brief study of Van't Hoff's work on oceanic salts should convince anyone that the determined facts for alloys and solutions do not justify such a statement. But even if the determined results of alloy and other systems did reveal the condition above outlined it would not constitute any valid reason for expecting such relations in silicate systems in the face of definite evidence to the contrary. There is now, indeed, a considerable body of experimental results on silicate systems and these furnish us with a much more reliable basis for the discussion of the crystallization of rocks than any supposed analogy with alloy and other systems, particularly if these are ill-chosen.

The accumulated data upon silicate systems offer us aid in two ways. They furnish us with definite facts concerning the behavior of certain silicates and their mixtures under varying conditions of temperature and pressure. They reveal the types of inter-relationship commonly displayed by silicates and the fundamental factors controlling these and thus permit us to deduce principles of more general applicability. It is our immediate purpose to consider what appear to be the most pertinent data, having in mind this dual service.

In the following pages a number of silicate systems that have been studied experimentally will be described and discussed. There will be no multiplication of systems and diagrams. Each is intended to bring out some fact or principle of importance in connection with the crystallization (particularly the fractional crystallization) of magmas and, insofar as possible, systems will be chosen that have crystalline phases closely related to rock minerals.

BINARY SYSTEM WITH EUTECTIC

The binary eutectic system is ordinarily considered the simplest. An example is afforded by anorthite $CaAl_2Si_2O_8$ and diopside $CaMgSi_2O_6$, the one an end member of the plagioclase group, the other an end member of the pyroxene group.[1] The relation revealed by experiment is shown in Fig. 5. The diagram shows the melting point of anorthite at 1550°C and of diopside at 1391°C. Each lowers the melting point

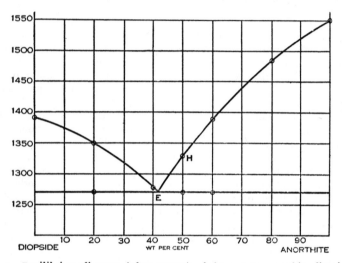

Fig. 5. Equilibrium diagram (after Bowen) of the system, anorthite-diopside.

of the other. The curves of melting (or freezing) point meet at 1270° and at the composition, 42 per cent anorthite-58 per cent diopside (E), which are therefore the eutectic temperature and composition for mixtures of these two compounds. A mixture of the composition 50 per cent anorthite-50 per cent diopside begins to crystallize at 1328° (H) with the separation of anorthite. As the temperature falls anorthite continues to separate and the liquid changes in composition along the curve HE, that is, becomes richer in diopside. At 1270° when the liquid has the composition E diopside begins to crystallize and thereafter diopside and anorthite separate in constant proportion, the liquid remaining of constant composition and the temperature remaining at 1270° until the whole mass has crystallized.

If, during the early stages of crystallization, some of the anorthite

---

[1] N. L. Bowen, *Amer. Jour. Sci.*, 40, 1915, p. 164.

crystals were removed or, say, settled to the bottom, no effect upon the course of crystallization would result therefrom. The composition of the last liquid to crystallize and the temperature of final crystallization would be exactly the same in the part to which the crystals had moved and in the part from which they had moved.

The melting of the same mixture of diopside and anorthite is exactly the reverse of the process described above. When the temperature 1270° is attained some liquid begins to form, and with further addition of heat the temperature remains constant until all of the diopside and most of the anorthite has melted, the liquid maintaining the composition E. As soon as the diopside has entirely disappeared the temperature will begin to rise and solution of anorthite will take place, the liquid changing in composition along the curve EH until at 1328° all the anorthite has dissolved.

In a mixture in which diopside is present in excess over the eutectic mixture the diopside separates first on cooling and dissolves last on heating in the same manner as does anorthite in the mixture described above. But in all cases the final liquid suffering crystallization upon cooling or the first liquid formed upon heating has the composition of the eutectic (E). Relative movement of crystals and liquid (crystal fractionation) could give rise upon complete crystallization to a mass locally enriched in either anorthite or diopside but to no other contrast of one part with another. Selective fusion of a crystalline mixture can, likewise, give no other contrast of one part with another.

BINARY SYSTEM WITH A COMPOUND HAVING A CONGRUENT MELTING POINT

Probably the next simplest type of binary system is one with a compound having a congruent melting point. It is not easy to find, among investigated silicate systems, a simple example uncomplicated by solid solution, inversion, or some other modifying influence. As simple as any is the binary system, sodium metasilicate-silica which shows the compound, sodium disilicate.[1] The equilibrium diagram is shown in Fig. 6. The diagram is divided into two parts, the line of separation lying at the composition of the compound, and a maximum occurs on the curve of melting points (liquidus) at the composition of the compound. When a maximum occurs at the composition of the compound the compound melts to a liquid of its own composition and is therefore said to melt congruently. Each of the two parts of the diagram, as divided at the composition of the compound, is a simple eutectic system. The crystallization of any mixture is exactly like that in any eutectic system, the mixtures on the one side of the compound proceeding to the one eutectic and those

1 G. W. Morey and N. L. Bowen, *Jour. Phys. Chem.*, 28, 1924, pp. 1167-79.

on the other side proceeding to the other eutectic with falling temperature. In view of the description of crystallization in a eutectic system given in the foregoing it is not necessary to describe the crystallization of any mixture. A mixture having less silica than the disilicate com-

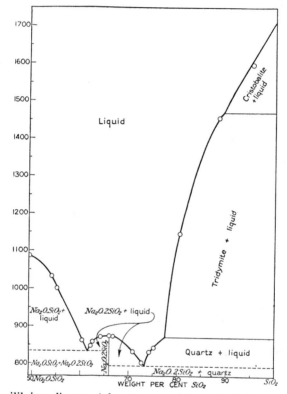

FIG. 6.   Equilibrium diagram (after Morey and Bowen) of the system, sodium metasilicate-silica.

pound must crystallize to disilicate and metasilicate. A mixture containing more silica than the disilicate compound must crystallize to disilicate and silica. By no process of crystal fractionation can a mixture containing less silica than the compound be induced to crystallize in such a way as to give rise to free silica.[1] We shall see that this is not always true

[1] The behavior of a crystalline mixture on heating is simply the reverse of the behavior of a liquid. The first liquid is formed at the eutectic of the subsystem in which the mixture lies and by no process of selective fusion can a liquid in the other subsystem be obtained from it.

of systems showing a compound; more explicitly stated, it is not true of a system in which a compound having an incongruent melting point occurs.

BINARY SYSTEM WITH A COMPOUND HAVING AN INCONGRUENT MELTING POINT

As an example of a system with a compound having an incongruent melting point we may take the system, forsterite (Mg_2SiO_4)-silica (SiO_2) which shows the compound, clino-enstatite ($MgSiO_3$). The equilibrium diagram is given in Fig. 7.[1] In this system the melting point

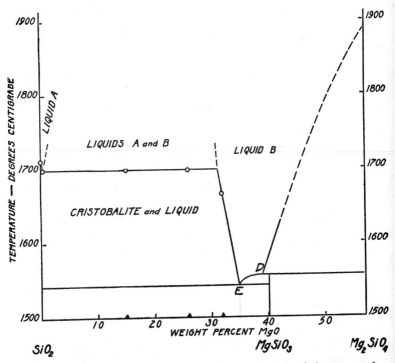

FIG. 7. Equilibrium diagram (after Bowen and Andersen) of the system, forsterite-silica (with additions by Greig in high silica compositions).

curve (liquidus) of forsterite falls from the melting point of forsterite at 1890° and is not intercepted by another curve until it passes beyond

1 N. L. Bowen and Olaf Andersen, *Amer. Jour. Sci.*, 37, 1914, pp. 487-500.

the composition of the compound clino-enstatite ($MgSiO_3$). There is thus no maximum on the liquidus at the composition of the compound, $MgSiO_3$, and when that compound is heated it melts with decomposition at $1557°$, giving forsterite and a liquid containing more silica than $MgSiO_3$. The compound, $MgSiO_3$, is therefore said to melt incongruently.

The crystallization of mixtures of Mg_2SiO_4 and $MgSiO_3$ takes place, when perfect equilibrium obtains, in the following manner: Forsterite crystallizes out first and increases in amount until the temperature $1557°$ is reached. At this temperature the liquid has the composition D and clino-enstatite begins to crystallize out, forsterite to redissolve or to react with liquid to produce more clino-enstatite, this process continuing until all the liquid is used up and the whole consists of forsterite and clino-enstatite. In pure $MgSiO_3$ the early separation of forsterite takes place in the same way, but at the reaction temperature ($1557°$) the last of the liquid and the last of the forsterite are used up at the same instant and the whole consists of clino-enstatite.

In compositions between $MgSiO_3$ and the point D, forsterite separates first as before, even although the liquid has an excess of silica over the composition $MgSiO_3$. At the reaction point it is completely resorbed, leaving clino-enstatite and some liquid. With further lowering of temperature clino-enstatite continues to separate until, at $1543°$, when the liquid has the composition E, cristobalite separates also and the whole crystallizes at this temperature, giving a mixture of clino-enstatite and cristobalite.

Such is the behavior of these mixtures when complete equilibrium is attained. However, with quick cooling, equilibrium is not attained and a preparation of composition $MgSiO_3$ will then crystallize to a mixture of clino-enstatite, silica, and forsterite. The same fact is true of all mixtures lying between Mg_2SiO_4 and D, the reason being that with quick cooling the liquid of composition D, instead of reacting with (redissolving) forsterite at $1557°$, simply crystallizes, as the temperature falls quickly below this point, to a mixture of $MgSiO_3$ and silica. Such mixtures containing both forsterite and silica are unstable but will persist indefinitely.

It is obvious that the same failure of the liquid D to react with forsterite would result if the forsterite were removed. Thus any mixture lying between forsterite and D might be crystallized in such a way that during the period of crystallization of forsterite there was relative movement of crystals and liquid. In such a case there would be an excess of forsterite in some parts and those parts would crystallize to forsterite and clino-enstatite. In other parts there would be an excess of liquid. The small amount of forsterite would there be used up by the liquid with concomitant formation of clino-enstatite and finally the liquid would crystallize to clino-enstatite and free silica.

By crystal fractionation in such mixtures, then, there may result not merely different proportions of mineral compounds in the different fractions but actually a different assemblage of compounds. Fractionation may bring it about that in any mixture, however rich in forsterite, the liquid may run down to the clino-enstatite-silica eutectic with precipitation of some free silica. The fundamental controlling factor is the incongruent melting of $MgSiO_3$.

It is important to note that melting of mixtures of this system is, in many cases, not simply the reversal of the crystallization. Thus, as we have seen, a liquid which is a mixture of Mg_2SiO_4 and $MgSiO_3$ (i.e., has a deficiency of SiO_2 with respect to the compound $MgSiO_3$) may be cooled in such a way as to give a final liquid containing excess SiO_2 and crystallizing only when the temperature has fallen to the eutectic $1542°$. But one can heat any crystalline mixture of Mg_2SiO_4 and $MgSiO_3$ for an indefinite period at a temperature somewhat above the eutectic $1542°$ without the formation of any liquid, indeed no liquid will form until the temperature is raised to the reaction point D ($1557°$), the temperature of complete consolidation of the perfect equilibrium type. Thus selective fusion is not simply the reverse of fractional crystallization. But if the liquid formed in the above case of selective fusion at $1557°$ be separated from crystals, allowed to crystallize and the product subjected to a temperature of $1542°$ some liquid will form by selective fusion at that temperature. Thus repeated selective fusion may give the same result as that obtained in a single course of fractional crystallization.

BINARY SYSTEM WITH MORE THAN ONE COMPOUND HAVING AN INCONGRUENT MELTING POINT

There may, of course, be several compounds in any binary system and any or all of them may melt congruently or again any or all of them may melt incongruently. Where there are two or more adjacent compounds that melt incongruently the possible effects of fractionation are correspondingly increased. An example is furnished by the system potassium metasilicate (K_2SiO_3)-water (H_2O) with the two compounds, hemihydrate and mono-hydrate, both of which melt incongruently.[1] The equilibrium diagram is given in Fig. 8. A mixture of the composition X begins to crystallize at $800°$ with the separation of K_2SiO_3, and this continues to separate until the temperature has fallen to $600°$ when the liquid has the composition B. At this temperature the liquid reacts with the crystals of K_2SiO_3 and converts a part of them to $K_2SiO_3 . \frac{1}{2}H_2O$, the temperature remaining constant until all the liquid is used up by this reaction. But if crystal fractionation had occurred, that is, if the

[1] G. W. Morey and C. N. Fenner, *Jour. Amer. Chem. Soc.*, 39, 1917, p. 1208.

crystals had accumulated in one part of the liquid and another part had been nearly freed of them, then in the latter part there would be some liquid of composition B in excess of that necessary to react with crystals. The remaining liquid would therefore continue its crystallization with the separation of $K_2SiO_3 . \frac{1}{2}H_2O$ and the temperature would fall to 380° when the liquid would have the composition C. At this temperature the liquid would react with the crystals and would convert a part of

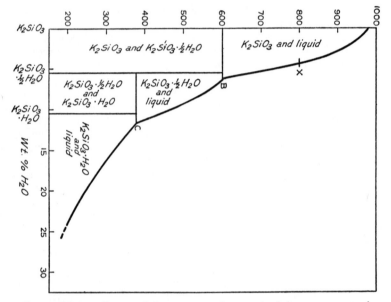

FIG. 8. Equilibrium diagram (after Morey and Fenner) of the system, potassium metasilicate-water (redrawn in wt. per cent).

them to $K_2SiO_3 . H_2O$, being itself used up in the process. But again, if there had been crystal fractionation in the meantime and a part of the liquid had been nearly freed of crystals, this liquid would continue to crystallize with separation of $K_2SiO_3 . H_2O$, a process which would continue at least until a temperature of 200° was reached.

The possibility of such results from fractionation depends entirely upon the incongruent melting of the two compounds. We have seen that a mass which, without fractionation, would be completely crystalline at 600° may, with adequate fractionation, become completely crystalline only locally at that temperature, whereas in other parts it would become completely crystalline at 380° and in yet other parts there would still be some liquid left at 200°. And although the study of the mixtures has not been extended to such low temperatures, there is no question that even at room temperatures there would be some liquid (aqueous solution) left over. The importance of crystal fractionation is therefore very great in such a system.[1] If there is a considerable number of compounds that melt incongruently the extent to which the final liquid may be offset in composition is, to all intents and purposes, limited only by the low-melting component. In this respect such a system approaches the continuous solid solution series without maximum or minimum. Indeed, such a solid solution system might, from the point of view of fractional crystallization, be conveniently regarded as a system with an infinite number of compounds having incongruent melting points. A system with a continuous series of solid solutions will now be considered.

BINARY SYSTEM SHOWING A COMPLETE SERIES OF SOLID SOLUTIONS, WITHOUT MAXIMUM OR MINIMUM MELTING TEMPERATURE

The most familiar example of a solid solution system without maximum or minimum is the plagioclase feldspar series with the end members anorthite ($CaAl_2Si_2O_8$) and albite ($NaAlSi_3O_8$).[2] The equilibrium diagram is shown in Fig. 9. The melting point of anorthite is at 1550° and that of albite at 1100°. They form a complete series of solid solutions whose melting intervals lie at temperatures intermediate between the melting points of the components. A mixture of composition Ab_1An_1 (A) begins to crystallize at 1450°, the first crystals having the composition about Ab_1An_5 (B). With further cooling and attainment of perfect equilibrium both liquid and crystals change their composition, the liquid along the curve ACE and the solid along the curve BDF. At 1400° the liquid has the composition C and the crystals the composition D. When the temperature has fallen to 1285° the crystals have attained the composition F (Ab_1An_1) and the liquid has just disappeared, the last minute quantity having the composition E.

This continuous change in the composition of the crystals with falling temperature involves not only the crystals separating at any instant but also those that had separated at earlier stages. The change in composition of these earlier crystals can be accomplished only by interchange of material with (reaction with) the liquid. If there is not sufficient time for this action the crystals will be layered or zoned, the inner zones being the more calcic. When the temperature has reached 1285° the zone then separating will have the composition F as before, but since the inner zones have remained more calcic the liquid has not been used

1 By heating a mixture of the crystalline hydrates it is impossible to obtain liquid at these low temperatures. The first liquid is obtained only at the temperature of complete crystallization of the perfect equilibrium type, not the fractional crystallization type. As noted before, here again it requires selective fusion oft repeated to give a result comparable with fractional crystallization.
2 N. L. Bowen, *Amer. Jour. Sci.*, 35, 1913, pp. 577-99.

up by reaction with the earlier zones. There will therefore still be some liquid of composition E remaining and this will continue to crystallize with falling temperature and give still more sodic zones.

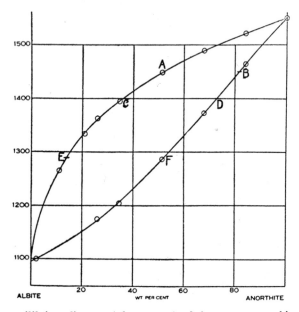

FIG. 9. Equilibrium diagram (after Bowen) of the system, anorthite-albite.

The resultant product is made up of crystals whose average composition is, of course, Ab_1An_1 but whose inner cores are much more calcic and outer rims correspondingly more sodic.

There is, theoretically, a condition under which this zoning may continue until the last deposited minute zone has the composition albite, the last minute quantity of liquid having also the composition of albite. This represents the extreme of fractionation. Less complete fractionation would be expected in any actual example because reaction between liquid and crystals could not fail entirely. It is plain, however, that as a result of fractionation by zoning a continuous offsetting of the composition of the liquid towards albite is accomplished and a great increase in the range of consolidation temperatures is brought about.

The fractionation might be effected in other ways, such as by relative motion of liquid and crystals. In such a case any part of the liquid which was continuously freed of crystals would have its temperature of consolidation correspondingly lowered.

In the melting of crystalline Ab_1An_1 no liquid will form, as the temperature is raised, until 1285° is reached. This is therefore the reverse of crystallization with perfect equilibrium but not the reverse of fractional crystallization. As we found in the case of a system showing a compound with an incongruent melting point, so in the present system it requires selective fusion oft repeated to give a result approaching that obtained in a single course of fractional crystallization.

BINARY SYSTEM SHOWING A COMPLETE SERIES OF SOLID SOLUTIONS WITH A MINIMUM MELTING TEMPERATURE

An example of a series of solid solutions with a minimum is furnished by the system gehlenite ($2CaO . Al_2O_3 . SiO_2$)-akermanite ($2CaO . MgO . 2SiO_2$). The equilibrium diagram is given in Fig. 10.[1] Crystallization

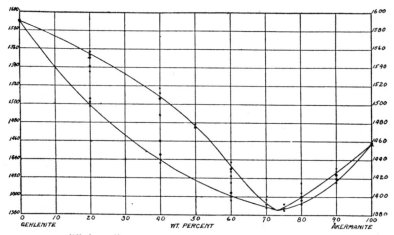

FIG. 10.   Equilibrium diagram (after Ferguson and Buddington) of the system, gehlenite-akermanite.

need not be discussed in detail. Any mixture of the system becomes completely crystalline at a temperature above the minimum if perfect equilibrium (complete reaction) between crystals and liquid obtains. On the other hand, if zoning (or any other kind of fractionation) occurs, any mixture may have its period of crystallization extended. The temperature of final crystallization may approach the minimum and the compo-

1 J. B. Ferguson and A. F. Buddington, *Amer. Jour. Sci.*, 50, 1920, p. 133.

sition of the final liquid may approach that of the minimum-melting mixture. However, since the minimum-melting mixture is of intermediate composition, there is no possibility of getting extreme compositions by fractionation. In this respect the system is like a eutectic system. The difference from a eutectic system lies, however, in the fact that the eutectic is always reached, whereas the minimum of the present system is attained or rather approached only with the aid of crystal fractionation.

BINARY SYSTEM SHOWING A COMPLETE SERIES OF SOLID SOLUTIONS WITH A MAXIMUM MELTING TEMPERATURE

No example of a series of solid solutions with a maximum is known among silicates and the phenomenon need not therefore be considered at length. Suffice it to say that during fractional crystallization the liquid tends to approach either the one pure component or the other, depending on whether the original mixture lay to one side or the other of the maximum.

BINARY SYSTEM WITH LIMITED SOLID SOLUTION SHOWING A EUTECTIC

No simple example of limited solid solution with a eutectic and uncomplicated by other phenomena has been investigated among silicates. The system anorthite $(CaAl_2Si_2O_8)$-nephelite $(Na_2Al_2Si_2O_8)$ furnishes an example, complicated to some extent by inversion, but this rather enhances its value as an illustration of the possible effects of fractional crystallization. The equilibrium diagram is shown in Fig. 11.[1]

A mixture of composition X begins to crystallize at $1415°$ and the liquid changes in composition along the curve AB with separation of crystals of carnegieite, the isometric form of $NaAlSiO_4$. When the temperature $1352°$ is reached the liquid has the composition B and the crystals the composition D. At this temperature liquid and crystals react, the carnegieite crystals being transformed to nephelite crystals of composition C (i.e., much richer in anorthite than the original carnegieite). The liquid B then continues to crystallize with falling temperature and changes in composition along the curve BG. The crystals change in composition along the curve CF and at $1340°$ all the liquid is used up. The mass then consists of homogeneous crystals of nephelite solid solution having the composition X.

On the other hand, if crystal fractionation occurred during the cooling, when the temperature $1352°$ was reached there might be no crystals to react with the liquid in some part of the mass. The liquid B would in these circumstances proceed to crystallize and if there was no further fractionation it would become completely crystalline at $1315°$ and would give a homogeneous solid solution of the composition B. If on the other

[1] N. L. Bowen, *Amer. Jour. Sci.*, 33, 1912, pp. 551-73.

hand there was further fractionation by zoning or otherwise, the liquid would proceed to the eutectic point (G, 1302°) and some crystallization of anorthite would then occur. Thus we have two contrasted results depending on the presence or absence of fractionation. A mixture may give a single, homogeneous nephelite phase (solid solution) in the absence of fractionation. The same mixture may give rise to nephelite solid solutions and excess anorthite under conditions permitting fractionation. Moreover the range of temperature through which liquid occurs is much increased by fractionation.

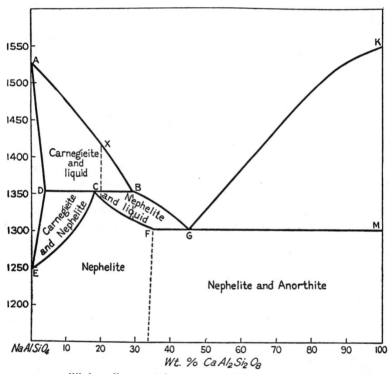

FIG. 11. Equilibrium diagram (after Bowen) of the system, anorthite-nephelite.

It is probable that albite and orthoclase constitute a system of the type just described, i.e., that they show limited solid solution with a eutectic.[1]

1 Strictly speaking this is impossible since orthoclase melts incongruently and the system is really ternary. The probable relation referred to above would, then, be that given by projection out of the ternary system.

BINARY SYSTEM WITH LIMITED SOLID SOLUTION
AND NO EUTECTIC

No example of a system with limited solid solution and without a eutectic has been investigated among silicates. For reasons that will be stated on a later page it may be considered probable that orthoclase[1] and anorthite show this relation and the hypothetical diagram illustrating it will be given at that point.

TERNARY SYSTEM WITHOUT COMPOUNDS OR SOLID SOLUTIONS

The type of ternary system that is commonly regarded as the simplest is that without compounds or solid solutions, in other words, with a single eutectic. No example altogether free from some complication is known, but perhaps the best approach to simplicity is given by the system, $CaSiO_3$ (wollastonite, pseudo-wollastonite)-$CaAl_2Si_2O_8$ (anorthite)-SiO_2.[2] The equilibrium diagram is given in Fig. 12. A liquid of composition X begins to crystallize about $1350°$ with separation of anorthite. The liquid changes in composition along the straight line XB as the temperature falls. When the composition B is attained the temperature is about $1200°$ and pseudo-wollastonite separates together with anorthite. The liquid now changes in composition along the boundary curve towards E, as the temperature falls. When the temperature reaches $1165°$ the composition is that of the eutectic E and silica (tridymite) separates together with anorthite and pseudo-wollastonite. Complete solidification occurs, the temperature remaining constant at $1165°$ and the composition of the liquid constant at E until it is used up.

Any liquid whose total composition is such that it lies in the tridymite field begins to crystallize with separation of tridymite and completes its crystallization at $1165°$, the detailed course being obtained by a similar simple geometrical construction. Likewise any liquid in the pseudo-wollastonite field completes its crystallization at $1165°$ with liquid of composition E.

If there is relative movement of crystals and liquid during the crystallization of any of these mixtures, all parts of the mass will still complete their crystallization at $1165°$ and the final liquid in all parts will have the composition E. Fractional crystallization can give rise to different relative proportions of the three crystalline phases and to no other result.

In the heating of any crystalline mixture some liquid of the composition E will form at $1165°$ and the whole course of fusion is the simple reversal of crystallization.

1 Neglecting the incongruent melting of orthoclase.
2 Rankin and Wright, *Amer. Jour. Sci.*, 39, 1915, p. 25.

TERNARY SYSTEMS WITH A COMPOUND OR COMPOUNDS
HAVING CONGRUENT MELTING POINTS

Every ternary system is made up of three fundamental binary systems and, of these binary systems, any or all may have one or more compounds. Besides these binary compounds there may be ternary com-

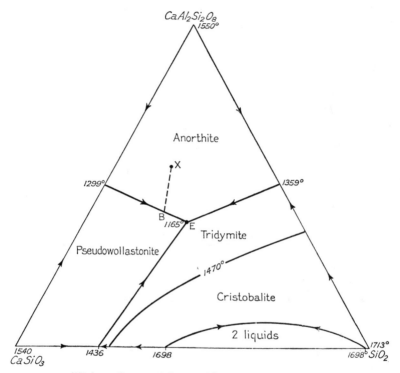

FIG. 12.   Equilibrium diagram (after Rankin and Wright) of the system, anorthite, wollastonite, silica (with additions from the work of Greig, *Amer. Jour. Sci.*, 13, 1927, p. 41).

pounds but, however many compounds there are, so long as all melt congruently no great complexity is ordinarily introduced. In fact such a system is compound rather than complex and merely consists of a number of subsidiary ternary eutectic systems placed side by side. This fact will be apparent from inspection of the several hypothetical diagrams given below in Fig. 13 (a), (b), (c), (d), (e), (f).
    Each subsidiary ternary eutectic system (as partitioned by the dotted

conjugation lines) of the various diagrams given in Fig. 13 may be regarded as a separate system analogous to the system shown in Fig. 12.

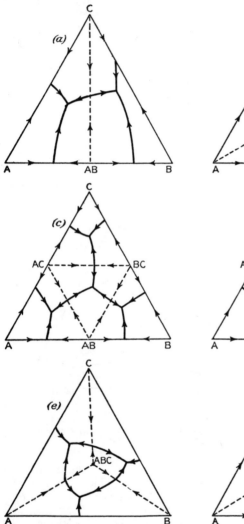
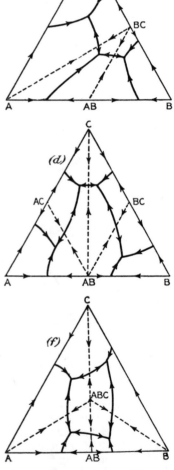

FIG. 13. Hypothetical diagrams of various simple ternary systems with compounds having congruent melting points.

Crystallization always proceeds towards the ternary eutectic of the subsystem in which the original mixture lies, and this eutectic is always attained. By no process of crystal fractionation can any part of the residual liquid acquire a composition which causes it to pass out of the subsystem in which the original mixture lay. Fractionation can not, therefore, bring about a diversity of crystalline products but can only produce a difference in the relative proportions of the same three minerals. Moreover, the temperature of consolidation of the final liquid must be the same in all parts of the mass.

TERNARY SYSTEM WITH A BINARY COMPOUND HAVING AN
INCONGRUENT MELTING POINT

We have already considered the binary system forsterite (Mg_2SiO_4)-silica (SiO_2), with its compound clino enstatite ($MgSiO_3$), having an incongruent melting point. The manner in which such a compound behaves in a ternary system is well illustrated by the system, anorthite ($CaAl_2Si_2O_8$)-forsterite (Mg_2SiO_4)-silica (SiO_2) investigated by Andersen.[1] The equilibrium diagram is given in Fig. 14. There is a small field in which spinel ($MgO \cdot Al_2O_3$) is the primary phase. Mixtures lying in this field and certain other mixtures whose crystallization will carry them into this field can not be treated as ternary, but for the present these need not concern us.

A mixture of composition P begins to crystallize at 1500° with separation of forsterite and the liquid changes along the straight line PQST. At 1375°, when the liquid has the composition T, clino-enstatite begins to separate and forsterite to be redissolved (or to be converted into clino-enstatite through the intervention of the liquid). While this action is going on the composition of the liquid changes along the boundary curve TR. When the temperature has fallen to 1260° the liquid has the composition R and anorthite begins to separate. The temperature then remains constant at 1260°, anorthite and clino-enstatite increasing in amount, liquid and forsterite decreasing in amount, until all the liquid disappears. The completely crystalline mass now consists of forsterite, anorthite and clino-enstatite.

If the original liquid had the composition Q, which lies on the anorthite-clino-enstatite join, the same course would be followed but at 1260° the liquid R and forsterite would be used up simultaneously and the crystalline mass would consist entirely of anorthite and lino-enstatite.

If the original liquid had the composition S, again the same course would be followed but at 1260° some liquid of the composition R would remain after the forsterite was exhausted. The liquid R would then con-

1 Olaf Andersen, *Amer. Jour. Sci.*, 39, 1915, pp. 407-54.

tinue its crystallization, with separation of clino-enstatite and anorthite, and would change in composition along RE. At 1222°, when the liquid has the composition E, tridymite (SiO_2) separates in addition to clino-

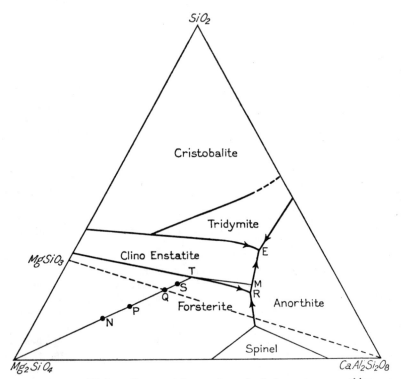

FIG. 14. Equilibrium diagram (after Andersen) of the system, anorthite-forsterite-silica.

enstatite and anorthite, and the temperature of the mass remains constant at 1222° until all the liquid disappears.

If the original mixture had the composition T the liquid would not follow the course TR but would immediately leave that curve and pass into the clino-enstatite field. As the result of crystallization of clino-enstatite the composition of the liquid would pass along the straight line TM. At 1250°, when the liquid has the composition M, anorthite begins to separate and the liquid changes along ME. At 1222° eutectic crystallization of clino-enstatite, anorthite and tridymite occurs as before.

It should be noted now that the liquids Q, S and T, whose crystalliza-tion has just been described, represent successive stages in the composition of the liquid left during the crystallization of the mixture P. If there is no crystal fractionation during the crystallization of a mixture of composition P, then of course all parts of the mass must at all times have the total composition P. On the other hand, if there is crystal fractiona-tion during the separation of forsterite, then a certain part of the mass might have the composition Q, another part the composition S and yet another part the composition T, representing increasing impoverishment in forsterite crystals. Complementary to these parts there would be others enriched in forsterite and having a composition such as N. The crystalli-zation of N is analogous to that of P except that there is a greater pro-portion of fosterite in the final product. The crystallization of Q, S and T has already been described. As a result of crystal fractionation, then, we may have different parts behaving in the contrasted manner already indi-cated in the description of the different behavior of the liquids P, Q, S and T. In short, those parts enriched in forsterite would be completely consolidated at 1260° (R) and would give forsterite, clino-enstatite and anorthite. Those parts sufficiently impoverished in forsterite would not be completely consolidated until a temperature of 1222° (E) was reached and would give clino-enstatite, anorthite and free silica. And of those parts which behaved in the latter manner, i.e., reached the eutectic, some would have followed a different course from that fol-lowed by others.

This flexibility introduced by fractional crystallization depends en-tirely on the fact that forsterite (except at the higher temperatures) bears a reaction relation to the liquid, that is, the liquid reacts with it to convert it to clino-enstatite. Crystal fractionation with respect to anorthite or clino-enstatite will not have an effect similar to that described for forsterite because anorthite and clino-enstatite do not, in this system, have a reaction relation to the liquid. They have what may be termed a purely subtraction relation. Once subtracted, the liquid is no longer concerned with them and is unaffected by their presence or absence.

It may be pointed out that any method of prevention of reaction be-sides actual relative movement of liquid and crystals will produce a like result. Thus if clino-enstatite forms as a layer, zone or corona about forsterite crystals, their reaction with liquid is prevented and the liquid will follow a course characteristic of the absence of forsterite and will pass on to the composition of the eutectic where crystallization of free silica (tridymite) occurs. A localized fractionation is thus possible and, in any small area within a mass, there may be forsterite crystals with a rim of clino-enstatite and a matrix of anorthite, clino-enstatite and tridymite.

Without discussing it in detail we may point out that in this system, as in a binary system with an incongruent-melting compound, the effect produced by heating is not the reverse of fractional crystallization. It requires repeated selective fusion to give an effect comparable with that of fractional crystallization.

TERNARY SYSTEM HAVING A TERNARY COMPOUND WITH AN INCONGRUENT MELTING POINT

An example of a ternary compound with an incongruent melting point is furnished by the system MgO-Al_2O_3-SiO_2, the compound being cordierite $2MgO \cdot 2Al_2O_3 \cdot 5SiO_2$. The equilibrium diagram is shown in Fig. 15.[1] It is an excellent diagram to illustrate the great prevalence of incongruent melting in silicates. There are four binary compounds and two of them, namely, mullite and clino-enstatite, melt incongruently. There is but one ternary compound, namely, cordierite, and it melts incongruently. As a consequence of these facts the relation of liquid to crystals characterized by reaction between them is exceedingly common during the crystallization of mixtures in this system. We shall not be concerned here with this relation in connection with the binary compounds, since that matter has just been discussed. We shall point out briefly, however, the course of crystallization in liquids exhibiting this relation principally in connection with the ternary compound.

A mixture of composition P begins to crystallize at 1600° with separation of mullite ($3Al_2O_3$ $2SiO_2$). At 1465° when the liquid has the composition Q, cordierite begins to crystallize and mullite to react with the liquid to produce more cordierite, the composition of the liquid changing along the curve QR. At 1425° when the liquid has the composition R tridymite (SiO_2) begins to separate and the temperature remains constant at 1425°, tridymite and cordierite increasing in quantity, liquid and mullite decreasing until all the liquid disappears. The crystalline product then consists of mullite, cordierite and tridymite.

However, crystal fractionation might occur, i.e., mullite might be protected from reaction with liquid either by its removal from a certain part of the mass or by the formation of a rim of cordierite about it. In these circumstances the liquid Q would not change along the boundary curve QR but would enter the cordierite field and change along QT with simple separation of cordierite. At 1420° tridymite would begin to separate and the liquid would change along TE, tridymite and cordierite separating. At 1345°, when the liquid has the composition E, clino-enstatite crystallizes in addition and the temperature remains constant until liquid disappears.

Thus as a result of fractionation a partly different crystalline pro-

1 G. A. Rankin and H. E. Merwin, *Amer. Jour. Sci.*, 45, 1918, with corrections after the work of Bowen and Greig, *Jour. Amer. Ceram. Soc.*, 7, 1924, p. 242, and Greig, *Amer. Jour. Sci.*, 13, 1927, p. 38.

duct has been obtained and the temperature at which final consolidation occurs has been considerably lowered. Again the fundamental controlling factor is the incongruent melting of a compound, in this case the ternary compound cordierite.

TERNARY SYSTEM WITH A BINARY SERIES OF SOLID SOLUTIONS

The simplest example, among silicates, of a ternary system with a binary series of solid solutions is furnished by diopside (CaMgSi$_2$O$_6$),

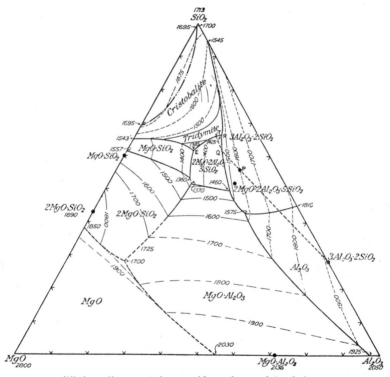

FIG. 15.   Equilibrium diagram (after Rankin and Merwin) of the system, MgO-Al$_2$O$_3$-SiO$_2$ with additions and corrections from the work of Bowen and Greig, *Jour. Amer. Ceram. Soc.*, 7, 1924, p. 242, and Greig, *Amer. Jour. Sci.*, 13, 1927, p. 41.

anorthite (CaAl$_2$Si$_2$O$_8$), albite (NaAlSi$_3$O$_8$).[1] The equilibrium diagram is shown in Fig. 16. There is but one boundary curve and it separates the field of the plagioclases from the field of diopside. Just as the tempera-

[1] N. L. Bowen, *Amer. Jour. Sci.*, 40, 1915, pp. 161-85.

ture falls continuously from the melting point of anorthite to that of albite in the binary system so there is a similar fall along the boundary curve from the anorthite-diopside eutectic to the albite-diopside eutectic, which is at practically pure albite.

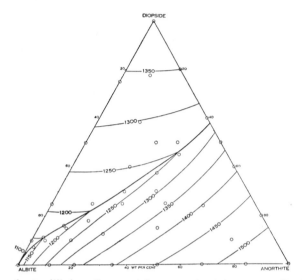

FIG. 16.   Equilibrium diagram (after BOWEN) of the system, diopside-anorthite-albite.

We may discuss the crystallization of mixtures lying in the plagioclase field with the aid of Fig. 17, which is to be regarded as representing Fig. 16 with some of the lines removed to avoid confusion. The mixture, Ab_1An_1 85 per cent-diopside 15 per cent (D, Fig. 17), begins to crystallize at 1375° with the separation of plagioclase of composition Ab_1An_4. As the temperature falls the plagioclase increases in amount and changes in composition until at 1300° the liquid has the composition P and plagioclase the composition Ab_1An_3. When the temperature has fallen to 1216° diopside begins to crystallize, the liquid then having the composition M and plagioclase the composition S (Ab_1An_2). With further lowering of temperature the liquid follows the boundary curve, both diopside and plagioclase crystallizing, and at 1200° the liquid is all used up, the last minute quantity of liquid having the composition H. The composition of the plagioclase is now Ab_1An_1 (F).

In the case of the liquid E ($Ab_{18}An_{82}$ 90 per cent-diopside 10 per cent) crystallization begins at 1480° with the separation of Ab_5An_{95} and the composition of the liquid follows the curve ERN. At 1245° diopside begins to crystallize. The composition of the plagioclase is now T $(Ab_{15}An_{85})$. As the temperature is lowered both plagioclase and diopside

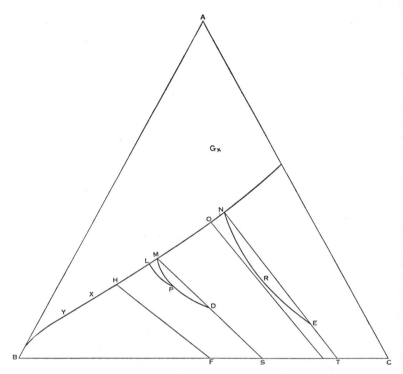

FIG. 17.   Diagram (after Bowen) illustrating the crystallization of mixtures in the plagioclase field. A = diopside; B = albite; C = anorthite.

crystallize until at 1237° all the liquid is used up. The composition of the final liquid is O and the feldspar has attained the composition $Ab_{18}An_{82}$.

It should be noted that the crystallization curves DPM and ERN apply to the liquids D and E respectively and to no other liquids. Thus the crystallization curve of the liquid P is not the curve PM but the new curve PL, that is, if we start with a liquid P free from crystals the composition of the liquid follows the course PL. Only when the liquid P contains in it the crystals formed during the change from D to P does the further course of the liquid coincide with PM. Moreover, the liquid

P, when originally free from crystals, becomes completely crystalline, not at 1200° (H), as before, but at a somewhat lower temperature.

Throughout the foregoing discussion of crystallization perfect equilibrium is assumed. The conditions are supposed to be such that crystals of plagioclase can change their composition through and through in response to the demands of equilibrium. It may be considered, however, that crystallization takes place in a quite different manner. When plagioclase of a certain composition has separated it may remain as such and become surrounded by layers of different composition deposited by the continually changing liquid. The liquid is in equilibrium at any instant only with the material crystallizing at that instant and not with crystals already formed. A plagioclase crystal, once separated, does not participate further in the equilibria. As far as any effect on the course followed by the liquid is concerned, the crystal may be considered absent. The course of crystallization of the liquid P, Fig. 17, in the absence of crystals, has been compared in the foregoing with that followed in the presence of crystals. If we examine also the liquid M, say, we find that if it crystallizes in the presence of crystals formed during the change in composition of the liquid from D to M, it then becomes completely crystalline at 1200° and the final liquid has the composition H. On the other hand, if the crystals referred to are separated from the liquid M, complete crystallization does not take place until the temperature 1170° is attained and the final liquid has the composition X, i.e., is very much richer in albite. If in this latter case a second removal of crystals took place when the liquid had the composition H, complete crystallization would not then take place until the temperature had fallen to 1125° and the final liquid has then the composition Y, exceedingly rich in albite.

If this separation of early crystals from liquid is a continuous process accomplished through zoning of the crystals, it is clear that the continual lowering of the temperature of final consolidation and offsetting in the composition of the liquid is limited only by the eutectic albite-diopside, 1085° and 97 per cent albite.

Thus, as we have noted for a compound with an incongruent melting point, so a solid solution series in a ternary system introduces a certain amount of flexibility of the course of crystallization, a possible variation in the temperature of final consolidation and in the composition of the final liquid.

Remelting of a crystalline mixture in this system does not begin at the very low temperature where there may still be some liquid if the same mixture were cooled from the liquid state. It requires selective fusion oft repeated to give an effect approaching that of fractional crystallization.

TERNARY SYSTEM WITH A SERIES OF BINARY SOLID SOLUTIONS THAT MELT INCONGRUENTLY

We have already considered the effect of the incongruent melting of clino-enstatite ($MgSiO_3$) in the binary system and in a ternary system without solid solution. If clino-enstatite forms solid solutions with another compound then at least some of these solid solutions (those

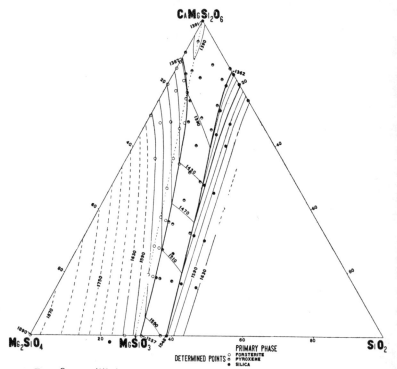

FIG. 18.   Equilibrium diagram (after Bowen) of the system, diopside-forsterite-silica.

close to clino-enstatite) must behave like clino-enstatite, i.e., must dissociate on melting, with formation of forsterite. It is true that the first liquid formed in the melting of any binary series of solid solutions has a different composition from the crystal phase but its composition lies in the binary system. In the case of solid solutions adjacent to clino-enstatite the liquid is not only different in composition from the crystal phase but, as a result of the separation of forsterite crystals, the liquid does not lie in the binary system. The melting and crystallization of such a series of solid solutions is illustrated by the system: forsterite (Mg_2SiO_4)-diopside $(CaMgSi_2O_6)$-silica (SiO_2).[1] The equilibrium diagram is given in Fig. 18. A skeleton diagram showing the boundary curves and the phases separating in the various fields is given in Fig. 19.

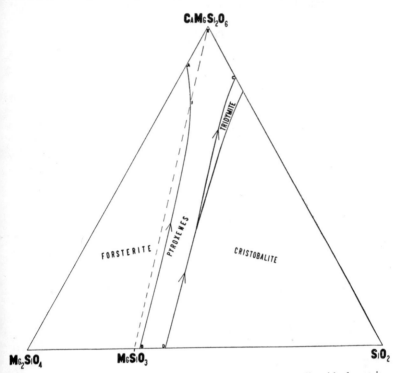

FIG. 19. Equilibrium diagram (after Bowen) of the system, diopside-forsterite-silica showing only the fields in which the various crystals appear as primary phases.

The course of crystallization of important mixtures may be discussed with the aid of Fig. 20 which is essentially the same as Figs. 18 and 19 except that slight distortion is introduced in order to avoid confusion of lines.

If a liquid of composition M (Fig. 20) is allowed to cool, forsterite begins to crystallize at the temperature of the isotherm through the

1 N. L. Bowen, *Amer. Jour. Sci.*, 38, 1914, pp. 207-64.

point M and continues to crystallize until the temperature of the point K on the boundary curve is reached. In the meantime the composition of the liquid has changed from M to K along the straight line AMK. Since liquid K is saturated with pyroxene that phase then begins to crys-

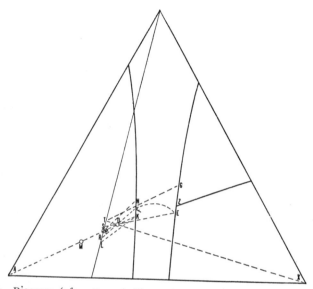

FIG. 20. Diagram (after Bowen) illustrating courses of crystallization in the system, diopside-forsterite-silica.

tallize and L represents the composition of the pyroxene. When the temperature is further lowered, pyroxene continues to crystallize and forsterite begins to redissolve or react with liquid to produce an increased quantity of pyroxene. The composition of the liquid now changes along the boundary curve and the composition of the pyroxene in the act of crystallization, as well as that of the pyroxene which has already separated (if there is perfect equilibrium), changes toward S. When the temperature of the point N is reached, the liquid finally disappears and the whole consists of pyroxene of composition P and forsterite. The proportion of pyroxene to forsterite is as MA:MP.

If the composition of the original liquid had been that of the pyroxene P, forsterite would crystallize first as before, and the whole course of crystallization would be precisely as given above. At the temperature of the point N, in this case, the last of the liquid and the last of the forsterite are used up simultaneously and the whole consists simply of pyroxene of composition P.

If the composition of the original liquid was that of the point D, forsterite would crystallize first as before, and crystallization would follow the same course as in the two preceding cases until the temperature of the point F is reached. At this temperature the last of the forsterite has dissolved and the mixture consists of liquid of composition F and pyroxene of composition R. The composition of the liquid now leaves the boundary curve and crosses the pyroxene field on the curve FE, pyroxene continuing to crystallize and changing in composition towards S. When the temperature of the point E is reached cristobalite begins to crystallize. At this temperature the liquid has the composition E and pyroxene the composition S. With further lowering of temperature the composition of the liquid changes along the boundary curve from E towards Z, cristobalite and pyroxene continue to crystallize and the pyroxene changes in composition towards T. At the temperature of the point Z cristobalite changes to tridymite, if perfect equilibrium is attained, and with further lowering of temperature tridymite and pyroxene continue to crystallize. When the temperature of the point G is reached the liquid finally disappears; the last minute quantity has the composition G, and the pyroxene has the composition T. The whole now consists of pyroxene of composition T and tridymite in the proportion, pyroxene:tridymite = DX:DT.

The crystallization of any mixture of a system such as the present may take place in a manner somewhat different from that outlined for the case of perfect equilibrium. When a crystal has separated it may never redissolve and when a mix-crystal separates its composition may not thereafter suffer any change. The crystals which are separating at any instant are at equilibrium with the liquid but those that have already separated may not be.

When crystallization takes place in this manner a liquid of composition M, P or D of Fig. 20 would behave as follows: Forsterite would crystallize out first and the composition of the liquid would change to K. At the temperature of the point K pyroxene of compositition L would begin to separate. In the case of complete equilibrium the re-solution of forsterite would now begin and the liquid would change along the boundary curve. In the case we are now considering, liquid and forsterite do not interact and the composition of the liquid crosses the pyroxene field and meets the boundary curve, pyroxene-tridymite, at a point lower than E, say G.[1] In the meantime the composition of the pyroxene separating has changed from L to T and there exist in the

---

1 The exact curve which indicates the change in the composition of the liquid is, unlike FE, such that the tangent to it at any point passes through the composition of the pyroxene in equilibrium with liquid of the composition and at the temperature represented by the point.

mixture zoned pyroxene crystals of all compositions varying from L to T. When the temperature is that of the point G and the composition of liquid is G, tridymite begins to crystallize and the composition of the liquid changes along the boundary curve, pyroxene-tridymite. Meantime the composition of the pyroxene separating changes from T towards pure diopside and final crystallization takes place only when the temperature is that of the eutectic, diopside-tridymite, when the remaining infinitesimal amount of liquid has the composition of this eutectic and the crystalline phases separating are tridymite and pure diopside.

It will be recalled that we started with any of the mixtures M, P or D. Any of these mixtures would, then, if crystallized in this manner, consist of forsterite, tridymite, and pyroxene varying in composition from L to pure diopside. The actual amount of pyroxene approaching L in composition would be relatively large; the amount approaching diopside, relatively very small; the amount of pure diopside infinitesimal.

This latter method of crystallization is, of course, a form of fractional crystallization. Any other method of crystal fractionation, such as relative movement of liquid and crystals, would have a similar effect on the temperature of final consolidation and the nature of the late crystalline phases in those parts of the mass impoverished in early crystals. However rich in forsterite the original may have been, there could be a final fraction from which free silica crystallizes and which consolidates at temperatures far below those at which forsterite-rich fractions crystallize.

CHAPTER V

THE REACTION PRINCIPLE

IT WILL be apparent from the discussion of crystallization in typical silicate systems that a relation of liquid to crystals, characterized by reaction between them, is exceedingly common during the normal course of crystallization. In the crystallization of any solid solution series such reaction between liquid and crystals must occur in greater or less amount, whether the solid solution series be one of the continuous types or the discontinuous types. Again, in the crystallization of any system exhibiting a compound with an incongruent melting point, reaction between liquid and crystals is a constant factor. We have seen that crystallization where reaction is involved is very different from crystallization in purely eutectic systems. In the eutectic systems crystals bear a simple subtraction relation to the liquid, and once subtracted they no longer take any part in the equilibrium. We have seen, too, that the reaction relation is of the utmost importance in connection with the problem of fractional crystallization in silicate systems; of such importance, indeed, that it has elsewhere been proposed to regard it as a fundamental principle.[1] It has been found convenient for the discussion of this principle to introduce certain terms whose significance will now be illustrated.

During the crystallization of the plagioclase feldspars there is, as shown in Fig. 9 and the discussion thereof, continual reaction between liquid and crystals. The change of composition of the crystals is a perfectly continuous one taking place by infinitesimal increments. Such a solid solution series may therefore be termed a *continuous reaction series*. This property of exhibiting a perfectly continuous variation of composition during crystallization is, however, just as definitely characteristic of each series of solid solutions in a system which shows two series with an hiatus. The fact is brought out by Fig. 11 and the discussion thereof, where it is shown that during crystallization the nephelite mix-crystals vary continuously in composition within certain limits. Within these limits, then, the nephelite solid solutions constitute a continuous reaction series. Thus any solid solution series, whether complete or limited, is a continuous reaction series.

---

[1] N. L. Bowen, "The Reaction Principle in Petrogenesis," *Jour. Geol.*, 30, 1922, pp. 177-98.

The essential feature of a continuous reaction series, the reaction relation of crystals and liquid, is retained when the series becomes a part of a more complex system. This is true even when the end members of the series bear a eutectic relation to the newly added component, as is well shown when diopside is added to plagioclase. The system has been discussed and the reaction relation emphasized in the explanation of Fig. 16.

In addition to the kind of reaction relation introduced by the existence of solid solution, another kind, of somewhat different character, has been described in connection with some of the foregoing systems. This is the kind introduced by the presence of a compound having an incongruent melting point. In the binary system $MgO-SiO_2$ the compound clino-enstatite $(MgSiO_3)$ has an incongruent melting point (Fig. 7). As a result of this fact certain mixtures precipitate forsterite (Mg_2SiO_4) first and this reacts with the liquid, in whole or in part, to produce clino-enstatite $(MgSiO_3)$. Forsterite and clino-enstatite may therefore be styled a *reaction pair*. By this is meant that crystals of the first compound react with the liquid to produce the second during the normal course of crystallization.

A reaction relation of this latter type may exist between three or more compounds and the compounds, arranged in proper order, may then be said to constitute a *discontinuous reaction series*. We have seen an example of this in the system $K_2SiO_3-H_2O$ illustrated in Fig. 8. Here the compound K_2SiO_3 reacts with liquid to produce $K_2SiO_3 . \frac{1}{2}H_2O$, and this in turn reacts with liquid to produce $K_2SiO_3 . H_2O$. The compounds K_2SiO_3, $K_2SiO_3 . \frac{1}{2}H_2O$ and $K_2SiO_3 . H_2O$ therefore constitute a discontinuous reaction series. It is plain that the distinction between the two types of reaction series lies in the fact that every gradation of composition is exhibited in the continuous series, whereas the change of composition in discontinuous series is by definite steps.

Just as the essential features of the continuous reaction series are retained when the series becomes part of a more complex system, so is this true of the reaction pair and the discontinuous reaction series. In the discussion of Fig. 14 it has been brought out how the reaction relation between forsterite and clino-enstatite is carried into a ternary system.

Another manner in which this reaction relation may be carried into a ternary system is illustrated in Figs. 18, 19 and 20 of the system: diopside-forsterite-silica. By the introduction of the component diopside, which forms a series of solid solutions with clino-enstatite, the latter compound appears in this system as a member of a continuous reaction series, while retaining its character as a member of the reaction pair, forsterite-clino-enstatite. Thus the pyroxenes of this system are at the same time members of both these reaction series. We have seen in the discussion of crystallization of mixtures of the system that both types of reaction are involved. The reaction of forsterite with liquid to give pyroxene and the reaction of more magnesian pyroxene with liquid to give less magnesian pyroxene go on at one and the same time.

The importance of reaction series, whether continuous or discontinuous, lies in the fact that a certain amount of flexibility in the crystallization of a mixture is introduced. The greater the reaction the sooner the liquid will be used up by reaction and the more the crystalline products will be confined to the earlier (higher temperature) members of the reaction series. The less the reaction the greater will be the temperature interval during which liquid occurs and the more the crystalline products will reveal the later (lower temperature) members of the reaction series in addition to the earlier members. In eutectic systems no such flexibility is possible. The change of composition of the liquid is always of the same magnitude for the eutectic is always attained and the crystalline products are always the same.

The flexibility introduced by the reaction series is not confined to a variability in the extent to which a certain goal will be approached by the liquid. In a binary system this is, to be sure, the only effect, but in a ternary system possible variety in the path followed in approaching this goal is introduced. This fact is illustrated by the alternative paths followed, while only plagioclase is crystallizing, in the system diopside-anorthite-albite and by similar effects in all the other ternary systems discussed, except, of course, the purely eutectic systems. In systems with a greater number of components a still greater flexibility in the course followed by the liquid will be introduced, provided, of course, that reaction series occur among the crystalline phases.

On account of general familiarity with eutectic systems there is an unconscious tendency to apply the principles pertaining to such systems to any kind of complex system. In an important particular the existence of the reaction series, continuous and discontinuous, causes the process of crystallization to depart from that obtaining in the eutectic system. In the crystallization of the plagioclase feldspars, when accomplished by simple cooling, a plagioclase always separates before any other plagioclase that is less calcic. There is no such thing as the separation of calcic plagioclase first from mixtures rich in calcic plagioclase, and of sodic plagioclase first (followed by calcic plagioclase) from mixtures rich in sodic plagioclase, as there would be in the eutectic system. And so with the reaction pair and the discontinuous reaction series the higher member of the series always separates before the lower, if at all. We do not have a condition in mixtures of forsterite and clino-enstatite such that forsterite separates first from mixtures rich in forsterite and clino-enstatite first in mixtures rich in clino-enstatite, as there would be in eutectic mixtures. On the contrary, forsterite, however small in amount, always separates first. And not only does forsterite begin to crystallize early, but it also ceases to crystallize early, an impossible condition in any eutectic system.

Thus the existence of reaction series tends to introduce a fixity in the order of crystallization, calcic plagioclase before sodic plagioclase, if at all; forsterite (olivine) before clino-enstatite (pyroxene), if at all; K_2SiO_3 before K_2SiO_2 . $\frac{1}{2}H_2O$ before K_2SiO_3 . H_2O.

A sufficient number of examples of the reaction relation have been given to illustrate the more important aspects of it. Moreover the examples have in most cases dealt with members of common, rock-forming groups and the prevalence of reaction series of one kind or the other among the rock-forming silicates is indicated by these few examples. The data are not at hand—and are not likely to be for some time—for a quantitative discussion of reaction series in mixtures corresponding to natural magmas. Nevertheless it is believed that much is to be gained from a qualitative consideration of this feature of rock-minerals.

It should be frankly stated that the existence of the reaction relation between two phases in a simple system is no guarantee of the persistence of an identical relation between them in a more complex system. In the case of the phases olivine and magnesian pyroxene, for which such a relation exists in the binary system MgO-SiO_2, it is not impossible that the relation might be modified in more complex systems. Actually, however, it is found that the relation persists in all the more complex systems examined, which fact renders it more likely, but by no means certain, that the reaction relation obtains in magmatic systems. The service rendered by experimental investigation, so long as it is confined to a limited number of components, must lie in its indicating where a reaction relation is to be expected. We are, moreover, instructed as to what we may expect in the way of indications of reaction and thus enabled to extend our inferences to phases not formed under laboratory conditions. This brings us to the question of the criteria of the reaction relation.

A criterion of the reaction series, common to both the continuous and discontinuous type, and serving to show their fundamental likeness, is simply the tendency of one mineral to grow around another as nucleus. In the case of the continuous series this is commonly known as zoning of mix-crystals and, in the discontinuous series, as the formation of reaction rims, coronas, etc. Thus we have plain evidence of this kind, from a wide range of rocks, that the plagioclases constitute a continuous reaction series and that pyroxene, amphibole, and mica form a discontinuous series.

Fortunately the continuous reaction series are easily picked out, for the mere existence of solid solution or variability of composition in a crystal phase is sufficient to establish that phase as a continuous reaction series. Their number is legion, all the important igneous rock minerals with the single exception of quartz being members of solid solution series. The detection of the discontinuous reaction series is not always so easy, and the element of judgment enters to some extent.

It has been stated above that pyroxene, amphibole and mica constitute a discontinuous reaction series, the conclusion being based on the fact that in certain rocks they show the corona relation. It is not necessarily to be inferred, however, that they do not have the reaction relation in rocks where the corona relation is not displayed. The development of one mineral of a discontinuous reaction series as a corona about another does not depend entirely upon the existence of the reaction relation. It appears to require comparatively rapid cooling in addition, though it is perhaps not the rapid cooling itself but conditions that attend it that are the fundamental control. In some uncompleted experiments with chemical salts having a reaction relation it has been found that under perfectly quiet conditions the later member of the reaction pair will precipitate as a layer about the earlier member, and reaction is shortly prevented. On the other hand if the solution is stirred the later member grows as separate crystals and one can see under the microscope these crystals growing larger while the crystals of the earlier member grow smaller (pass into solution). The cause of the difference between stirred and unstirred masses is no doubt the fact that stirring brings in small crystals of the later member from the colder marginal parts of the solution where crystallization is farther advanced and these crystals act as nuclei for the growth of crystals of the later member. In the absence of these nuclei the later member precipitates rapidly from the layer of solution immediately surrounding (and saturated with) the earlier member, which layer is supersaturated with the later member.

Applying these observations to crystallization in silicates we find that the likelihood of corona formation should increase with rapid cooling, for under such conditions there is less opportunity for nuclei of a later member of a reaction series to be brought in (by convection currents, gravity or otherwise) from adjacent parts of the crystallizing mass. Moreover, even in the presence of such nuclei, rapid cooling may not allow sufficient time for the diffusion of material dissolving from crystals of the earlier member and its deposition upon nearby crystals of the later member. *Per contra*, slow cooling tends to promote both these factors and therefore to give separate crystals of the later member rather than zones of the later member about the earlier which should, however, be rounded during this slow cooling. We thus see that the corona relation is not to be expected universally even with a pair of minerals which always have a reaction relation to each other.

There is nothing in the existence of a reaction relation between two crystal species to prevent the occurrence of both as perfectly formed crystals with no evidence of a resorption of the earlier member of the reaction pair. A liquid of the system, anorthite-forsterite-silica, such as

P, Fig. 14, will precipitate forsterite in such amount that the remaining liquid will attain a composition on the boundary curve olivine-pyroxene. The liquid is now saturated with pyroxene and with adequate opportunity for reaction it will precipitate pyroxene and dissolve olivine, thus suffering a change of composition represented by motion along the boundary curve. But under certain conditions of rapid cooling the liquid may ignore the olivine crystals and pass immediately into the pyroxene field with formation of crystals of that phase. That the corona relation should be found fairly commonly is sufficient to establish a reaction relation between the two minerals concerned.

The existence of reaction series may sometimes be inferred from a general survey of an igneous rock sequence even when there is no other evidence of it, though usually there is confirmatory evidence furnished by coronas. To illustrate the kind of general survey referred to we reproduce a table given by Harker showing the mineral relations in the sequence at Garabal Hill.[1]

TABLE I

MINERAL AND ROCK SEQUENCE AT GARABAL HILL (*after Harker*)

	1. Iron Ore	2. Olivine	3. Augite (diallage)	4. Brown Hornblende	5. Green Hornblende	6. Biotite	7. Plagioclase	8. Orthoclase	9. Quartz
A. Olivine-diallage rock	+	+	+	+	—	—	—	—	—
B. Biotite-diorite	+	—	—	—	+	+	+	—	—
C. Hornblende-biotite-granite	+	—	—	—	+	+	+	+	+
D. Porphyritic biotite-granite	+	—	—	—	—	+	+	+	+
E. Eurite vein	+	—	—	—	—	—	+	+	+

It will be noted that the minerals *appear* in a certain order, as they might in a system where simple eutectic relations prevailed, but they also *disappear* in a similar order, a feature that is altogether foreign to a eutectic system. In a eutectic system no mineral ever disappears.[2] The first-formed mineral is simply joined by another, the pair by a third, and so on until all the minerals appear together in a final eutectic product. Very different from this is the condition actually found, namely, the

1 *The Natural History of Igneous Rocks*, 1909, p. 131.
2 A case of simple inversion without change of composition would require to be excluded from this statement, but it has little importance in rocks.

disappearance of minerals in the order in which they appear which is of the very essence of the reaction series.

Upon examination in detail it is found that 2, 3, and 4 disappear in B, 5 disappears in D, and 6 in E. From this we conclude that these phases bear a reaction (not a mere subtraction) relation to the liquid and that, as a result of the reactions, phases appearing later are formed. We arrive at the definite conclusion that 4, 5, and 6 constitute a reaction series and at the same time note indications that they are but a part of a series containing more members.

TABLE II

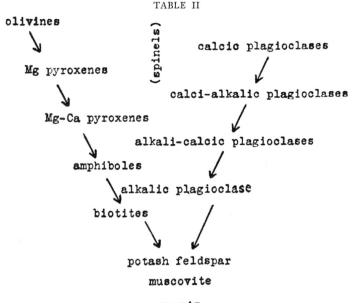

By piecing together the information to be obtained from the examination of such sequences and from observation of the structural relations of the minerals, a conclusion as to the reaction series in rocks is to be arrived at. Without going into further detail as to the evidence, an attempt is made below to arrange the minerals of the ordinary subalkaline rocks as reaction series. The matter is really too complex to be presented in such simple form. Nevertheless the simplicity, while somewhat misleading, may prove of service in presenting the subject in concrete form.

Beginning at the upper end of the series in the more basic mixtures we have at first two distinct reaction series, the continuous series of the plagioclases and the discontinuous series, olivines-pyroxenes-amphiboles, etc. As we descend in these series, however, they become less distinct, in the aluminous pyroxenes and amphiboles a certain amount of interlocking begins and they finally merge into a single series.[1] This is expressed diagrammatically by a convergence of the series, with a dovetailing of the mineral names at first, and finally a joining of the two series by the arrows converging upon potash feldspar. Just where the two series merge completely is more or less a question, but it is given closely enough for our present purpose in the figure.

It is to be noted that the minerals given at the end of the series after their convergence are on a somewhat different basis from the others. It can not, of course, be said that any liquid reacts with biotite and alkalic plagioclase to give potash feldspar. On the contrary these last minerals (principally potash feldspar and quartz) are those that form from the liquid, if any, which is left over at this final stage and are thus the result of the failure of this liquid to be used up in the reactions which produce biotite and alkalic plagioclase carrying some potash. The manner in which reaction controls the presence or absence of potash feldspar as such is more fully discussed on later pages, especially in connection with Fig. 28.

That the series, olivines-pyroxenes-amphiboles-biotites, constitute a reaction series is well attested in many rock varieties. By this is meant that liquid reacts with olivines to produce pyroxenes, with pyroxenes to produce amphiboles, and with amphiboles to produce biotites. In the increasing water content of the series it is related to the series K_2SiO_3-K_2SiO_3 . $\frac{1}{2}H_2O$-K_2SiO_3 . H_2O. The impression seems to have been gained by some petrologists that the postulated reaction relation between, say, pyroxene and hornblende carries with it the implication that all hornblende is secondary after pyroxene. Nothing is farther from the fact. Just as there are many liquids of the system, anorthite-forsterite-silica that precipitate pyroxene as a primary phase without any previous separation of olivine so many magmas may precipitate hornblende directly without any previous precipitation of pyroxene from that particular magma. To be sure it is part of the general thesis of this study that magmas which precipitate hornblende are, in general, derived from magmas that have precipitated pyroxene but that is a different matter to be discussed later.

The continuous reaction series of the plagioclases is perhaps the best-understood series of rock minerals. This is fortunate, for the series

[1] The two series are bridged at the very outset by spinel but this has, on the whole, no great practical importance.

happens to be of particular importance in that it runs through a wide range of conditions and compositions in rock series. We simply have a continual enrichment of the liquid in alkaline feldspar, with the separation of the potash variety of alkaline feldspar as a separate phase when it has exceeded its solubility in the plagioclase mixture. With the formation of potash feldspar in the one series and of biotite in the other, the two series are now so intimately intermingled as to constitute a single series.

There is little of the nature of eutectic crystallization in the crystallization series given in the foregoing. At early stages, and as between the two series, there is some suggestion of the eutectic relation in that a member of one series lowers the melting "point" of a member of the other series. Moreover, the one or the other begins to separate first according to whether the one or the other is present in excess over certain fixed proportions. There the analogy with eutectic crystallization ends for the simple reason that there is no eutectic, no inevitable end-point where final solidification must take place when the liquid has attained a certain composition. The minerals have a reaction relation to the liquid, not a mere subtraction relation. Each separated mineral tends always to change into a later member of the reaction series. This change of composition is effected by reaction with the liquid, and according to the opportunity for reaction the liquid is entirely used up, in some cases sooner, in others later, and only then is solidification complete.

Thus we see that rock series can not be partitioned off into such divisions as gabbro, diorite, etc., each having a eutectic of its own. All of these belong to a single crystallization series, to a single polycomponent system, which is dominated by reaction series.

CHAPTER VI

THE FRACTIONAL CRYSTALLIZATION OF
BASALTIC MAGMA

GENERAL CONSIDERATIONS

WE SHALL now endeavor to discuss the fractional crystalliza-tion of basaltic magma. By way of anticipation it may be stated now that we shall first develop the thesis that normal subalkaline series of rocks could be developed from basaltic magma by fractional crystallization. In doing so, much use will be made of the data of investigated systems, but since the investigated systems are always much simpler than magmas, it is not possible to use directly the actual quantitative values of concentrations, temperatures, etc., of the experi-mental results. The principal service of these must be rather to point the road. We find in a certain investigated system a definite course of crystallization and definite possibilities of differentiation through frac-tional crystallization. We turn then to an actual magma as near as may be to some liquid of the investigated system and ask ourselves whether the crystallization of such a magma presents any parallelism with that of the investigated liquid. To answer this question we may consider what indications there may be in physico-chemical theory as to the expected extent of departure from the simple system. We then turn to the evidence of the course of crystallization of the magma, as determined from rocks, and see to what extent there is parallelism with the simple liquid and to what extent the departure is of the expected kind. If the correspondence with our expectations appears to be sufficiently good we may then proceed to deduce the results of *fractional* crystallization of the natural magma, using the evidence from the simple liquid with such modifications as may be appropriate in the light of physico-chemical theory. The deduced course of fractionation is then to be checked against actual rock series and if the rocks are of the anticipated kind, there is considerable like-lihood that they have been formed by fractional crystallization. There is a little in this of the nature of circuitous reasoning but it is the kind of circuitous reasoning upon which all scientific generalizations are based. It is the common procedure of science, given indications that a certain general relation is true, to *assume* that it is true, to push deductions to their ultimate consequences in all directions and to make the degree of correspondence of observation with deduction the measure of the probable truth of the original assumption. This does not mean of course that one should not entertain alternative hypotheses. Nevertheless the alternatives must be checked in precisely the same manner, and it is but a poor recommendation for an hypothesis that it can be checked against observation to such a limited extent that it is difficult to prove wrong.

Of all the hypotheses of differentiation of magmas none except the hypothesis of fractional crystallization can be checked against observation in any detail. It is therefore the only one that can be regarded as having any sound scientific basis.

THE EARLY SEPARATION OF BOTH PLAGIOCLASE AND PYROXENE

Before the fractional crystallization of basaltic magma can be considered it is necessary to state what is meant by basaltic magma. In a broad way it may be said that any magma is basaltic which, on rapid crystallization, gives rise to a rock having intermediate plagioclase and clino-pyroxene as its principal constituents. We shall consider first a magma which gives rise to these minerals almost exclusively. The nearest approach to such a magma that has been investigated experimentally is afforded by liquids of the plagioclase-diopside system. The equilibrium diagram of this system has been given in Fig. 16 and with the aid of Figs. 16 and 17 the crystallization of typical mixtures has been discussed. The salient features in the present connection are the following. In mixtures of labradorite and diopside containing roughly equal amounts of these (60 per cent plagioclase when it has the composition Ab_1An_1 and 55 per cent when it has the composition Ab_1An_2) plagioclase and diopside begin to separate from the liquid simultaneously (at 1230°C for the former mixture and 1250°C for the latter). The first plagioclase separating is highly calcic and the progress of fractional crystallization is such as to continually enrich the liquid in albite.

These facts and figures can not, of course, be applied directly to the mixture of plagioclase with the more complex pyroxene of the natural rock. Summary rejection of the results as wholly inapplicable is one course but there is another, viz., enquiry as to the extent and nature of the modifications likely to be introduced by the difference in composition. We may begin this inquiry by considering the effect of the addition of another component to the three already present. To represent the composition it is necessary to resort to a tetrahedron whose base may be the albite-anorthite-diopside triangle and whose apex will then be the point representing the other component. No matter what the other component may be there will rise into the tetrahedron, from the boundary curve on the base, a boundary surface which separates those liquids that precipitate diopside first from those that precipitate plagioclase first. How far this boundary surface will rise into the tetrahedron without being cut off by other boundary surfaces is a matter which depends on the specific properties of the fourth component.

Now let us make the specific assumption that the fourth component is $FeSiO_3$ and construct a tetrahedron illustrating this case (Fig. 21). Since

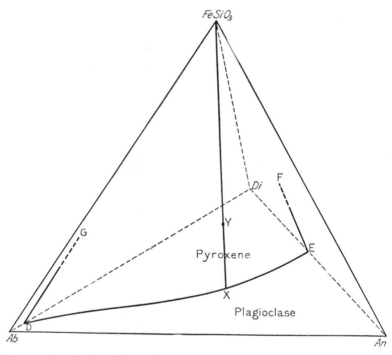

FIG. 21. Tetrahedron illustrating the composition of a quaternary liquid (Y) which approaches basaltic composition.

this may be taken into solid solution in diopside the boundary surface rising into the tetrahedron will separate the plagioclase field from the field of the more complex pyroxene, ferriferous diopside. Now let us join the point on the boundary curve on the base corresponding with the composition, Ab_1An_1 60 per cent-diopside 40 per cent (X), with the $FeSiO_3$ apex and take a point (Y) 20 per cent of the distance along this join. This point will represent the composition of a mixture containing 20 per cent $FeSiO_3$ and 80 per cent of the mixture (X). In other words the mixture (Y) will have the composition

$$
\left.\begin{array}{ll}
FeSiO_3 & 20 \\
CaMgSi_2O_6 & 32 \\
Ab_1An_1 & 48
\end{array}\right\} \text{pyroxene}
$$

It must be regarded as highly probable that the boundary surface (FEDG) rising into the tetrahedron from the boundary curve on the base would pass close to the point (Y) representing the composition of the mixture we have described. The point might lie on the plagioclase side of the surface or on the pyroxene side, nevertheless none but a very determined doubter would consider it likely to lie far on either side. In other words it would probably require the crystallization of only a quite small amount of either plagioclase or pyroxene from this liquid to bring it to a composition from which both plagioclase and pyroxene would separate together.

This quaternary liquid does not, of course, correspond exactly with a natural basaltic liquid, but it is a surprisingly close approach. Under I of Table III is given the oxide composition of the quaternary liquid and under II the average composition of Deccan traps as calculated by

TABLE III

COMPARISON OF A SIMPLE QUATERNARY MIXTURE AND AVERAGE DECCAN TRAP

	I	II
SiO_2	53.7	50.6
Al_2O_3	13.2	13.6
FeO	10.9	12.8
MgO	6.1	5.5
CaO	13.1	9.5
Na_2O	3.0	2.6
K_2O	0.0	0.7

I—Chemical composition of the quaternary mixture containing Ab_1An_1 48 per cent, diopside 32 per cent, and $FeSiO_3$ 20 per cent.
II—Average chemical composition of Deccan traps (after Washington). All iron stated as FeO. TiO_2, H_2O and minor constituents omitted.

Washington from his several analyses.[1] It plainly requires but a very slight further modification of the quaternary liquid to render it typically basaltic. This further modification is not to be regarded as a greater extrapolation from investigated liquids. Rather is it a more complicated extrapolation and therefore not amenable to graphic presentation. We may therefore regard it as highly probable that typical basaltic liquid such as II is not far from a composition which is saturated with both plagioclase and pyroxene.

1 H. S. Washington, *Bull. Geol. Soc. Amer.*, 33, 1922, p. 797.

So much for the probabilities in the light of physico-chemical considerations. What, now, are the indications to be derived from direct observation of the course of crystallization in basaltic rocks? Here we must consider the opinion recently expressed by Fenner that, in the crystallization of basaltic magma, all of the plagioclase crystallizes out first and is followed by pyroxene.[1] If this is correct we have, as Fenner points out, radical departure from the investigated ternary liquids at the very outset. Instead of enrichment of the liquid in albite by fractional crystallization we would have enrichment in pyroxene and the possibility of development of late salic differentiates from basaltic magma would be completely nullified. This opinion as to the course of crystallization of basaltic magma has been expressed by others, but it is a very surprising one to come from Fenner. In 1910 he made a special study of the subject and reached a very different conclusion. He was then much puzzled by the apparent indication of the ophitic texture that plagioclase had separated out completely and was followed by pyroxene. Such a behavior violated the physico-chemical concepts then entertained by him, and he set out to get all the evidence available in the rocks themselves. The conclusion reached was thus not derived from any theoretical reasoning but was based on careful observations. He expressed his conclusion as follows: "No matter to what degree glass may be present in the slide, plagioclase and diopside appear side by side. It is evident that almost from the beginning of crystallization these two constituents were being eliminated simultaneously."[2] Apparently, then, the Watchung basalts, of which Fenner was speaking, fully justify the conclusion reached from theoretical considerations that basaltic magmas are ordinarily not far from a composition which is saturated with both plagioclase and pyroxene. These basalts are included by Washington among plateau basalts. Deccan traps and plateau basalts in general are stressed in this section because they are considered to represent parental magma, a feature more fully discussed on later pages.

Regarding the highly glassy basalts as quenched products which reveal the early stages of crystallization, as did Fenner, we may cite other examples which prove that in basalt, plagioclase and pyroxene separate together from a very early period of crystallization. Adams and Gibson have recently measured the compressibility of a basaltic tachylite. The actual sample was collected by Daly in Hawaii and was chosen because it represented the nearest available approach to a basaltic glass which was at the same time in compact form. Adams and Gibson had no interest in the course of crystallization of basalt. They merely placed on record the nature of the material studied. It was stated by them as follows:

1 C. N. Fenner, *Jour. Geol.*, 34, 1926, p. 756.
2 *Amer. Jour. Sci.*, 29, 1910, p. 220.

"The total amount of crystalline material is about 3 per cent, and consists mainly of plagioclase with a little olivine and pyroxene."[1] Plainly when the proportion of all crystals was as little as 3 per cent pyroxene was represented.

We may refer, also, to the observations of Washington on the Deccan and other plateau basalts. From these observations we must reach the same conclusion though a different conclusion has been stated. The actual observations as given by Washington are as follows: "Augite is also very abundant in the more holocrystalline types but becomes less so as the glass content increases and seems to be quite absent in one or two *highly glassy* specimens."[2] Even if "seems to be quite absent" is read "is absent" this observation would prove merely that, for these particular basalts, there is a very brief early period during which plagioclase is separating alone and that throughout the major portion of the crystallization period the two were crystallizing together. The conclusion is scarcely justified that the *whole* course of crystallization is towards a liquid consisting of augite and magnetite.

Let us now consider other lines of evidence concerning the course of crystallization of basaltic magma. The ophitic texture is often regarded as indicating that plagioclase crystallized first and pyroxene afterwards. Of that point, too, Fenner made an investigation in 1910. Concerning this view he states: "It is only necessary to devote a little study to the question to determine that this is emphatically not the case."[3] He then describes his observations proving mutual interference of the crystals of the two minerals throughout the period of their growth.

It would appear then that, while a cursory examination gives the impression that ophitic texture in basalts is the result of late crystallization of pyroxene, a more detailed examination indicates simultaneous crystallization with plagioclase or even continuance of the outer rims of plagioclase after the pyroxene. On this point we may quote the authors of the Mull memoir, who say: "It is noteworthy that the *ophitic* augites of the Mull *Plateau Type* often completed their growth well within the crystallization-period of the associated feldspar."[4]

The ophitic texture of basalts is probably to be taken as the result of simultaneous crystallization of two minerals the one of which, viz., plagioclase, has a greater specific tendency to idiomorphism under conditions of rapid growth. Practically the same idea is expressed by the Mull authors (p. 240), who state, "The relatively early date of the

1 *Proc. Nat. Acad. Sci.*, 12, 1926, p. 276.
2 H. S. Washington, *op. cit.*, p. 769.
3 Fenner, *op. cit.*, p. 225.
4 "Tertiary and Post-Tertiary Geology of Mull, Loch Aline and Oban," *Mem. Geol. Surv. Scot.*, 1924, p. 16. Italics are mine.

augite is obscured by its almost complete refusal to exhibit crystal-boundaries."

When the crystallization is followed step by step as it is revealed in basalts with varying proportions of glass no ambiguity is possible. It has never been shown in any basalt that *all or practically all* the plagioclase has crystallized and left pyroxene substance alone or practically alone as a glass base. Fenner's studies of 1910 prove this fact for the New Jersey basalts and Lacroix's work on the basalts of Reunion agrees entirely. In largely glassy types he finds that the earliest microlites, as well as any phenocrysts there may be, are of both augite and labradorite. Brown spherulites of a pyroxenic nature may be abundant when the total amount of crystallization is quite moderate. To be sure, when the crystallization has advanced practically to completion an ophitic texture results and Lacroix concludes *from this texture* that pyroxene crystallized later than plagioclase, a conclusion which does not appear to be supported by his own observations of the stages of development of crystallization in the partly glassy types.[1] If we accept Fenner's conclusion of 1910 that the ophitic texture does not indicate late crystallization of pyroxene there is no conflict between the observations on these Reunion basalts.

While it is believed that the evidence of these quenched glass phases utterly negatives the conclusion that fractional crystallization of basaltic magma would give a highly pyroxenic late differentiate, we shall not place principal reliance on these observations. They are always open to the possible objection that these basalts crystallized from a strongly undercooled state and that the course of crystallization observed is not that in more slowly cooled masses. This objection is not, however, likely to favor the conception that the plagioclase separates out completely at an early stage in the more slowly cooled masses, for the more slowly a basaltic liquid is cooled the more its texture approaches the gabbroid which indicates essentially contemporaneous crystallization of pyroxene and plagioclase. Indeed, if we accept unreservedly the conclusion that the partly glassy basalts prove the storing up of pyroxene in the residual liquid, it is this condition, if any, that is to be regarded as aberrant and probably due to rapid crystallization from an undercooled state, whereas it is with slow crystallization that we are concerned in discussing differentiation. There are, however, as we have seen, the strongest reasons for doubting the validity of this conclusion even for the quickly chilled varieties.

We may, then, state it as the conclusion of this section that in the crystallization of basaltic magma plagioclase and pyroxene crystallize out together at a very early period.

[1] Lacroix, *Compt. rend.*, 154, p. 252, 1912.

In the preceding section we have given some attention to the crystallization of basaltic magma, especially to the sensibly simultaneous separation of feldspar and pyroxene at early stages. In the very earliest stages the femic material separating from a great many basalts is olivine. This fact does not, of course, weaken the conclusion that there is no general enrichment of the residual liquid in femic constituents. However, an effect upon the residual liquid is produced which is different from that produced when only pyroxene and plagioclase separate, and to this effect attention is now directed.

In certain investigated systems it has been found that the olivine, forsterite, has a reaction relation to the liquid. It separates from the liquid in amounts greater than its actual stoichiometric proportion, later to react with the liquid to be partially or wholly converted to pyroxene. Upon the extent of this reaction depends, to a very considerable degree, the future course of the liquid and the products of its crystallization, especially the presence or absence of free silica among the late crystals. If necessary, the reader may refresh his memory by reference to Figs. 7, 14, 18, 19 and 20, and the discussion thereof. The question now to be considered is whether the olivine of rocks and especially basaltic rocks bears a similar reaction relation to pyroxene and to magmatic liquids.

The objection has been raised that this reaction takes place at 1557° in the system MgO-SiO_2 and since rocks crystallize at temperatures far below this it can have no application to magmas.[1] The objection is, of course, utterly absurd. Even if there were no experimental data on the subject, the elementary principles of phase equilibrium show that an incongruent melting point is affected by the presence of other substances in the same manner as an ordinary (congruent) melting point. But quite apart from theoretical considerations it has been shown experimentally that this reaction temperature is lowered some 300° by the presence of the appropriate amount of anorthite (see Fig. 14).

Among the criteria of the reaction relation is the formation of reaction rims or coronas. The formation of coronas of pyroxene about olivine in rocks is too well recognized to require discussion here. Another indication of the reaction relation of a mineral is the fact that it crystallizes out at an early stage, then ceases to crystallize and is wholly unrepresented among the later crystallization products of the magma. Olivine meets this requirement eminently.

Another characteristic of a mineral having a reaction relation is that it may separate in an amount greater than its actual stoichiometric proportion in the mixture. There is no necessity that a reaction mineral should always do so. That it should sometimes do so is a sufficient indi-

1 J. W. Evans, *Trans. Faraday Soc.*, No. 60, Vol. 20, pt. 3, 1925, p. 474.

cation, for a failure of excess separation can be accomplished in the identical liquid by different conditions of cooling. Fortunately there is a means of verifying the excess separation of olivine in a large number of olivine basalts. The normative proportion of olivine is perhaps not an exact value but is certainly a very close approximation to what may be called the stoichiometric proportion of olivine. It has been repeatedly noted in olivine basalts, especially where these have a very fine-grained or partially hyaline base, that olivine has separated in amounts far in excess of the normative olivine. The fact has been commented upon by Cross and by Washington in connection with their investigation of Hawaiian and other basalts,[1] by Lacroix, especially in his studies of basalts of Madagascar and Reunion,[2] and by Tyrrell.[3]

The evidence that the reaction relation of olivine, demonstrated in investigated systems, persists in the more complex natural magmas is thus entirely satisfactory. Some petrologists have liked to think that high pressure or the presence of volatile components destroys the relation, but this brief review of the actual relations displayed in rocks does not justify their hopes. Curiously enough, other petrologists, admitting that the relation exists in magmas but wishing to minimize the importance of experimental work, insist that the existence of the reaction relation is caused by the presence of volatiles. No better examples could be found of the tendency to confer omnipotence upon the volatile components. The facts show that we are dealing with a dry-melt relation, no doubt modified, possibly enhanced, but certainly not destroyed by the presence of volatiles. The reaction relation of olivine to liquid carries with it an inevitable consequence. The early separation of olivine in excess of its actual stoichiometric proportions necessitates the late formation of free silica, if for any reason, such as relative motion of crystals and liquid, the olivine fails of complete reaction.

This fact is of great general significance in the genesis of rock types. A very simple example illustrating its consequences is afforded by the Palisade diabase sill of New Jersey. In this igneous body there is a layer near the base much enriched in olivine crystals, while the rest of the mass, except at the chilled contacts, is free from olivine. Lewis interprets this olivine-rich layer as formed by the accumulation of sunken olivine crystals from the overlying mass.[4] If the olivine has the reaction

1 Whitman Cross, *U.S. Geol. Surv. Prof. Paper* 88, 1915, p. 55; H. S. Washington, *Amer. Jour. Sci.*, 5, 1923, pp. 469-70.
2 A. Lacroix, *Mineralogie de Madagascar*, III, 1923, p. 46; Analysis B and foot-note 2; also *Compt. rend.*, 177, 1923, p. 663.
3 G. W. Tyrrell, "Petrography of Jan Mayen," *Trans. Roy. Soc. Edin.*, 54, 1926, p. 762.
4 J. Volney Lewis, *Geol. Surv. New Jersey*, 1907, Pt. IV, pp. 109 ff.

relation to liquid that it is found to display in investigated systems then its removal from the upper parts of the sill should have induced the formation of quartz in those parts. Quartz does occur as a constituent of interstitial micropegmatite throughout the main body of the sill. Fenner, while accepting the sinking of olivine, denies that it has anything to do with the formation of quartz in the rest of the mass.[1] Now it is quite probable that this Palisade diabase magma would have developed some quartz even if there had been no sinking of olivine, for the norm of the chilled contact phase (presumably representing the original magma as intruded) contains a very little quartz,[2] and normative minerals are on the whole the most siliceous possible. Still the magma was not so siliceous that it could not develop olivine as an early mineral and it therefore presents a parallel with those liquids of investigated systems which, while containing more than enough silica to form magnesium metasilicate, yet lie in the forsterite field and precipitate forsterite as an early phase (e.g., S of Fig. 14). In these simple liquids removal of the forsterite augments the amount of free silica formed. If the olivine of the Palisade magma had the same reaction relation to the liquid, then its sinking must have augmented the amount of quartz. Lewis points out that in the quickly chilled facies at the upper margin, from which olivine crystals did not have the opportunity to settle, the olivine crystals have a reaction rim of enstatite.[3] The reaction relation of olivine is further demonstrated by a fact brought out by all studies of the rock and admitted by Fenner, viz., that olivine began and *ceased* to crystallize at an early stage. No crystalline phase can cease to separate at an early stage unless it has a reaction relation. In the absence of such a relation it must be represented among the final crystals. These observations remove the whole matter from the realm of the debatable. The sinking of olivine crystals from the main mass of the diabase and consequent lack of opportunity for the liquid to react with olivine must have constituted a source of quartz.

In this connection it would perhaps be well to reproduce a table of the mineral constitution of the different facies of the diabase as given by Lewis (p. 123). If the micropegmatite interstices with their free quartz are secondary, as Fenner contends, it is a very remarkable fact that the secondary action studiously avoids the olivine diabase proper and the phases of the diabase associated with it which contain smaller amounts of olivine. The table demonstrates the complementary nature of olivine and quartz, a fact which is rationally explained by the reaction relation of olivine and liquid but is wholly unaccounted for by any kind of sec-

---

1 Fenner, *Jour. Geol.*, 34, 1926, p. 749.
2 Lewis, *op. cit.*, p. 121, Analyses IV and V; and Washington's Tables, 1917, Analyses 21, p. 640, and 27, p. 642.
3 Lewis, *op. cit.*, pp. 115, 127.

ondary action. Attack of the salic residuum upon surrounding minerals is no evidence of its secondary nature. The residuum formed by fractional crystallization should have a tendency to attack earlier minerals when these minerals are members of reaction series, and it is the failure of completion of this reaction that constitutes fractional crystallization. While the Table is before us we may refer to another feature of this

TABLE IV

MINERAL CONSTITUTION OF THE PALISADE INTRUSIVE DIABASE

	I	II	III	IV	V	VI	VII
Quartz	19	7	..	..	..	..	..
Feldspar	44	42	37	30	20	38	26
Augite	27	34	59	63	73	46	56
Biotite	3	..	..	..	1	1	1
Olivine	..	..	1	5	4	13	16
Ores	6	17	3	2	2	2	1
Apatite	1	..	..	..	..	..	..

I—Homestead, Pennsylvania R.R. tunnels, 400 feet from west end.
II—Marion Station, Jersey City, coarse-grained rock 420 feet east of platform.
III—Englewood Cliffs, immediately below the olivine-diabase.
IV—Englewood Cliffs, immediately above the olivine-diabase.
V—Weehawken, apparently intruded into the olivine-diabase, in roadside near West Shore ferry.
VI—Weehawken, road near West Shore ferry, olivine-diabase.
VII—Englewood Cliffs, olivine-diabase.

differentiated mass. The portions of the mass intimately associated with the olivine diabase are enriched in augite. There has plainly been quite notable sinking of augite crystals as well as olivine. The augite could not collect entirely in a definite layer because it did not cease to crystallize at an early stage as did olivine. The crystallization of augite continued until quite late stages and its settling was gradually brought to an end by the crowding of plagioclase and augite crystals. Fenner contends that separation "ceased abruptly" but does not point to any feature of the mass which indicates this, indeed he describes very clearly a process which would bring about a gradual cessation of sinking of crystals. But the probability that this action was slowly brought to an end at a certain stage in the Palisade mass does not indicate that it would cease at the same stage in a large and much more slowly cooled mass. The very fact that any effect of this kind is unknown in most small bodies, and that one must go to considerable masses like the Palisade sill to find it, justifies the assumption that the same action may take place to a greater extent in the really large masses of which the lower portions are inaccessible. But, even if it were true that in any mass of basaltic magma, however large, crystal settling would be brought to an end at precisely the same stage as it was in the Palisade sill this fact would not prove that the Palisade mass has exhausted the possibilities of crystallization differentiation. Crystal fractionation does not depend solely, nor perhaps even principally, upon the gravitative separation of crystals. The zoning of crystals is a very important process, indeed was so in the Palisade body, and the squeezing out of liquid residues may, perhaps, be the most important of all.

The example just given of a case of differentiation (Palisade sill) in which the early separation of olivine has augmented the late development of quartz has been shown to present a certain parallelism with investigated liquids which have a slight excess of silica and yet may precipitate olivine in early stages. We shall turn now to the case of liquids that are definitely more basic, having a deficiency of silica, in order to show that they too are analogous to investigated liquids in that the early separation of olivine may take place in excess of its stoichiometric amount and thus may transfer the liquid residue at this stage to the class containing an excess of silica. This character of the liquid residue will be retained if for any reason there is failure of complete reaction with the excess olivine.

The excess silica of such liquids is ordinarily manifested as quartz occurring as a late crystallization, in most cases as a constituent of micropegmatite. This interstitial acid material, appearing so frequently in basic rocks, has, largely by reason of its strong contrast with the rest of the rock, been considered by many investigators to be the result of contamination of the magma by wholly extraneous salic rock. There is no objection to such action as a possible means of production of salic interstices and the factors controlling it will be discussed in another chapter. Nevertheless, contamination is by no means essential to the production of the salic residue. We shall discuss this problem with special reference to the basaltic lavas of the Brito-Arctic region, particularly the Hebridean area, because the problem has been brought up in that connection.

The authors of the Mull Memoir divide the igneous rocks of their area into "magma types," the distinction between types being "based upon composition alone"[1] and having no reference to the state of crystallization. Among their various basalts they recognize a Plateau type and a Non-Porphyritic Central type to which they assign the compositions shown in the Table opposite.

The development of all their more salic types from the Non-Porphyritic Central type they are able to accomplish by a straightforward process of crystallization differentiation because the latter normally has salic intersertal matter. Of the passage from the Plateau type (which

[1] "Tertiary and Post-Tertiary Geology of Mull, Loch Aline and Oban," *Mem. Geol. Surv. Scot.*, 1924, p. 13.

they regard as probably parental) to the Non-Porphyritic Central type they are not so sure. On this point they state:

"If it be admitted that the Plateau magma-type holds a parental position in Mull petrology, a difficulty is at once manifest. In its most typical representatives, analcite and natrolite, rather than quartz, seem to be the last products of consolidation. How then was the passage brought about from the Plateau magma-type to the unstable Non-Porphyritic Central type?

"A possible answer is that diopside, spinel, and silica might result during crystallization as an alternative to aluminous augite, or that a magnesian olivine, magnetite, and silica might develop instead of a ferriferous olivine; but Mull does not seem to supply evidence bearing upon such matters. Possibly the change from the one magma-type to the other was due in part to assimilation, as Professor Daly has argued in comparable cases. There is again no direct evidence bearing upon this point; all that can be said is that if assimilation has been of importance in modifying the Mull Magma it must have been accomplished at a high temperature under conditions admitting of *complete admixture of melted sediment and original magma*.[1] There is no inherent impossibility in this conception."

They thus leave the matter open, but in the chemistry of these types there is a very definite reason for choosing between the method of crystallization-differentiation and the method of assimilation as a means of passage from the more basic to the more salic of the two.

TABLE V

COMPARISON OF HEBRIDEAN MAGMA-TYPES

	Plateau Type	Non-Porphyritic Central Type
SiO_2	45	50
Al_2O_3	15	13
$FeO+$ / Fe_2O_3	13	13
MgO	8	5
CaO	9	10
Na_2O	2.5	2.8
K_2O	0.5	1.2

To change the Plateau magma to Non-Porphyritic Central magma by assimilation, material of a certain type must be added. To make the same change by fractional crystallization material of a certain type must be

[1] Italics are mine, and are used to emphasize the fact that for this example, the relation between the two magmas and the supposed sediment should be a purely additive one.

subtracted. When these two materials are compared the choice of processes is readily made. By a graphic construction we may ascertain the possible compositions of the added and subtracted substances. In Fig. 22 the compositions of the two magmas are plotted as in the ordinary

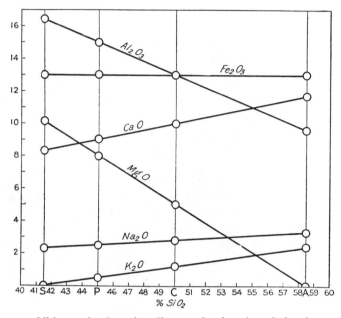

FIG. 22. Addition and subtraction diagram showing the relation between the Plateau Magma type (P) and the Non-Porphyritic Central Magma type (C) of Mull.

variation diagram, the Plateau magma at P, the Central magma at C. The two points indicating the proportions of each oxide are joined by straight lines and these lines are produced in both directions. When this is done it is found that the curve of K_2O falls to zero at about 41.5 per cent SiO_2 and the curve of MgO falls to zero at 58.5 per cent SiO_2. Rock substance having 58.5 per cent SiO_2 (A) is therefore the most siliceous that could be added, and a crystal aggregate having 41.5 per cent SiO_2 (S) the least siliceous material that could be subtracted to produce a change of the required kind. The proportions of all the other oxides in these two materials are fixed by their ordinates above these two points S and A. We may therefore express the composition of these as in Table VI.

TABLE VI

COMPOSITION OF ADDED AND SUBTRACTED MATERIAL REQUIRED TO CHANGE
PLATEAU TYPE TO NON-PORPHYRITIC CENTRAL TYPE

	A	S	S (Calculated to minerals)		Per cent
SiO_2	58	41.5	Plagioclase ($Ab_{37}An_{63}$)		53
Al_2O_3	9	16	Olivine $\begin{cases} \text{Forsterite } 16 \\ \text{Fayalite } 11 \end{cases}$		27
FeO	13	13			
CaO	12	10	Diopside $\begin{cases} CaSiO_3 \ 3.8 \\ MgSiO_3 \ 2.2 \\ FeSiO_3 \ 1.5 \end{cases}$		7.5
MgO	0	8.5			
Na_2O	3.5	2.25			
K_2O	2	0	Magnetite		4.5

NOTE: The summations are not 100 because H_2O, TiO_2 and all minor constituents are omitted. In making the calculation to minerals the iron has been assumed to be 10 per cent FeO and 3 per cent Fe_2O_3, which corresponds approximately with the distribution in the rocks under consideration.

The most siliceous possible added material (A) has a very remarkable composition. It is very doubtful whether any mixture of known igneous and sedimentary rocks could approach such a composition. Moreover, even granting such a mixture as possible, it would require to be assimilated to the extent of 62.5 per cent of the original magma in order to produce the required change. Any less siliceous material would have to be assimilated in still greater amount. One can avoid the necessity of adding so much by assuming that the MgO curve is not known accurately enough to fix its intersection with the axis of abscissae exactly at 58.5 per cent SiO_2. If the intersection were really at a higher silica percentage, the amount of this material required to be added would, of course, be less, but it would be of still more peculiar composition, for CaO, Na_2O and K_2O must continue to mount, FeO must maintain its high value, and Al_2O_3 must continue to fall. Moreover, even a mass containing 70 per cent SiO_2 must be added to the extent of 25 per cent of the original liquid. Plainly solution of foreign rock is not a satisfactory means of passing from the one magma to the other.

Compare this result with the straightforward solution of the problem given by supposing that the change was effected by subtraction of crystals. The material S corresponds accurately with the mineral composition calculated for it in Table VI. It is a mixture of a basic plagioclase with olivine, some pyroxene and a little magnetite. All of these are minerals known to have crystallized from the Plateau magma. Moreover, the mixture emphasizes those minerals which, according to direct observation, were emphasized in the earlier stages of crystallization. The authors of the Mull Memoir state (pp. 136-7): "Abundant olivine is in many cases the only micro-porphyritic constituent. . . . In some cases, there is a tendency towards a glomero-porphyritic grouping of the feldspars either alone or with olivine." In other words this material required to be subtracted corresponds quite satisfactorily with the kind of crystalline material that may reasonably be supposed to have been subtracted. We need not insist on the actual figures calculated; indeed, consideration of TiO_2 and of aluminous pyroxene would change the figures somewhat, but that the one magma-type differs from the other in terns of actual early minerals is not open to question. It is especially to be noted that over one-half of the Plateau magma must be crystallized before the required change is effected.

In the comparison of the two Mull magma-types given above there would appear to be sufficient reason for the statement that the passage from one to the other was the result of the separation of crystals of olivine and basic plagioclase. The one type is characterized, when crystalline, by the presence of olivine and, ordinarily, the absence of a salic residuum, the other by the absence of olivine and the presence of a salic residuum, and there are some intermediate varieties that show both. The free silica of the salic residuum is the result of the separation of olivine. In a number of investigated systems we have seen that the early crystallization of olivine, if fractionation occurs, will bring about the formation of a late residuum containing free silica. The determining factor is the continuance of the crystallization of olivine until the composition of the liquid is such that it can be expressed only as metasilicate with free silica. Plainly the passage from one magma-type to the other is the result of a strictly parallel relation, in fact the Plateau magma-type, when not cooled too rapidly, may itself show the salic residuum developed during zonal fractionation within the single rock. This rock will therefore have both olivine and free quartz, the latter in the salic residuum, a form of crystallization of the Plateau magma which is exemplified in the Early Basic Cone Sheets.[1] The whole story rests upon the fact that olivine bears a reaction relation to the liquid and if, as a result of fractionation, reaction is incomplete, the liquid will remain enriched in free silica. Thus crystal fractionation may bring about not only different relative proportions of minerals but, when the reaction relation enters, it may also induce the formation of different assemblages of minerals, and the passage from the olivine-bearing rocks of Mull to rocks with a quartzose residuum is a straightforward result of fractional crystallization, just as definitely as is the passage from those carrying but little of the quartzose residuum (Intermediate magma-types) to those made up principally of that material (Acid magma-types).

1 Mull Memoir, p. 241.

THE GENERAL TREND OF THE FRACTIONAL CRYSTALLIZATION OF BASALTIC MAGMA AND THE FORMATION OF BIOTITE

The first section of this chapter brought together, partly from extrapolation of investigated systems but more from examination of rocks themselves, the evidence that plagioclase and pyroxene separate together from basaltic magma at a very early stage. We have still to consider the general direction of crystallization subsequent to these early stages. In some measure this has been done in treating the development of free silica at late stages as a consequence of early separation of olivine. This is, of course, only a small part of the story. The question now to be raised is whether the residual liquid is continually enriched in alkaline feldspar in a manner analogous to the enrichment in albite, found in investigated systems, particularly the diopside-anorthite-albite system (Fig. 16).

The only suggestion alternative to the enrichment of the liquid in alkaline feldspar that has been made is that enrichment in iron compounds occurs. The pyroxene of basalts has been considered to be so different from pure diopside, especially in having a notable content of iron oxides, that crystallization in a system involving only pure diopside has been regarded as having little significance in connection with the crystallization of basalt. The crystallization temperature of ferriferous pyroxene is said to be so low that pyroxene rather than alkalic feldspar would be stored up in the liquid during crystallization.

On the experimental side nothing of a reliable nature is known of the crystallization temperatures of ferriferous pyroxenes nor of the effect of their presence in a system. We must therefore turn to the evidence of rocks. In reality we have done a good deal in this direction when considering the early stages of the crystallization of basalt as revealed in partly glassy types. There we found reason for agreeing with Fenner's conclusions of 1910 that there was no storing up of ferriferous pyroxenic material in the residual liquid. To be sure there are some basalts in which the glassy residuum is of a brown color and this glass has been assumed, on the basis of its color, to be rich in iron. No one has shown that this glass is richer in iron than the whole rock; indeed, there is little reason to doubt that we are there dealing with rapid, non-fractional crystallization and that this brown glass has sensibly the same composition as the original liquid and, of course, as the whole rock. In any case such rapidly cooled rocks are not reliable indications of the course of crystallization in slowly cooled magmas which are the only magmas of importance in connection with the problem of crystallization differentiation.

No support is to be found in more slowly cooled rocks for the belief that there is enrichment of ferriferous pyroxene in residual liquids. A certain rock from Mull has an important bearing on this question. The rock as a whole has about 7.5 per cent total iron oxides whereas early crystals (phenocrysts) of pyroxene contained in it have 29.5 per cent of those oxides. There is certainly no tendency towards enrichment in iron in this case. On the contrary there is abundant evidence in innumerable rocks of enrichment of the residual liquid in alkaline feldspar. We shall consider one example in detail and incidentally we shall find a source of free silica in addition to that brought about by the early formation of excess olivine.

The course of fractional crystallization of basaltic magma may be studied by considering an example in which fractionation has occurred in an individual rock as a result of zoning of crystals. It is then not a question of comparing two rocks of more or less problematical relationship, but the whole story is read in a single rock. A noritic gabbro described by Asklund[1] will be chosen, partly because it has received such careful chemical and petrographic study at his hands, and partly because it has been considered by him to disprove the hypothesis that a basic magma could give an acid residuum by fractional crystallization. The

TABLE VII

ANALYSIS, NORM AND MODE OF STAVSJÖ NORITIC GABBRO

	I		II		III	
			Norm		Actual Mineral Composition	
					wt. per cent	
SiO_2	52.79	Or	8.90			
TiO_2	1.12	Ab	26.20			
Al_2O_3	13.79	An	19.18	Quartz	2.53	
Fe_2O_3	1.91			Microcline	1.39	
FeO	8.13			$An_{51}Ab_{49}$	19.53	
MnO	0.09	Di	18.48	$An_{31}Ab_{69}$	19.17	
MgO	8.34	Hy	16.31	$An_{25}Ab_{75}$	3.96	
CaO	8.84	Ol	5.26	Hypersthene + Hypersthene		
Na_2O	3.12			Augite	27.09	
K_2O	1.48	Mt	2.78	Diallage-like Augite	18.24	
P_2O_5	0.29	Il	2.13	Biotite	6.57	
$H_2O +$	0.20	Ap	0.67	Magnetite	3.19	
$H_2O -$	0.12			Apatite	0.54	
	100.22					

chemical composition of this noritic gabbro is given under I and the norm under II of Table VII. As Asklund's work shows, it is easy to go astray in the study of crystallization by placing too much reliance upon the norm, yet we are on safe ground when we make the statement, based on the norm, that this magma could have crystallized to a rock consisting

1 B. Asklund, "Granites and Associated Basic Rocks of the Stavsjö Area," *Sveriges Geol. Undersök. Årsbok*, 17, 1923, No. 6.

almost exclusively of plagioclase and hypersthene-augite. There is no more orthoclase than could have formed a homogeneous mix-crystal with the plagioclase though it is just upon the limit. There need have been no olivine, or at most very little, for the pyroxene would have been aluminous and some of the alumina assigned to anorthite would actually have occurred in the pyroxene. Some SiO_2 and $CaSiO_3$ allotted to anorthite in the norm would have been combined with the molecules allotted to olivine and have existed as pyroxene. Some little of the soda allotted to albite would also have formed pyroxene molecules and there would thus have been a little more silica for the conversion of olivine to pyroxene. There need have been no quartz, no biotite, no potash feldspar as such, nothing but homogeneous plagioclase and hypersthene-augite with a little titano-magnetite, and according to the hypothesis of fractional crystallization there would have been nothing but these minerals had the magma been crystallized very rapidly. Actually the rock formed had the mineral composition given under III. The cause of this mineral composition was fractional crystallization taking place within the rock itself as a result of zoning of crystals. This fact is brought out by a consideration of the course of crystallization as revealed by Asklund's careful microscopic analysis.[1] The determined course is here given in a graphic form (Fig. 23) rather than by quotation, for this form facilitates the discussion of fractional crystallization.

We note that, because the early plagioclase contained much less of the soda and potash feldspar molecules than the total feldspar composition of the rock, there was a continual storing of these molecules in the residual liquid. We note also that most of the ferromagnesian material had crystallized at a relatively early stage. The remaining material of that nature, finding itself in a strongly alkalic medium, crystallized as mica (instead of as pyroxene) together with the more sodic plagioclase and the microcline, and as a result of the formation of biotite there was left some free silica to crystallize as quartz. Without this fractionation the liquid would never have become alkaline enough to precipitate microcline and to develop biotite and there would consequently have been no quartz. The material of all of these would merely have gone into the plagioclase and pyroxene.

After giving in the descriptive part of his paper the clear outline of the course of crystallization of the noritic gabbro and picturing the early crystallization of pyroxene and basic plagioclase with final crystallization of the salic residuum (alkaline feldspars, biotite and quartz), it is scarcely credible that in the theoretical part Asklund is able to convince

[1] *op. cit.*, p. 69.

himself, by assuming the separation of normative minerals, that the noritic gabbro could not crystallize in such a way as to give a salic residuum.[1] Because there is not enough silica to make all the bases into metasilicates in the norm he believes that the separation of hypersthene would render the liquid less siliceous. But the actual modal mineral

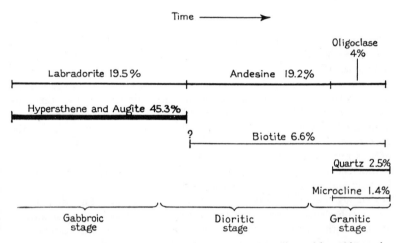

FIG. 23. Graph of periods of crystallization of Asklund's noritic gabbro. The length of the line for each type of mineral shows the length of its period of crystallization and together with the thickness of the line gives a rough indication of its quantity.

hypersthene is never a metasilicate. It is always somewhat more basic and in the present instance it (and also the diallage) must have a slightly lower SiO_2 percentage than the rock, otherwise the total composition of the rock could not be what it is, for there is a small but definite content of highly salic minerals. A silica content of about 51 per cent is common in hypersthene from such rocks. Asklund's actual determination of the course of crystallization by direct observations is indisputable; indeed, the rock is typical of a large class of basic rocks that show granophyric or micropegmatitic interstices in which this material is almost universally recognized as the last crystallized residuum. Fenner, as we have noted, argues that these interstices are secondary and formed by some sort of action taking place during the last stages of consolidation of the rock.[2] In the Asklund rock they are plainly no more secondary than the

1 Asklund, *op. cit.*, pp. 93-4.
2 C. N. Fenner, *Jour. Geol.*, 34, 1926, p. 754.

sodic rim of the plagioclase and are the result of the same crystal fractionation. The fact that the micropegmatite may in some cases constitute a separate mass occurring independently as an intrusive, which may be quenched to a glass at its margins, completely negatives the notion that there is anything of a secondary nature about it as it ordinarily occurs in igneous rocks.

Another notion of the origin of micropegmatite interstices is disposed of definitely as far as the Asklund rock is concerned. On account of the strong contrast between such interstices and the basic rock in which they are found, some writers, and notably Daly, have concluded that the basic magma has been contaminated with wholly extraneous salic material; in other words that it has assimilated or dissolved a salic rock. Now there is no objection to this as one possible means of production of salic interstices, but the described noritic gabbro shows that contamination is quite unnecessary. The total composition of the rock is such that there is no excess silica. Indeed there is a deficiency of silica below that necessary to convert all of the bases into perfectly stable and mutually compatible silicate minerals. This fact is shown by the occurrence of olivine in the norm. The rock is undersilicated, not oversilicated, and yet there develops an interstitial material of highly salic character. It is the result of crystal fractionation.

The basaltic material occurring as dikes and sills in the Cobalt and adjacent regions of Northern Ontario is often undersaturated yet contains micropegmatite interstices.[1] Many other examples of such rocks might be mentioned. The development of free quartz at a late stage is the result of those reactions which lead to the formation in the liquid of the molecules characteristic of mica and to the precipitation of some ferromagnesian and alkalic material in the form of biotite. It is an action furnishing a source of free SiO_2 quite independent of any previous formation of olivine but it may also occur in conjunction with the olivine effect.

The continual enrichment of the liquid in alkaline feldspars, demonstrated in the noritic gabbro and other rocks just discussed, may, of course, be effected by any method of fractionation other than the zonal method displayed by them. There is also a continual enrichment in the volatile constituents such as water, CO_2, S, Cl, etc. From this liquid are precipitated the minerals of the salic differentiate, including alkaline feldspars, quartz, and biotite. The chemical characters of these minerals give direct evidence of a number of equilibrium reactions in the liquid. The most important of these reactions involve the breakdown of part of the polysilicate molecules as follows:

1 N. L. Bowen, *Jour. Geol.*, 18, 1910, p. 660. W. H. Collins, *Geol. Surv. Can. Mem.* 95, 1917, p. 87. See also norms of analyses in Washington's Tables, p. 605, Nos. 2 and 4.

$$KAlSi_3O_8 \leftrightarrows KAlSiO_4 + 2SiO_2$$
$$NaAlSi_3O_8 \leftrightarrows NaAlSiO_4 + 2SiO_2$$

There must also exist such equilibria as the following:

$$NaAlSiO_4 + H_2O \leftrightarrows HAlSiO_4 + NaOH$$
$$KAlSiO_4 + H_2O \leftrightarrows HAlSiO_4 + KOH$$

and, doubtless,

$$NaOH + HCl \leftrightarrows NaCl + H_2O$$
$$2NaOH + H_2S \leftrightarrows Na_2S + 2H_2O$$
$$2NaOH + CO_2 \leftrightarrows Na_2CO_3 + H_2O$$

with similar reactions for the corresponding potash compounds, besides very complicated equilibria between the molecules S, SO_2, SO_3, C, CO, CO_2, H_2S, H, H_2O, O, HCl, Cl, etc.

Though we understand very little about the exact form in which iron and magnesia enter into the micas, it appears that there is the same tendency to partial breakdown from the more siliceous metasilicate to the less siliceous orthosilicate.

There is no doubt that the increased concentration of water in the magma at this stage exerts a strong influence in promoting this breakdown of the polysilicate molecules of the alkalis and the metasilicates of iron and magnesia into the orthosilicate molecules with setting free of silica, an action which may be compared with hydrolysis. Niggli[1] ascribes such action to water in discussing the rocks of Electric Peak and Sepulchre Mountains described by Iddings.[2] In some of the deep-seated rocks of Electric Peak quartz and biotite occur, whereas they are absent in surface rocks of the same composition at Sepulchre Mountains. This is ascribed to loss of water by the surface rocks.

It is not to be imagined that any of these reactions in the magma begin abruptly at any special stage in the history of the magma. For any given concentration of the molecules $KAlSi_3O_8$ and $NaAlSi_3O_8$, however small, there is a certain corresponding concentration of $KAlSiO_4$, $NaAlSiO_4$, and SiO_2. During the *slow* crystallization of the basaltic magma, with the continual increase in the concentration of $KAlSi_3O_8$, $NaAlSi_3O_8$, and the promoting agent water, there is a corresponding increase in the concentrations of $KAlSiO_4$, $NaAlSiO_4$, SiO_2, and others, until finally some of these exceed their saturation limit. SiO_2 then separates as quartz; $KAlSiO_4$ with $HAlSiO_4$, certain complex ferromagnesian molecules, and a limited amount of $NaAlSiO_4$ separate as a solid solution making up the mineral biotite. The molecules which separate are not necessarily the most concentrated; certain others may be much more concentrated, but correspondingly more soluble. Neither is

1 "Die gasförmigen Mineralizatoren im Magma," *Geologische Rundschau*, Band 3, 1912, p. 479.
2 *U.S. Geol. Survey*, 12th Ann. Rept., 1, 1891, p. 657; Iddings, *Igneous Rocks*, I, pp. 152-3.

it necessary that the molecules separate in the stoichiometric proportions represented by the reactions given. They are formed in the liquid in these proportions, but the extent of their separation from the liquid is determined by their solubility relations. The relative amounts of quartz and biotite contained in a given rock have, therefore, no necessary relation to the proportions of the various molecules indicated in the foregoing reactions. Indeed, quartz, the one product of these reactions, usually is greatly in excess of biotite in salic differentiates, but this may be principally due to the fact that much of the quartz has arisen from other causes. There is a relative storing up in the liquid of the molecules, other than SiO_2, which result from these reactions. The molecules stored up are, then, principally the very soluble and fusible compounds of the alkalis listed among the products.

In the courses of crystallization just discussed there appears to be adequate reason for belief in the development of granitic material as a late differentiate of basaltic and a satisfactory accounting for the mineralogy of the late differentiate.

SOME RELATIONS INVOLVED IN THE SEPARATION OF HORNBLENDE

We have now examined two methods whereby free quartz may be developed through fractional crystallization of basaltic magma. We have seen that through the early crystallization of olivine there may be a storing-up of silica in the liquid which goes on at the same time as the storing up of alkaline feldspar from the feldspar fractionation and thus gives a residuum rich in alkaline feldspar with some quartz. Either accompanying this or occurring independently of it there may be a reaction between ferromagnesian material and potash feldspar in the presence of water at late stages (when the concentration of water and potash have reached appropriate values) with resultant formation of biotite and setting free of quartz. In addition to these processes depending on the formation of olivine at an early stage and of biotite at a late stage there is a third, taking place dominantly at intermediate stages, namely, the formation of hornblende. It may occur together with one or both of the other processes, or it may be omitted altogether, a flexibility for which we shall attempt to assign reasons on a later page.

The formation of hornblende will be discussed with special reference to a series of rocks with hornblendic members described by Asklund. These rocks are chosen because Asklund has studied them so minutely and because he has concluded that the series could not be formed by fractional crystallization. The series of rocks has for its most basic member the noritic gabbro whose fractional crystallization we have already discussed for the case where fractionation is the result of zoning of crystals within the rock itself. The series passes from this member through hornblende diorites to quartz diorites and on to granites. Asklund attacks the problem of fractional crystallization by attempting to determine the nature of the material that must have been subtracted from (crystallized from) the noritic gabbro to give quartz diorite. This is a proper mode of approach but the problem is insoluble unless a fundamental assumption is made, viz., that the relative proportion of the subtracted material is known or that the percentage of some oxide in the subtracted material is known. These are not independent of each other; assuming the one fixes the other. Asklund chose to assume that there was no K_2O in the subtracted material and in so doing, of course, fixed the relative amount of that material and the percentage of all the oxides. Making this assumption Asklund deduced quartz diorite from noritic gabbro and thus determined the nature of the material that must be removed from the noritic gabbro to give the quartz diorite. The actual calculation was carried out in molecular proportions and the result was expressed as normative minerals.[1] It is given below.

Sal:	Ab	1.83	Fem:	Di	18.81
	An	10.56		Ol	12.34
	Ne	6.11		Mt	2.00
				Il	0.10
	Sal	18.50		Fem	33.25

Sal + Fem 51.75

Quantitative system : Rossweinose

Although Asklund's work shows plainly that he realizes the difference between normative and modal minerals he nevertheless goes on to discuss this result in the following terms:

"Evidently the condition for fractional crystallization must be that the $MgO + FeO$ in the femic crystallization-phase crystallized as orthosilicates, and also that CaO and Na_2O were capable of crystallizing as silicates poorer in silica than pyroxene or albite. The silicates poor in silica which can take up MgO and FeO can only be olivine or biotite. The rapid settling-out of small quantities of olivine would give the remaining magma a composition so rich in silica that CaO and Na_2O could not crystallize as silica-poorer silicates than pyroxene and albite. And, naturally, it would be diopsidic pyroxene, and not Ca-orthosilicate, that would crystallize out. Consequently, the whole of the $MgO + FeO$ cannot be taken up by olivine, and thus the formation of such mixtures of minerals as those just calculated is out of the question. Hence, finally, we may state that the crystallization of the norite cannot give remaining magma-solutions so rich in silica as quartz-diorites and granites."[2]

1 Asklund, *op. cit.*, p. 95.
2 *idem*, p. 97.

If one prefers to make such calculations with the aid of normative minerals there is only one proper procedure to follow, having obtained a result expressed in such minerals. Magmas do not crystallize as normative minerals, they crystallize as modal minerals, and if one is testing a process which may have resulted from crystallization it is necessary to consider modal minerals. In other words one should ask the question, What modal minerals does such a normative composition represent? When this question is asked in the present instance a possible solution of the problem is immediately apparent. We find that material rich in hornblende could have an abundance of the same very basic normative minerals, as evidence of which we reproduce below the analysis and the normative composition of a hornblende from a hornblende gabbro of Ivrea, Piedmont.[1] The gabbro is a member of a norite-diorite series analogous to the Stavsjö rocks.

		Sal		Fem	
SiO_2	39.58				
Al_2O_3	14.91	An	36.70	Di	2.26
Fe_2O_3	4.01	Lc	3.05	Ol	34.64
FeO	10.67	Ne	13.35	Cs	2.92
MgO	13.06			Mt	5.80
CaO	11.76				
Na_2O	2.87				
K_2O	0.62				
H_2O	2.79				

Moreover, hornblendes usually recalculate to such basic molecules, as evidence of which we may refer to the norms of 10 hornblendites given in Chapter XIV, Table XV.

In considering the possibility that these rocks are related as crystal-differentiates it would appear to be rather premature to dismiss the problem without asking oneself whether a hornblende-rich differentiate might not represent the complementary material. This is especially true when one of the two analyzed rocks, the quartz diorite, is a hornblendic rock and is a member of a series of diorites some of which are much more basic than the particularly acid quartz diorite that was chosen for analysis. "These more basic rocks continuously pass into more acid, granodioritic rocks" and it was "this granodioritic type of the quartz-diorites" that "was analyzed."[2]

We may now question somewhat more closely the nature of the material that might have been subtracted from the noritic gabbro to give the quartz diorite, confining ourselves entirely to crystalline phases known to have formed in the described rock series. We may obtain a perfectly general solution of the problem graphically and the graph is especially useful in making clear the significance of any assumptions

1 Frank R. Van Horn, ,T.M.P.M., 17, 1898, p. 410.
2 Asklund, op. cit., pp. 35 and 36.

that are made in obtaining a particular solution. We plot (Fig. 24) the compositions of the two rocks as in the ordinary variation diagram and produce the straight lines indicating the change of each oxide in the direction away from quartz diorite, i.e., to lower SiO_2 percentage. Any material whose composition is represented by points on these lines (vertically above each other, of course) has a composition such that, if it were subtracted from the noritic gabbro, the quartz diorite would result. We note that the K_2O line falls to zero at 45 per cent SiO_2 and this is therefore the most basic possible material, for any more basic would

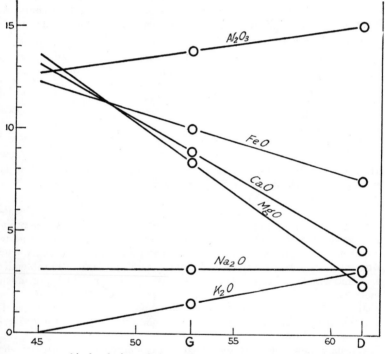

Fig. 24. Graphical solution of the nature of the material subtracted from the Stavsjö noritic gabbro (G) to give the quartz diorite (D).

require to have a negative content of K_2O. Absence of K_2O in the subtracted material was, it will be recalled, the assumption made by Asklund and he therefore obtained the particular solution of the problem given by the material having 45 per cent SiO_2. His result was the most basic possible material and it is not surprising that he found an abundance of basic molecules in its normative composition. But if we are to make an appropriate test of the hypothesis of fractional crystallization we must make a reasonable assumption as to the nature of the subtracted material. It requires but a moment's thought to convince one that the assumption of no potash is wholly unreasonable, for the analyzed quartz diorite is of granodiorite affinities and actually has arrived at a stage of crystallization where some potash feldspar has separated as such. It is entirely out of the question to assume that there was no potash in the crystal complex previously separated, regarding this as made up of known minerals of the rock series, plagioclase, pyroxene and hornblende. We must therefore choose a composition with some K_2O, how much there is no means of deciding. If we chose only 0.5 per cent K_2O the composition of the subtracted material would be as given below in Table VIII with the calculated normative composition beside it.

TABLE VIII

COMPOSITION AND NORM OF A POSSIBLE SUBTRACTED CRYSTAL COMPLEX REQUIRED TO CHANGE THE STAVSJÖ NORITIC GABBRO TO QUARTZ DIORITE.

SiO_2	47.50					
TiO_2	.47					
Al_2O_3	13.01					
Fe_2O_3	2.36					
FeO	9.17	Sal : Or	3.34	Fem : Di	30.09
MnO	.07	Ab	15.20	Ol	21.34
MgO	11.91	An	20.29	Mt	3.25
CaO	11.61	Ne	5.68	Il	0.91
Na_2O	3.07					
K_2O	.51					
P_2O_5	.16					
	99.84					

This result makes very clear the extreme consequences of the assumption made by Asklund. By permitting a very moderate and wholly reasonable potash content in the subtracted material we obtain a composition which, as its norm shows, could quite readily be made up of plagioclase, pyroxene and hornblende, all recognized early crystal phases in the rock series. We thus see that, while the material supposed subtracted by Asklund is, in view of the chemical nature of hornblende, by no means the impossible material he considers it to be, it is nevertheless unnecessary to imagine a crystal complex as rich in hornblende as his result would require. A much more moderate content of hornblende is indicated by the solution given above and, of course, still other solutions will be obtained, if somewhat greater potash content of the subtracted crystals is assumed, with consequent further decrease in the very basic normative molecules that go to make up modal hornblende. No unique solution of the problem is possible but it is plain that the quartz diorite can be derived from the noritic gabbro by the subtraction of known early crystals of the described rock series. It is to be noted, too, that the formation of hornblende at some stage of the crystallization of noritic gabbro would augment the amount of free quartz and therefore the possible quantity of a quartzose differentiate. This setting free of quartz is a plain inference from the very basic nature of hornblende which we have just discussed, but it is also an observed fact in the rock series under discussion. After its complete consolidation the noritic gabbro has locally been transformed into a hornblende-bearing rock and the reactions between plagioclase and hypersthene which have accomplished the formation of hornblende have likewise resulted in the setting free of quartz.[1] When the formation of hornblende took place during the magmatic stage, as it did in the diorites, there can be no question that it would have a like result, indeed the hornblende of the diorites has a tendency to include some of the quartz as small grains.[2]

We may now ask the question why the magma of the noritic gabbro could crystallize under certain conditions as the minerals of the noritic gabbro and under other conditions gave the series of rocks, diorites, quartz diorites and granites. In both cases the final goal aimed at during crystallization was substantially the same. The late-crystallizing residuum of the noritic gabbro had the composition of a biotite granite and granites were the late members of the rock series. The intermediate stages were, however, different, that is, the liquid followed different courses in reaching that goal. In several investigated systems we have seen that liquid may follow alternative courses and that the determining factor is the amount of reaction between liquid and past crystals of reaction series. There is much reason for believing that the same factor is of fundamental importance in determining what course may be followed by a crystallizing natural magma. The crystallization of the noritic gabbro itself was characterized by high fractionation (low reaction) in the plagioclase reaction series, the fractionation being brought about by zoning. The product was the characteristic gabbroid rock with granophyric interstices, occurring very commonly in nature, and it would appear to be in general the result of rather rapid crystallization of a mass of basaltic magma of moderate dimensions. The association gabbro-granophyre, in which some of the granophyre occurs as a separate body, may reasonably be supposed to result from a similar process during which gravitative settling and squeezing out of residual

1 Asklund, *op. cit.*, pp. 21-5.
2 *idem*, p. 35.

liquid, one or both, operated to separate the granophyric liquid. The evidence points, then, to high fractionation in the plagioclase reaction series, effected dominantly by zoning, as the principal factor in the production of the gabbro-granophyre association. The liquid passes through every gradation of composition between gabbro and granophyre, the product of crystallization at intermediate stages, say the dioritic, being represented by certain layers of the crystals (see Fig. 23) and not by a separate rock mass. The discontinuity in the crystallization series, that is sometimes claimed, has no real existence, though a physical discontinuity between gabbro and granophyre rock masses may result from factors superimposed upon and in some measure supplementing the zonal fractionation. The dioritic stage of the gabbro-granophyre association is represented (normally only in certain crystal layers) by augite diorite, whereas the similar stage is represented in a truly dioritic sequence as a hornblende diorite, normally occurring as a separate rock. Since hornblende appears to contain essential water the simple and obvious explanation of the difference is that the hornblendic sequence is formed from magmas richer in volatile components. This is, however, in all probability not the correct explanation, or at best only a partial one, for there is, in the magmas of the two sequences, a characteristic difference in the content of the ordinary oxide constituents. This difference is to be referred, not necessarily to a contrast in the original magmas, but probably to a different placing of the emphasis in crystal fractionation as we have already suggested. Further discussion of this method of control over the development of the hornblendic sequence must be deferred until after a consideration of the chemistry of the liquid lines of descent in sub-alkaline rock series.

ADDENDUM

On pages 64-66 of this chapter it was concluded from theoretical considerations that plagioclase and pyroxene should separate together from basaltic magma at very early stages in its crystallization. On pages 67-69 the evidence of the crystallization of basalts in nature was examined and it was concluded that they do behave in the above manner. When this material was in page proof, information of an entirely different character became available. Greig and Shepherd showed experimentally, with two Hawaiian basalts and a Maryland diabase, that feldspathic and femic constituents begin to crystallize nearly simultaneously. Three different lines of evidence are thus in thorough agreement.

CHAPTER VII

THE LIQUID LINES OF DESCENT AND VARIATION DIAGRAMS

GENERAL CONSIDERATIONS

A NUMBER of methods of plotting the chemical composition of associated igneous rocks have been devised. The principal purpose of a diagram is, of course, to offer a quick and graphic means of apprehending the chemical relations between the rocks. Regarded in this light the great majority of proposed diagrams have been dismal failures for it is commonly easier to see the relations of the rocks in the table of analyses than in the diagram. An outstanding exception is the diagram now most commonly used, in which the weight percentage of each oxide is plotted on rectangular coordinates against the percentage of one of the oxides, usually silica. A particular advantage is the facility with which one may determine the nature of the material that must be added to or subtracted from any rock mass to obtain another. We have already given examples of its use for this purpose.

Harker has found variation diagrams particularly serviceable in distinguishing between changes brought about by differentiation as distinct from those brought about by addition of foreign matter. In his hands this type of diagram has thus shown itself capable of acquiring a philosophic significance that is impossible in most other types of diagram.

In considering the characteristic curves of variation of the oxides in the usual diagrams for rock series even Harker has, on the whole, a tendency to look at the matter from the descriptive rather than the genetic viewpoint. Thus he says of the alumina curve that it declines at the high silica end because of the coming-in of the alkalic feldspars (which are relatively poor in alumina) and of quartz.[1] From the genetic viewpoint the question is, Why do the alkali feldspars and quartz come in and thus give a falling alumina curve? We shall find that the answer to this and to similar questions is forthcoming when variation diagrams are examined with the problem ever in mind of trying to decipher from

[1] *Natural History of the Igneous Rocks*, p. 122.

them the changing composition of the liquid and of eliminating from them such complications as are imported by the fact that not all rocks correspond in composition with a possible liquid. When this is done we find in variation diagrams strong corroborative evidence of crystallization-differentiation.

We shall now devote some space to the consideration of variation diagrams with the object of demonstrating the truth of the above statement. It will first be necessary to consider what the theory of crystallization-differentiation leads one to expect.

FACTORS GOVERNING THE BULK COMPOSITION OF A ROCK

The liquid in any given example must have followed some definite course of crystallization. If in any way a means is provided in nature for separating a portion of the liquid at any definite stage and of forming a rock therefrom without further opportunity for variation, a stage in the course followed by the liquid is revealed. If a series of rocks which meet this requirement is available the general course followed by the liquid is revealed. But a series of rocks can at best only approach this condition as an ideal. The magma from which any given rock sample formed must ordinarily have arrived in the position where it congealed to that rock with some crystals suspended in it. If these crystals represent all of the material which separated from some other liquid of the series while it was changing to the liquid in which they are now suspended the total composition will, of course, still represent a liquid of the series. If the mass has lost any of these crystals, or gained any, it will depart from the composition of any liquid even though the proportions of the different kinds of crystals relative to each other have been unchanged. Again, if there has been any relative movement of different kinds of crystals, then some kind of crystal will be represented in the total composition in greater relative amount than is possible in a mass formed by the direct congelation of any past liquid. And, of course, some other kind of crystal will be represented in lesser relative amount than is possible in a mass formed by the direct congelation of any past liquid. We may illustrate these facts with the aid of an equilibrium diagram for which we may take, say, the plagioclase-diopside diagram (see Fig. 17). During the crystallization of plagioclase and diopside from the liquid N all possible compositions of the changing liquid are represented by points along the nearly straight boundary curve, NMB, and if the compositions of the liquids are plotted after the manner of the ordinary variation diagram they lie along smooth curves (nearly straight lines). If the separating plagioclase and diopside remained precisely where formed, then the composition of the uniform final product would be that of the original mass, and therefore a possible liquid of the series. Or if a portion of the liquid were separated at any time it would, of course, represent a possible liquid of the series. Again, if the separated liquid contained only a very small amount of "past" crystals it would approximate to a possible liquid, and even if it contained a considerable amount of "past" crystals, so long as these were approximately in the proportions in which they were produced from a "past" liquid, the total composition would approximate that of the "past" liquid. But if there was considerable relative movement of plagioclase and diopside crystals in the liquid then a certain part of the mass would be enriched in diopside crystals and another part in plagioclase crystals. The total composition of each of these parts would be represented respectively by such points as G and D.[1] They are thus not possible liquids of the series and the points representing their compositions will not lie on the smooth curves of the variation diagram representing the change of composition of liquid. Nor will they lie on any kind of smooth curve. There will be a scattering of points.

This variation from any possible liquid may be great or small and the mere analysis of a rock gives no clue to the amount of departure or even to its direction. But on the assumption that, usually, the departure is small it is often permissible to indulge in a certain amount of "smoothing" of the curves as is ordinarily done in constructing the variation diagram of a series of rocks. The curves so smoothed thus represent more or less closely the change of composition of the liquid during crystallization.

We may now consider what indications there may be in the nature of the rock as to its probable approach to the composition of a liquid. If it is an entirely glassy rock the matter is not open to question and again if it is a dense, aphanitic rock there is ordinarily little doubt of its essential identity with the liquid from which it formed. A porphyritic rock may depart very little from a possible liquid or it may depart very markedly, and of a panidiomorphic granular rock the same is true.

RELATIVE SIGNIFICANCE OF THE DIFFERENT CLASSES OF ROCKS

Of the different classes of rocks the effusive are more likely to approach possible liquids. They are prone to have a minimum of past crystals, in part because as these increase in amount the liquid becomes less freely eruptible, but more because they are probably derived in general from the uppermost part of a crystallizing magma-mass or are the filter-pressed liquid from such a mass. We would thus expect the

1 It may be of interest to note that this part of the discussion takes a quite similar form to a discussion of the same factors by Tsuboi though it was written before Tsuboi's paper was seen. cf. *Jour. Imp. Univ. Tokyo*, II, pt. 2, 1925, pp. 78-9.

points representing the composition of the rocks of a volcanic field to lie particularly close to smooth curves, which curves would represent, approximately, the changing composition of the liquid. That the composition of rocks of a volcanic field do tend to be readily represented by smooth curves appears to be a fact. Thus we find Harker remarking of the volcanic rocks of Lassen Peak that "the variation-diagram, though constructed from a large number of analyses, requires very little smoothing."[1] His diagram is given in Fig. 25. The same is true of the eruptive rocks of the Whangerei-Bay of Islands Area of New Zealand,[2]

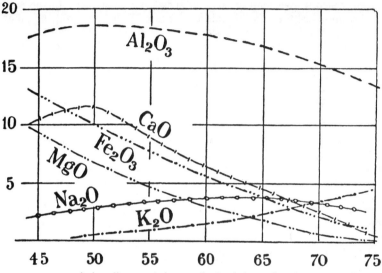

FIG. 25. Variation diagram (after Harker) of the rocks of Lassen Peak, California.

of the dominantly volcanic rocks of the Katmai region of Alaska,[3] and of other areas. If only non-porphyritic lavas were chosen in making such plots a still better approach to the smooth curves of changing composition of liquids would usually be obtained.

Plutonic masses of the type known as subjacent might rather frequently show a notable approach to the composition of successive liquids in the various associated rock types which they display, for they may represent to a considerable extent the same sort of material as lavas.

1 *Natural History of the Igneous Rocks*, p. 125.
2 J. A. Bartrum, *Verhandl. Geologisch-Mijnbouwkundig Genootschap Voor Nederland en Kolonien*, VIII, 1925, p. 13.
3 C. N. Fenner, "The Katmai Magmatic Province," *Jour. Geol.*, 34, 1926, pp. 673-772.

On the other hand, in bodies of the kind known as "floored," where erosion has exposed every layer of the mass, the types occurring may include those in the formation of which crystal accumulation has played some part and, according to the degree of sorting, the departure of any rock type from a composition represented by a member of the series of liquids may be very great. It is not surprising, therefore, that when Fenner plots the composition of the analyzed rocks of the Palisade sill, in which crystal accumulation has certainly occurred, he finds it impossible to draw smooth curves through the points representing each oxide.[1] The same is true of his plot of all Mull analyses[2] and in part for the same reason, as we shall attempt to show on a later page. In the Mull rocks, however, there is an additional cause of the scattering of points in that two different lines of descent are represented. The Mull authors have clearly distinguished between these two lines of descent in both their field and chemical relations and have plotted them on two separate diagrams. When so plotted each line of descent gives smooth curves.[3] In plotting them all on one diagram Fenner has lightly set aside the results of a magnificent example of petrologic study.

THEORETICAL SHAPES OF THE CURVES OF VARIATION OF THE LIQUID

Having considered what types of rocks are most likely to reveal the changing composition of the liquid, we may now proceed to enquire into the probable course of the curve of each oxide. A beginning may be made with the plagioclase-diopside diagram. This has already been done in part by Fenner. Taking a certain mixture of this system he plots the changing composition of the liquid as it crystallizes, with the result given in Fig. 26. We note that there is a discontinuity on the curves corresponding with the point at which the boundary curve of the equilibrium diagram is encountered but that thereafter the curves are nearly rectilinear. Now if we had the diagram of a series of rocks we would look at some corresponding point to find a discontinuity, that is, we would look to some point where a single separating mineral was joined by a second or a pair by a third. In the Mull diagram for the normal series (Fig. 27) such a point is readily located. The first courses of the curves we have found to correspond with the separation of olivine and plagioclase (see Fig. 22 and discussion thereof). These courses change when the lime-bearing augitic pyroxene appears and they change in the appropriate direction. The curve of lime, which has hitherto risen a little, now begins to fall because the crystal assemblage now separating has more lime than the liquid. Alumina now decreases less rapidly because there is less plagioclase in the separating assem-

1 Fenner, *op. cit.*, p. 713.
2 *idem*, p. 712.
3 Mull Memoir, Fig. 2, p. 14, and Fig. 4, p. 27.

blage. Iron, which has remained nearly constant, now begins to decrease, because the augite first separating is rich in iron.[1] After these changes have occurred the curves assume a nearly rectilinear character. Now in

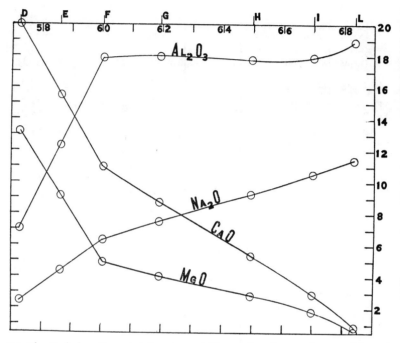

Fig. 26.   Variation diagram (after Fenner) illustrating change of composition of liquid during crystallization of a liquid in the system, diopside-anorthite-albite.

the Mull diagram these curves represent the simultaneous crystallization of augite and calcic plagioclase and consequent storing up of alkaline feldspar, both sodic and potassic, in the residual liquid. This fact is as well demonstrated by direct observation in the Mull series as any fact of igneous geology is likely to be. It is not surprising, therefore, that corresponding curves of the Mull diagram have the same general slope (in so far as the same oxides are present in both) as do the curves of the diagram of the investigated system (Fig. 26) where the shape of the curves is again determined by the separation of plagioclase and pyroxene (diopside). The curve of CaO falls rapidly, that of MgO

1 See: Analyses of phenocrystic uniaxial augite, Mull Memoir, p. 34.

less rapidly, while that of Na_2O rises and Al_2O_3 stays nearly constant in both diagrams.

Unfortunately no silicate system of more than three components has ever been completely investigated so it is not possible to demonstrate,

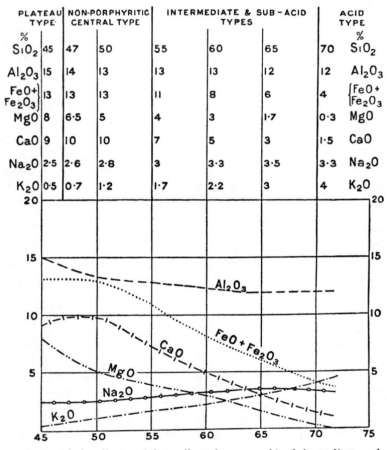

	PLATEAU TYPE	NON-PORPHYRITIC CENTRAL TYPE		INTERMEDIATE & SUB-ACID TYPES				ACID TYPE
% SiO_2	45	47	50	55	60	65	70	% SiO_2
Al_2O_3	15	14	13	13	13	12	12	Al_2O_3
$FeO+Fe_2O_3$	13	13	13	11	8	6	4	$FeO+Fe_2O_3$
MgO	8	6·5	5	4	3	1·7	0·3	MgO
CaO	9	10	10	7	5	3	1·5	CaO
Na_2O	2·5	2·6	2·8	3	3·3	3·5	3·3	Na_2O
K_2O	0·5	0·7	1·2	1·7	2·2	3	4	K_2O

FIG. 27. Variation diagram (after Bailey, Thomas, et al.) of the Mull Normal Magma series.

from experimental determinations, the course of the curves of other oxides. It is, however, possible to make some very useful deductions from theoretical considerations. We shall examine first the course of the curves of the various oxides concerned in the crystallization of ternary mixtures of anorthite, albite, and orthoclase. In Fig. 28 this system is represented as of three components. Strictly speaking this is not correct because orthoclase melts incongruently and can itself be represented

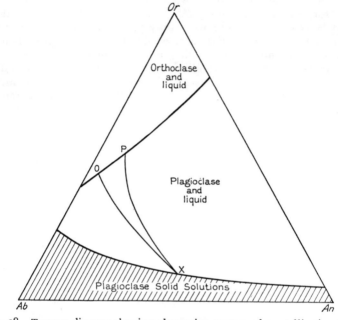

Fig. 28. Ternary diagram showing alternative courses of crystallization in the feldspar mixture X.

only as a two-component system. The relations shown must therefore be regarded as a projection from a quaternary system upon the ternary plane. We have represented on this diagram the limits of solid solution of orthoclase in plagioclase and have placed a boundary curve separating the field of plagioclase from that of orthoclase. For our present purpose it does not matter whether these lines have been exactly placed or not. We merely know that the general relations must be as indicated. Let us now consider the crystallization of a mixture which lies on the line indicating the limit of solid solution of orthoclase in plagioclase. We may take the mixture X which has the composition, anorthite 45, albite 45, orthoclase 10. The crystallization of such a mixture may occur in different ways depending upon the extent of fractionation or the converse, reaction. With very quick chilling, such that crystallization took place from a somewhat undercooled condition, the result would be homogeneous mix-crystals or solid solutions having the total composition of the original liquid. In this case there would be no change of composition of liquid during crystallization. The crystallization would be non-fractional. Again, with very slow cooling and no opportunity for relative motion of crystals, the early crystals, rich in anorthite, would be continually made over into crystals having more soda and potash and as a final product there would again be homogeneous mix-crystals of the uniform composition of the original liquid. The course of the liquid in this perfect equilibrium type of crystallization would be rather strongly curved, such as XP, the boundary curve of the orthoclase field would be barely attained and no orthoclase would separate as a distinct phase.

On the other hand if the cooling was at such a rate that zoning of the crystals was accomplished, or if fractionation was effected in any other way (such as by sinking of crystals or filter-pressing of liquid) the course of the liquid would be considerably different. It would be represented by a curve such as XO, which is drawn so as to represent rather strong fractionation. The orthoclase boundary curve would be reached and at O some crystallization of orthoclase as a separate phase would begin. Now the fractionation of the plagioclase would be practically independent of the presence of other phases so that in this contrasted behavior we see why a dolerite, say, may crystallize in such a way as to contain all the orthoclase molecules in the plagioclase, or again may crystallize in such a way as to contain free orthoclase in the last-crystallizing residuum.

However, at the present time we are particularly interested in the variation of the oxides as indicated by the course of the liquid. In Fig. 29 and in the full curves there is plotted, after the method of the ordinary variation diagram, the change of the various oxides corresponding with the curve XO. The curves for the oxides have, then, the shape that is necessitated by rather strong crystal fractionation in the feldspar series. We note that CaO and Al_2O_3 fall rapidly with increasing silica and in particular that K_2O increases at an increasing rate, that is, the curve is concave upward. The total increase of K_2O is some three- or four-fold. On the other hand Na_2O increases at a decreasing rate, that is, the curve is convex upward and the total increase is quite small (not more than 25 per cent). There is, moreover, a very distinct tendency for the Na_2O curve to pass through a flat maximum and then to decrease, the K_2O curve, meantime, crossing it. Now these are well recognized common tendencies of the curves of Na_2O and K_2O in sub-alkaline series of rocks even though they may represent somewhat different lines of descent. We direct attention in particular to the Lassen Peak and the Mull diagrams already given (Figs. 25 and 27), the one line of descent emphasizing hornblende, the other characterized by its essential absence, but in both of them the Na_2O and K_2O curves have these characters. We may take it therefore that the characteristic shapes of these curves are the result of fractional crystallization. Daly has made the mistake of sup-

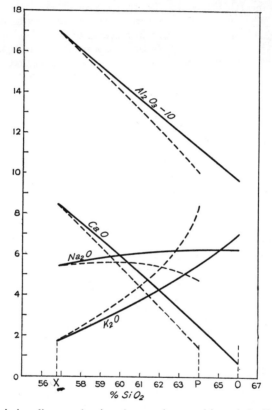

FIG. 29. Variation diagram showing change of composition of the liquid during the crystallization of the liquid X of Fig. 28.

posing that during fractional crystallization Na_2O should increase in about the same ratio at K_2O and has regarded the actual facts as inimical to the hypothesis of fractional crystallization.[1] Analysis of the problem

1 *Jour. Geol.*, 20, 120, 1918.

makes it plain that the facts are precisely as one would expect from fractional crystallization.

An important effect of fractional crystallization, namely, possible variety of courses of crystallization, is well illustrated by the diagram (Fig. 28). We have shown on the figure the two strongly contrasted courses XO and XP, the former characterized by strong fractionation (limited reaction), the latter by lack of fractionation (complete reaction) of the feldspar series. Between the two there is every possible gradation. The curve XP, however, represents the course of a liquid when fractionation is limited and the reaching of the orthoclase boundary curve is barely accomplished. If liquid approaching the composition P were separated from past crystals it would of course precipitate some orthoclase and in it orthoclase begins to separate at a time when the plagioclase then separating is still rather strongly calcic, whereas with strong fractionation (XO) orthoclase appears only when the plagioclase has become particularly sodic.

The variation of the oxides corresponding with XP is shown on Fig. 29 as dotted curves. We see that the dotted curves of CaO and Al_2O_3 are not significantly different from the full curves. The dotted curves of Na_2O and K_2O are, however, changed considerably, having in more marked degree the characters already pointed out in the full curves. The dotted Na_2O curve soon passes through its maximum and then falls considerably. The mounting of the dotted K_2O curve is correspondingly more rapid.

In drawing the curves of Fig. 29 the actual percentage of each oxide has been calculated for various points along the curved courses of crystallization of Fig. 28. It may be pointed out now that the general shape of these curves could be read from the triangular diagram (Fig. 28) by mere inspection. Since albite is the only compound containing Na_2O, lines indicating constant Na_2O content are one set of the ordinary coordinates of the diagram, viz., that set parallel to the side of the triangle opposite to the albite corner. The curves XO and XP, indicating the changing composition of the liquid, at first cut across this set of lines in the direction of increasing albite (or Na_2O), then become parallel to them and then again cut them but now in the direction of decreasing albite (or Na_2O). The Na_2O content of the liquid thus increases, passes through a maximum and then diminishes as crystallization proceeds. It is, of course, necessary in addition to show that this is a course of increasing SiO_2 so that the character described will persist when plotted against SiO_2. Anorthite has the least SiO_2 of the three components. Orthoclase and albite have so nearly the same amount of SiO_2 that the lines of constant SiO_2 are nearly parallel to the orthoclase-albite side of the triangle. Plainly the curved courses of crystallization cut across these at all times in the direction of increasing SiO_2. By mere inspection of Fig. 28, then, it could be stated that the curve of Na_2O, when plotted against SiO_2, would rise, pass through a maximum, and then fall. The general course of each of the other curves of the oxides can be read in a similar manner from the triangular crystallization diagram. One feature is particularly worthy of note. Since anorthite is the only compound containing CaO the lines of constant CaO content are parallel to the side opposite the anorthite corner, i.e., the albite-orthoclase side. We have already noted that the lines of constant SiO_2 content are nearly parallel to that side. Lines of constant CaO and of constant SiO_2 are thus nearly parallel to each other. Therefore, any course of crystallization in mixtures of these three, be it curved, tortuous or even discontinuous, will appear as a nearly straight line on a plot of CaO against SiO_2 after the manner of the ordinary variation diagram. A like condition is true of Al_2O_3 and for like reasons. It is not surprising, therefore, that in Fig. 29 we found nearly linear curves for CaO and Al_2O_3 whether the course of crystallization was strongly curved, as a result of strong reaction, or but slightly curved, as a result of weak reaction.

We may now inquire into the effects produced on these curves by the presence of other substances. One method of regarding the matter is to consider these substances as mere diluents. The other substances present together with plagioclase in the more basic rocks, such as basalt, constitute approximately 50 per cent so that the low-silica ends of the curves of those oxides that are confined almost exclusively to the plagioclase will move about half way towards the axis of SiO_2. Thus Al_2O_3, Na_2O and K_2O will be reduced about one-half at the low-silica end, but not so CaO which is present in other constituents in prominent amounts. Feldspar becomes an increasingly important constituent of the more salic members, quartz coming in as a diluent but usually only to the extent of about 25 per cent. The high-silica ends of the curves are thus moved down but not as much as the low-silica ends and, in particular, since the diluent is silica, the curves are stretched out to higher silica percentages. When the curves of variation of oxides (except CaO) in the pure feldspar liquids are thus modified they remain of the same general shape but less strongly curved and swung from their original course in such a way that their low-silica ends are much lowered, their high-silica ends not so much. They would thus present a satisfactory degree of correspondence with the curves of typical rock series. Alumina thus falls less rapidly than lime and not at a parallel rate as in the pure feldspar liquids. It is particularly noteworthy that the curves corresponding with an intermediate degree of reaction, when they are thus modified, would give a roughly constant amount of Na_2O and a rather rapidly increasing amount of K_2O. The curves corresponding with a high degree of fractionation will show soda mounting considerably and thus not so great a contrast in the rates for Na_2O and K_2O. Now we have seen that the association, gabbro-granophyre, is ordinarily representative of a high degree of fractionation in the feldspar series and it is not surprising therefore that granophyres of this association have a tendency to run higher in soda than the general average of granites. Thus we find that the average of 12 "granophyres" in Washington's Tables (1917) shows Na_2O 4.16 and K_2O 2.77, and of 7 "micropegmatites" Na_2O 4.94 and K_2O 3.86. The Na_2O therefore tends to be in excess of K_2O. The average of all "granites" according to Daly shows the reverse condition, with Na_2O 3.28 and K_2O 4.07.

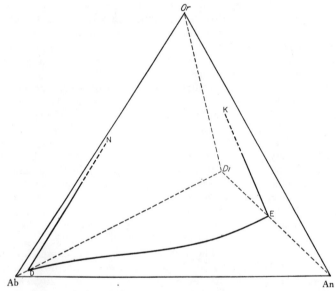

FIG. 30. The tetrahedron, diopside-anorthite-albite-orthoclase with the hypothetical boundary surface, diopside-plagioclase (NDEK).

This method of regarding other constituents as modifying the curves of the feldspar oxides merely as diluents may seem to veer too much towards the descriptive rather than the genetic method of regarding the matter. We shall therefore consider the question in another way. The determined three-component diagrams often make it possible to deduce with much assurance the course of crystallization in related but somewhat more complex systems. The albite-anorthite-diopside triangle is known and we may construct a tetrahedron with that triangle as base and orthoclase as the opposite apex. Into this tetrahedron there must rise, from the plagioclase-diopside boundary curve on the base, a plagioclase-diopside boundary surface. These general relations are shown in Fig. 30. Of the position assumed by the surface KEDN as it rises into the tetrahedron we need have, for our present purpose, no precise knowledge. Consider now the crystallization of a mixture with a little orthoclase, but dominantly of plagioclase and diopside and in such proportion that it lies near the boundary surface. Only a small amount of crystallization of either plagioclase or diopside, as the case may be, will take place from such a liquid before the boundary surface is reached, whereupon plagioclase and pyroxene (diopside) will separate together.[1] The general course which the liquid will follow on this boundary surface as crystallization proceeds is readily deduced. It will vary with the degree of fractionation in the plagioclase series and will follow different curved courses dependent upon this factor. In Fig. 31

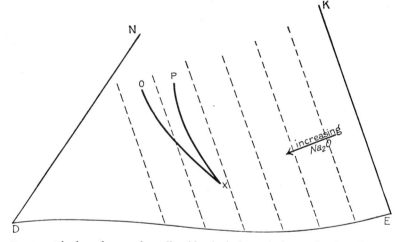

FIG. 31. The boundary surface, diopside-plagioclase of Fig. 30 showing alternative courses of crystallization during the simultaneous separation of plagioclase and diopside. The dotted lines are lines of constant Na₂O content.

we have taken *this boundary surface* out of the tetrahedron and represented it on the plane of the paper. On it are shown alternative courses that may be followed by the liquid. We also show lines of constant Na₂O content which are, of course, the traces of planes parallel to the

1 We have already stated reasons for believing that basaltic magma is normally of such a nature. See pp. 64-9.

face of the tetrahedron opposite to the albite apex. The curves XO and XP represent respectively high and low degrees of fractionation. It is plain that the same tendencies we have already pointed out are true of the potash and soda curves. The course of crystallization tends to parallel the lines of constant Na_2O so that Na_2O increases but moderately and tends to pass through a maximum. Moreover the course of crystallization swings towards the orthoclase apex so that K_2O in-

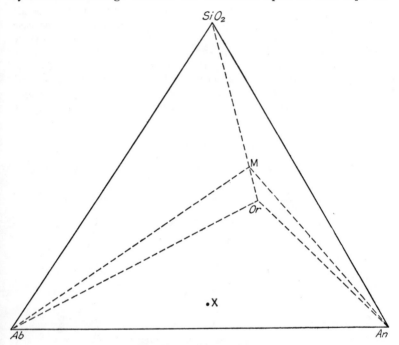

FIG. 32. The tetrahedron, anorthite-albite-orthoclase-silica. The point X lies in the plane M-An-Ab.

creases at an increasing rate. By constructing such a tetrahedron one will find too that, if it can be assumed that the boundary surface is approximately normal to the base, the planes of constant Al_2O_3 will be roughly parallel to it, in other words, any kind of motion in this boundary surface will induce but little change of Al_2O_3. We thus find that by regarding diopside as a component of the mixture and as crystallizing together with plagioclase we reach the same conclusion as to the shape of the Na_2O and K_2O curves as we did from the simple feldspar diagram or from it as modified by regarding pyroxene merely as a diluent.

A tetrahedron with the albite-anorthite-orthoclase triangle as base and SiO_2 as the apex may now be considered. A point (X, Fig. 32) close to the albite-anorthite edge but within the tetrahedron will represent a mixture consisting mainly of plagioclase. A plane determined by this point and the albite-anorthite edge (which plane cuts the opposite [Or-SiO_2] edge of the tetrahedron at M) will represent approximately the plane in which the course of the liquid lies. Strictly speaking, since there is some potash feldspar in the plagioclase, the course of the liquid will turn upward from this plane somewhat but consideration of movement in this plane is sufficiently accurate for our present purpose. The plane is taken out of the tetrahedron and represented in Fig. 33. On it

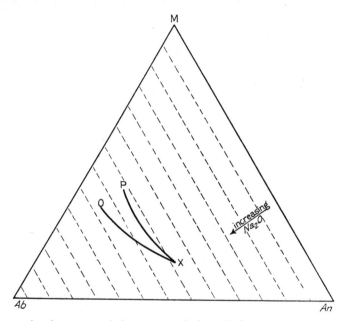

FIG. 33.   The plane M-An-Ab from the tetrahedron of Fig. 32 showing alternative courses of crystallization of the liquid X. The dotted lines are lines of constant Na_2O content.

are shown the traces of the planes of constant soda (planes parallel to the face of the tetrahedron opposite to the albite apex). The curves indicating the courses of crystallization with little and with high fractionation are also shown and it is plain that there is still the same tendency for the liquid to move in a course approximately parallel to the lines of equal Na_2O and to pass through a maximum for that oxide. Moreover, since any motion in this plane has a component towards silica, there is not as great a tendency as in the simple ternary system for K_2O to increase at an extremely high rate with respect to silica, in the event of rather free reaction (limited fractionation). Such rapid rise of the K_2O curve as is indicated in the dotted curve of Fig. 29 is therefore not to be expected in an actual rock series.

Combining the information derived from these two tetrahedra we find that, whether at an early stage when diopside is the principal constituent separating with feldspar or at a later stage when SiO_2 is the principal constituent present in addition to feldspar, the curves of Na_2O and K_2O have the characteristic shapes they are found to have for rock series.

We have considered in the foregoing what one would expect of the curves of the oxides that go to make up the feldspars, if these curves are determined by fractional crystallization. A satisfactory agreement has been found with characteristic curves of these oxides in rock series. There is, however, another reaction series of great importance in rocks and the crystallization of this will give curves of characteristic shape. This series is the femic reaction series olivine, pyroxene, hornblende and mica, each member of which is itself a solid solution series or reaction series. Unfortunately our knowledge of these minerals as reaction series is very limited. That the order in which they are stated above is the proper order for the reaction series of a normal crystallization sequence is almost certainly correct but of the details of change of composition within each solid solution series very little is known. Since the shapes of the curves of various oxides are to a considerable extent interdependent it is necessary to consider the probable effects of the femic series.

As a matter of fact, one of the systems used in deducing the shapes of the curves of the feldspar oxides involved the pyroxenic mineral, diopside, and some indications are to be obtained therefrom as to the probable shape of the MgO curve. We revert, therefore, to the tetrahedron, albite-anorthite-diopside-orthoclase (Fig. 30) and in particular to the boundary surface, plagioclase-diopside which as we have seen must rise from the boundary curve plagioclase-diopside. This is the same boundary surface already used as an indicator of the shape of other curves. On this surface are shown in Fig. 34 lines of constant MgO content which are the traces of planes parallel to the face of the tetrahedron opposite to the diopside apex.[1] The curve indicating the course of the liquid upon crystallization of plagioclase and diopside (with some

---

1 These lines as well as all other lines representing constancy of any oxide have been drawn on the deduced boundary surface at the correct slope with the aid of a hollow tetrahedron having transparent removable faces.

plagioclase fractionation) cuts directly across the lines of constant MgO at first and then swings so that it cuts them more and more obliquely, or, in other words, becomes more nearly parallel to them. The curve of variation of MgO in the changing liquid should therefore

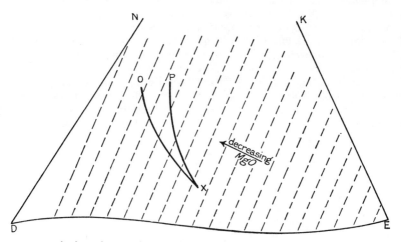

FIG. 34. The boundary surface of Fig. 30 showing alternative courses of crystallization during the simultaneous separation of plagioclase and diopside. The dotted lines are lines of constant MgO content.

fall steeply at first and less steeply later, being thus convex upward. It is to be noted that this shape of the MgO curve is the result, purely and simply, of feldspar fractionation, the pyroxene separating being pure diopside throughout. Even without any fractionation in the femic series, then, the deduced MgO curve has the characteristic shape it is found to have in the variation diagrams of typical rock series. If there is fractionation in the pyroxene series, of the same nature as that demonstrated experimentally in the clino-enstatite-diopside series (Fig. 20), the convexity of the MgO curve would be increased thereby, for the crystals first subtracted would be higher in MgO, i.e., the rate of subtraction of MgO from the liquid would be higher at first. That the early crystals separating from rocks are rich in MgO has been demonstrated by Vogt's statistical studies of rocks, not only for pyroxenes but for other femic minerals as well.[1]

Not only the sign of the slopes but also the characteristic shapes of the curves of K_2O, Na_2O, MgO and apparently CaO and Al_2O_3 of the

[1] J. H. L. Vogt, *Vidensk. Selsk. Skr. I. Mat.-Naturv. Kl.*, 1924, No. 15.

typical sub-alkaline rock series are such as to offer much support
for the hypothesis that these series have come into being through fractional crystallization. Of course, the systems that have been used to
illustrate the correspondence of actual and expected curves are much
simpler than actual magmas. Some modification of these curves is to be
expected in the presence of other oxides, the principal of which in
magmas are the oxides of iron. Of their effect we know very little from
experimental work but it is surely not very unreasonable to suppose that
the presence of iron modifies these relations somewhat, without destroy-
ing them, and that the shapes of the curves may still be taken as indicating the control of crystallization. The alternative is to suppose that
iron completely wipes out these relations, that the course of crystallization in the presence of iron oxides is completely changed and that
differentiation by some totally different process, such as liquid immiscibility or gaseous transfer, then endows the variation curves with a remarkable resemblance to the curves that would be given by crystallization-differentiation in the absence of iron oxides. There can be no
question as to the choice to be made between these alternatives even on
purely general grounds. But we do not have to base our choice on purely
general considerations. We may examine the evidence of rocks themselves as to the course of crystallization in the presence of these oxides
of iron. In urging the invalidity of the deductions based on systems
containing no iron it is sometimes contended that the iron compounds,
being of low-melting "point," would be emphasized in the last-crystallizing residuum.[1] Nevertheless, this contention can not be supported by
appeal to the evidence of rocks themselves. In the noritic gabbro of
Asklund, in the basaltic rocks of Mull, in Fenner's Katmai andesites, as
we shall see, and in any number of similar rocks the last crystallizing
residuum is of a granitic nature and is nearly as well purged of iron
oxides as it is of lime and magnesia. Crystallization-differentiation, thus
epitomized in the zonal fractionation of these individual rocks, is certainly not of such a nature as to emphasize iron in the late differentiates.
It may very well be that as compared with silicates of lime and magnesia
those of iron are of low melting point,[2] but it is plain that by entering
into combination and solid solution in the femic minerals, so-called, iron
silicates are constrained to act in much the same way as the lime and
magnesia silicates of these minerals in so far as their relation to the
feldspar crystallization series is concerned.

1 See Fenner, *op. cit.*, pp. 756-7.
2 It should be pointed out that while the melting point of hedenbergite may be lower than that
of diopside this relation can not be regarded as established by the evidence cited by Fenner (*op.
cit.*, p. 762). He quotes a melting point determined by a school of investigators who, at the time
of the determination, used a method quite incapable of giving a reliable result. One can find
support in their results of that date for a melting point of almost any silicate at about the same
temperature. This is true even of anorthite which melts 400° higher.

POSSIBLE EFFECTS OF THE SEPARATION OF HORNBLENDE

All of the deductions made in the foregoing as to the characteristic shapes of curves of the oxides apply, strictly speaking, only to the period of simultaneous separation of plagioclase and pyroxene. In the gabbro-granophyre association it is quite common to find pyroxene continuing to crystallize even into the granophyre stage.[1] In this sequence it may be expected that the deduced curves would show notable correspondence with the actual curves. In a very important modification of the sub-alkaline sequence, indeed the most important, hornblende comes in as a prominent constituent of the medio-silicic rocks and to its separation some consideration has already been given. We may now inquire into the factors which determine that hornblende shall separate and the effect of its separation upon the course of the curves. It has already been pointed out that the presence of water may be one of the factors influencing the formation of hornblende but that it is probably not the sole control, for there is a difference in the relative proportions of the other oxides as well. The character of this difference will now be pointed out.

In the Mull diagram of the sub-alkaline sequence (Fig. 27) the early courses of the curves depict a decrease of alumina, a rise of lime and other relations with which we need not now be concerned. If we examine the same curves in the variation diagrams of a typical series of rocks where hornblende is prominently represented in intermediate members, we find that towards the basic end the alumina curve rises with increasing silica. Subsequently it passes through a maximum and then falls. This relation is shown in the variation diagram of the Lassen Peak rocks (Fig. 25), and other hornblendic series are quite similar.

If we seek a cause of this difference in the alumina curve in the Mull sequence, on the one hand, and the same curve in a hornblendic series, on the other, we must consider which mineral is, at this stage, most potent in its influence upon the alumina curve. This is, plainly, basic plagioclase. If basic plagioclase crystallizes from the liquid and its reaction with liquid is limited, the alumina curve should fall slowly. On the other hand if it separates but reacts with liquid to produce more sodic plagioclase the alumina curve may rise, for this means a decidedly less net subtraction of alumina. It is, then, to a difference in the degree of reaction, at early stages, between liquid and the already separated plagioclase crystals that this difference in behavior of the alumina curve is probably to be attributed. The cause of the contrasted fractionation in the plagioclase reaction series is to be sought, in general, in the size of the differentiating mass and its effect on the rate of cooling. A rapid, but not too rapid, rate of cooling is conducive to a strong development

1 Mull Memoir, p. 21.

of zoning and therefore a high degree of fractionation in the plagioclase series. A slower rate of cooling permits much reaction between liquid and crystals with equalization of composition of the layers of the crystals and its consequent effect upon the course of the liquid. The slower rate of cooling permits, to be sure, the entrance of other factors in the production of fractionation, such as the sinking of the crystals, but this action is more readily affected by, say, convection in the case of plagioclase, for that mineral, unlike the femic minerals, is ordinarily not significantly different in density from the liquid from which it separates, particularly the more basic liquids. Ample opportunity for considerable reaction in the plagioclase series will therefore frequently be offered in a slowly cooled mass at least in early stages. We may expect a tendency towards the hornblendic line of descent when the differentiation has occurred in a large mass and towards a gabbro-granophyre association in a smaller mass.

The different courses which the liquids follow in consequence of the contrasted fractionation are depicted in the contrasted variation diagrams. When the course is of one kind the liquid soon attains such a composition that hornblende crystallizes from it. When the course is of the other kind the liquid never attains a composition that permits the formation of hornblende, unless it be at a very late, highly siliceous stage when an alkaline amphibole, not common hornblende, may form.

On account of the more restricted feldspar fractionation at early stages the Al_2O_3 content comes to be about 18 per cent in the sub-basic and medio-silicic members of the hornblendic sequence, whereas it is only about 13 per cent in the corresponding members of the gabbro-granophyre sequence. All of this is consistent with the same *general* shape of the early courses of the other curves, since it is merely a question of slight variation in feldspar fractionation, upon which the shape of the curves largely depends. But, with the appearance of hornblende, it is quite reasonable to expect a change of direction of the curves. Theory demands some change in the course followed by the liquid upon the appearance of a new crystalline phase but specifies nothing as to its magnitude. The change may be very small. Alteration of direction, and that a large one, has already been noted as occurring in the curves of the Mull diagram at the point where the separating assemblage changes from plagioclase and olivine to plagioclase and augite. In the case of a change of the separating assemblage from plagioclase and pyroxene to plagioclase and hornblende it is quite probable that the effect on the course of the curves would be very small. The reason for this statement will be apparent from the following considerations.

The curves of the various oxides have a certain direction at any given silica percentage because the crystal assemblage separating[1] from the liquid of that silica percentage has a composition which lies on the tangents to the curves at that silica value. A change of the crystal assemblage to any other assemblage which could be represented by other points along these tangents would not alter the direction of the curves in any way. Likewise a change in the crystal assemblage such that the new assemblage could be represented by points departing but little from these tangents would alter the direction of the curves but little. Now it is more than probable that a change of the separating assemblage from plagioclase plus pyroxene to plagioclase plus hornblende would fall in the latter class, for the total composition of the one assemblage may be very close to that of the other. Therefore only a small change of direction of the curves is a reasonable expectation. In the hornblendic line of descent, the change of direction of the Al_2O_3 curve from a rising to a falling slope may be due to the appearance of hornblende. Coincident with this change of the Al_2O_3 curve there must be a change in each of the other curves but these are evidently small. It is to be remembered that our information on any change of direction of the curves is obtainable only from analyses of a series of rocks that show a greater or less departure from the composition of the changing liquid. We can not hope, therefore, to locate any of them accurately at the present time but when a large number of analyses of the aphanitic or glassy groundmass of lavas of individual rock associations is available much more may be done. It is probable, indeed, that some of the convexity of the curve of MgO, and perhaps also of Fe_2O_3 and CaO, is an averaged-out discontinuity of slope coinciding with the point of change of direction of the Al_2O_3 curve, itself not accurately located in any given series of rocks and, of course, subject to variation in different series.

It is plain that the most we can hope to do at present is what has already been done, namely, to draw smooth curves which average the points for each oxide. When this is done an approximate picture of differentiation is obtained which, in the light of the facts we have just considered, can scarcely be regarded as anything but crystallization-differentiation. A notable departure of any rock of a series from the liquid line of descent may be produced by the fact that its composition has been affected by crystal accumulation. If there is reason to suspect any type of having this character it is well to leave it out in constructing a variation diagram.

VARIATION OF THE LIQUID LINE OF DESCENT

There is, in addition, a very great flexibility in the liquid lines of descent themselves, produced by varying degrees of fractionation. There

---

[1] By this is meant of course net separation or subtraction, which is the resultant of simple subtraction together with reaction.

may be great fractionation of one reaction series accompanied by moderate fractionation of the other. There may be strong fractionation in the one reaction series at an early stage which gives place to moderate fractionation of the same series at a later stage. The possible complexity is, therefore, very great and yet for the great bulk of rocks there is a sufficient approach to a "normal line of descent" that the foregoing detailed analysis is rendered possible.

At this point it is desirable to recall certain features of the deduced curved courses of crystallization pictured in Fig. 28. There are two extreme conditions with all intermediate possibilities; the one extreme gives a strongly curved course and the variation diagram (Fig. 29) shows strong bending of the corresponding curves of the oxides, especially K_2O and Na_2O. These curves converge rapidly as crystallization proceeds. The curve of Na_2O passes through a rather distinct maximum and is crossed by the curve of K_2O. When the later liquids are rich in potash the courses of Na_2O and K_2O are thus of relatively strong curvature.

At the other extreme is the condition in which there is very little curvature in the course of crystallization. In the variation diagram the curves of Na_2O and K_2O depart but little from straight lines, the maximum on the curve of Na_2O is ill-defined and it may not be crossed by the curve of K_2O. When the later liquids are relatively rich in Na_2O the curves thus tend to approach straight lines. There is no reason in the nature of things why these conditions should exist. They are solely the result of the control of crystallization. If the various liquids were formed as a result of other factors there is no reason why the curve of K_2O should not pass through a maximum and that of Na_2O through a minimum. If the controlling factors were the combination of all physico-chemical processes known to man and some others not yet known, it would be very remarkable if the curves did not show several maxima and minima and perhaps cross each other several times.

THE KATMAI ROCK SERIES

Among rock series the condition first described, high potash in the late differentiates and relatively strong curvature of the courses of the oxides, is represented by the general diagram of pitchstones given on a later page (Fig. 38). The second condition, relatively high soda in the late differentiates and little curvature of the courses of the oxides, appears to be represented by the series of rocks from Katmai recently described and analyzed by Fenner. Of these he gives a variation diagram on which the courses of the oxides are represented as straight lines.[1] His diagram is here reproduced as Fig. 35. The rocks are dominantly

1 C. N. Fenner, *Jour. Geol.*, 34, 1926, p. 700.

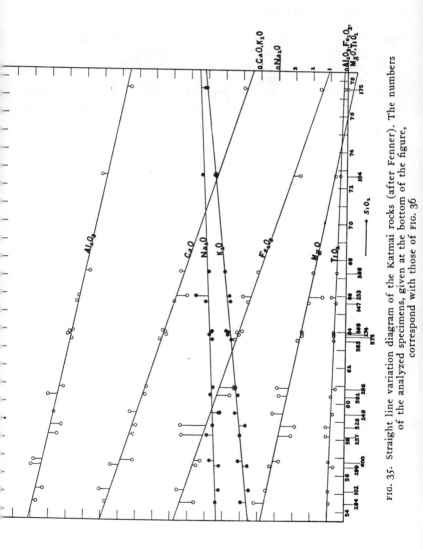

FIG. 35. Straight line variation diagram of the Katmai rocks (after Fenner). The numbers of the analyzed specimens, given at the bottom of the figure, correspond with those of FIG. 36

lavas and in accordance with the deduced tendency of such rocks there is but moderate scattering of the points. The most siliceous rock is a rhyolitic pumice with Na_2O somewhat higher than K_2O.

Without immediately raising the question whether the variation is really rectilinear, let us consider the slopes of the lines. With increasing silica, the curves of lime, magnesia, iron and alumina descend, while those of soda and potash rise. This is all in accord with the tendency of the separation of femic minerals and calcic plagioclase to bring about a continual increase in the amount of alkali feldspars and silica. More than this, the relative slopes of the curves are nicely adjusted to each other in such a way as to conform absolutely with this condition. There is not merely a decrease in this oxide and an increase in that, but the quantitative values of these changes are such as to express a changing composition of the liquid of the nature of an enrichment in alkali feldspar and quartz. This fact may be verified by a glance at the table of normative compositions (Fenner, p. 677). Fenner does not appear to have stated any conclusion as to the course of crystallization in his rocks but the facts for such rocks are quite well established. We may quote Iddings on the course of crystallization in nearly identical rocks of Electric Peak, Yellowstone Park, . . . "the earlier crystallizations were of hypersthene and augite and of part of the hornblende while the greater part of the hornblende and biotite followed. The labradorite commenced to crystallize early, while the alkali plagioclase, orthoclase and quartz closed the series."[1] Examination of Fenner's rocks and especially a determination of the nature of the glassy groundmass of the andesites, referred to later, would lead one to accept this general course for the Katmai series. It is plain that the course of these lines is an expression of the course of crystallization and that the contrast between successive members might be due to crystallization-differentiation.

We may now raise the question whether the straight lines Fenner has drawn really represent the variation as well as curves which depart somewhat from these lines. Necessity for departure from a straight line is most obvious in the MgO curve. As drawn by Fenner it runs off the diagram toward the higher SiO_2 percentages, which is of course an impossible condition. The curve of MgO must therefore be concave upward at the high silica end and, once this decision is made, it is plain from mere inspection that a curve which is concave upward throughout its length would average all the points better than a straight line.

The content of Na_2O increases definitely and fairly regularly from 54 per cent SiO_2 to 66 per cent SiO_2, but from 66 to 78 per cent SiO_2 (an equal increment of SiO_2) it does not increase further; indeed, a proper

1 J. P. Iddings, "The Mineral Composition and Geological Occurrence of certain Igneous Rocks in the Yellowstone National Park," *Phil. Soc. Wash. Bull.*, 11, 1890, p. 196.

averaging of the points would indicate a slight decrease. The highest determined Na_2O content (4.59 per cent) is in a rock with 65.59 per cent SiO_2. A curve of Na_2O which is convex upward and passes through a flat maximum would therefore average all the points better than any straight line. A maximum on the Na_2O curve at approximately the same SiO_2 value is a nearly universal condition in variation diagrams.

Three points at the high silica end and three points at the low silica end all lie above the straight line drawn by Fenner for CaO. Of the intermediate points only 3 lie above the line and 8 below it. It is plain that a curve which is concave upward, lying below the straight line in the middle portions and above it at both ends would average the points better than any straight line.

Averaging of the points for Al_2O_3 can be slightly, but very slightly, improved by substituting for the straight line of Fenner a curve which is convex upward as is this curve in similar rock series. The departure from the rectilinear condition is, to be sure, usually quite small but, when it is realized that the curves of MgO, of Na_2O, of CaO and probably of Al_2O_3 have precisely the same character as those deduced by Harker for the Lassen Peak rocks, additional confidence is to be placed in their curved character.[1] Aurousseau[2] finds curves for these rocks which show excellent agreement with Harker's, with a slight difference in the case of CaO. Furthermore, all of these curvatures are of the kind that one is led to expect from the application of physico-chemical principles to the process of fractional crystallization.

We find, then, that even when all the points determined by Fenner are given equal weight in fixing the curves, these curves deviate from straight lines in the expected manner. At the same time it can not be too strongly emphasized that in any test of the hypothesis of fractional crystallization the points do not have equal weight. We have already seen that the only points that might be expected to give smooth curves would be those for rocks belonging to or closely approaching the liquid line of descent. The rocks giving this condition best would be the glassy or aphanitic rocks and those which might depart most from this condition would be holocrystalline phanerites and porphyritic rocks with large phenocrysts. Greatest weight should therefore be given to rocks that are almost wholly aphanitic to glassy, and increasingly less weight to those with an increasing proportion of coarse crystals. Now when Fenner drew his straight-line diagram he found, of course, a deviation of some of the points from these lines. He singles out for special mention Nos. 257 and 526 as giving points which show greatest departure.[3]

1 Indeed, Harker finds the same types of curves for sub-alkaline rocks in general and offers "generalized variation diagrams" depicting these. *Natural History of the Igneous Rocks*, pp. 150, 151.
2 M. Aurousseau in Day and Allen, "The Volcanic Activity and Hot Springs of Lassen Peak," *Carnegie Inst. Wash., Pub. No.* 360, p. 38.
3 C. N. Fenner, *op. cit.*, p. 704.

Dr. Fenner has kindly permitted me to examine his specimens and slides, and of all the rocks analyzed No. 526 is certainly the very best and No. 257 among the best for the fixing of points lying on the liquid lines of descent. Specimen No. 526 is of a uniform aphanite nearly free from phenocrysts and No. 257 has but a small proportion of phenocrysts. Another analysis whose points show notable departure from several of the straight lines is that of specimen No. 253 and this again is among the best for determining the form of the curves in any test of the hypothesis of fractional crystallization. On the other side of the picture we have a group of four analyses clustering close to 64 per cent SiO$_2$ (Nos. 583, 575, 274 and 568) which show great agreement with each other in the position of the points for all their oxides and have thus had a predominant influence in determining the position of the central portions of the curves and their straight-line character, when all points are assigned equal weights. Yet their petrographic character is such as to show that they are among the very worst from the point of view of giving reliable information as to the composition of a liquid in the line of descent by fractional crystallization, which is the question we are testing. These specimens have nearly, if not quite, 50 per cent of phenocrysts about 2 mm in diameter lying in a glassy groundmass.[1] They *may*, therefore, depart widely from the composition of any liquid. That they *do* depart notably from such composition is shown by their disagreement with the nearby analysis, No. 253, which, as we have seen, must give a very close approximation to a liquid composition.

A study of specimens and slides of the 18 analyzed rocks shows that only 5 have a character which insures that they have a composition close to that of a liquid. These include the three already mentioned, Nos. 257, 526 and 253 and in addition No. 102 with 55 per cent SiO$_2$ and No. 175, the rhyolitic pumice with about 77 per cent SiO$_2$. In Fig. 36 all of Fenner's determined points have been plotted as circles but the points referring to the five above analyses are distinguished as double circles. Curves have then been drawn whose shape is determined entirely by these points representing the composition of the five rocks so chosen. These curves (Fig. 36) give the best information available as to the liquid line of descent in the Katmai rock series. One sees at a glance that the curves have the characteristic shape of those of the typical variation diagram of sub-alkaline series.[2] There is nothing unusual, therefore, about the Katmai rock series and the variation of the liquid, like that in

1 In two of these the glass is sufficiently unchanged that its refractive index may be determined and it is found to be close to 1.49. This value lies within the range for rhyolitic obsidians according to Tilley's determinations (*Min. Mag.*, 19, 1922, p. 279). We may therefore take it that the groundmass of these andesites is of rhyolitic composition though not as salic as the rhyolitic pumice of Katmai (n = 1.480 — 1.485, Fenner, p. 694).
2 The curve of iron oxides has not usually been drawn in the form determined by these points but in at least one diagram, that of the Mull normal series, it is so determined (see Fig. 27).

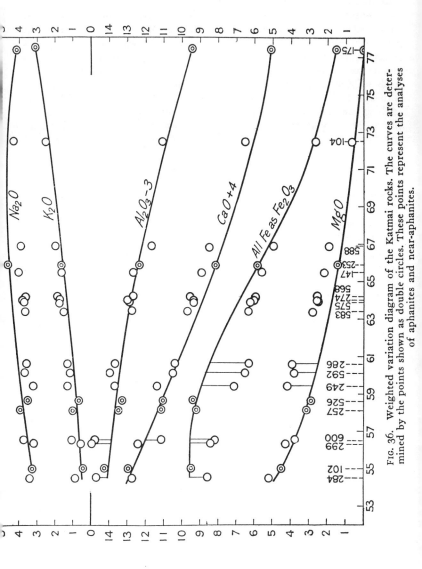

FIG. 36. Weighted variation diagram of the Katmai rocks. The curves are determined by the points shown as double circles. These points represent the analyses of aphanites and near-aphanites.

other series, is of the kind which theory demands in fractional crystallization.

A large proportion of the points in Fig. 36 show notable departure from the curves of the liquid line of descent. The rocks to which these points correspond always show a character which is consistent with a deviation from the liquid lines. Two of them are coarsely crystalline phanerites which may depart notably from a liquid but which permit no study of the causes of the departure. The other rocks are porphyritic with aphanitic or glassy groundmass and the nature of the phenocrysts enables one to deduce the possible kind of departure from liquid composition. The phenocrysts are of two classes, both present in essential amounts, plagioclase and a femic mineral, hornblende or pyroxene which latter may be of two varieties, augite and hypersthene. These are the early crystals of the rock series and their bulk composition must lie on a tangent to the curves at points towards the low-silica ends of the curves. According to the proportion of these crystals the composition of the groundmass of the rock must lie on the curves an appropriate distance on the high-silica side of the composition of the rock as a whole.[1] Any line which joins the groundmass composition with early crystal composition must therefore be a chord of the curve of liquid composition. Points representing the composition of rocks consisting of an aphanitic groundmass and early crystals will therefore lie on this chord, i.e., they will lie on the concave side of the curve of liquid composition.[2] It is noteworthy that in Fig. 36 practically all the points that deviate from the curves of liquid compositions lie on the concave side of the curves. When curves are constructed which give equal weight to all the points and when two-thirds of the rocks analyzed are these porphyritic types it is only natural that the curves should vary toward straight lines from the true curves of liquid descent and should have lost nearly all, but not quite all, their characteristic curvature.

It is frankly admitted by Fenner that no known process, not even his favored process of gaseous transfer, will account for the straight-line variation of liquid composition which is implicit in his discussion of the Katmai rock series and there is apparently still less capability of explaining deviation of points from the curves. Curves of liquid variation constructed in the manner of Fig. 36 are susceptible of a rational explana-

---

1 The rhyolitic groundmass of the andesites which have 64 per cent SiO_2 furnishes an example.
2 The possible directions of deviation of points from the liquid lines of descent are, of course, not exhausted in this discussion. The deviation is of the kind noted only when both kinds of early crystals (femic and feldspathic) are prominently represented among the phenocrysts, in other words when there has not been strong sorting. With much sorting points may be widely scattered on either side of the curves, not, however, in a wholly random way but in a manner determined by the strong preponderance of some one type of phenocryst. Sorting has probably been a prominent factor in the formation of rocks in Alaskan regions adjacent to Katmai, to be mentioned on a later page.

tion in terms of fractional crystallization as is also the manner of deviation of points from these curves.

Another possible explanation of the manner of deviation of points on this diagram is by hybridism. But the lack of aphanites of the compositions corresponding with the aberrant types shows that such hybridism as may have occurred in the formation of the analyzed types was not in any important degree a simple addition of one liquid to another or a simple solution of rock in a liquid, although Fenner offers evidence of some action of that kind. If hybridism occurred in significant amount it was of the reactive type discussed in a later chapter which insures that the liquid portion remains on the normal lines of descent, a fact which the petrographic character of the analyzed specimens attests.

Against the hypothesis of fractional crystallization Fenner urges that the Katmai rocks show hornblende in some types and in others the hypersthene-plagioclase assemblage which takes the place of hornblende, and that they should therefore belong to two different lines of descent. There is some evidence, however, to indicate that only the hornblendic line of descent is represented. The facts suggest that the total composition of any rock was largely, if not entirely, determined by differentiation in a magma chamber where the hornblendic line of descent was followed. Liquids prepared in this chamber may, however, have been moved to higher levels and there have crystallized in whole or in part to the hypersthene mineral assemblage, but when this happened the part of the liquid so crystallizing had reached the end of its career as a producer of further liquids by fractional crystallization. The evidence that this is so we find in Fenner's statement that hypersthene is confined to the effusive rocks, which suggests its crystallization, in the Katmai rocks at least, under surface or near surface conditions and confirms the opinion that no new line of descent is there initiated by the appearance of hypersthene. In addition Fenner himself notes (p. 692) that in the pyroxenic types there is evidence that hornblende was formerly present. We thus find that the diverse mineralogy displayed is quite consistent with the determination of the bulk composition of any type by a single line of descent (hornblendic) and the general agreement of the points of the Katmai diagram with smooth curves is probably an expression of the fact that the rocks did correspond closely with a single line of descent. The data are inadequate for a definite decision on this point. During the crystallization of any large body of magma it is inevitable that the liquid follow a somewhat different course in different parts of the mass. It is possible that in the intrusive and extrusive derivatives of the mass this flexibility of the liquid descent may be revealed if the derivatives come from different parts of the mass. The problem could be solved by a sufficient number of analyses of the aphanitic groundmass of rocks of the series. It will never be solved by determination of the bulk composition of coarse granular or of porphyritic rocks.[1]

Fenner gives ten other analyses of rocks from areas adjacent to Katmai and when he attempts to plot these on the straight-line diagram he finds the widest divergence (Fig. 8, p. 707). Believing that the straight-line character is the result of some process which must give such variation, he is thus forced to appeal to a different process for the rocks of adjacent areas. These nearby rocks are of the same general types and one is naturally reluctant to accept a different process. No detailed petrographic descriptions of the rocks are given, but if one considers them in the light of fractional crystallization and sets aside coarse, equigranular rocks it is probable that with the aid of the curves of liquid descent of Fig. 36 one could predict which of the rocks might be altogether aphanitic, which of them could be only partly aphanitic with phenocrysts in addition and at the same time the probable nature and abundance of these phenocrysts. Only if these predictions failed of realization in the rocks would it be necessary to appeal to a different liquid line of descent and even then due consideration should be given to the possibility that this is a variant of the liquid line of descent produced by crystallization. In these neighboring rocks the deviations are of the nature of a greater scattering of the points which indicates notable crystal sorting as a factor in their formation.

GENERALIZED VARIATION DIAGRAM AND ITS SIGNIFICANCE

In the foregoing study of the general significance of variation diagrams only those of sub-alkaline series have been considered. Even in these there is some flexibility of the liquid lines of descent the principles of which have been discussed with the aid of Figs. 28-34. Alkaline series represent still wider variation, to which some study will be given on a later page.

The general character of variation diagrams is of much significance in certain broader problems of petrogenesis. If granite and basalt were due to a splitting of dioritic magma then the series basalt-diorite-granite would give rectilinear curves for the oxides. The characteristically curvilinear shape is thus in itself sufficient to rule out that hypothesis even apart from the detailed consideration of the signs of the curvature which we have seen to point indubitably to crystallization as the controlling factor. The curves of oxides in the average composition of rocks of the types, basalt, andesite, dacite and rhyolite show the deduced shape (see Fig. 37). In such averages the effects of variation in the liquid line of descent and also the effects of crystal accumulation are averaged out, leaving the curves of characteristic shape.

1 See also Tsuboi, *Jour. Imp. Univ. Tokyo*, II, Vol. 1, pt. 2, 1925, p. 81,

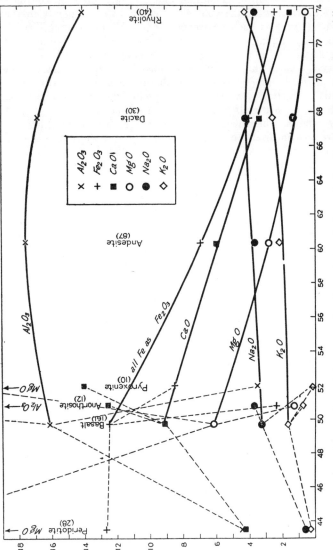

FIG. 37. Variation diagram for Daly's average basalt, andesite, dacite and rhyolite (the full curves). In addition three dotted lines radiate from each oxide point for basalt towards the corresponding oxide point for peridotite, pyroxenite and anorthosite (Daly's averages). These dotted lines show the random relation of the sorted rocks to the rocks of the average liquid line of descent (full curves). The number in parenthesis with each rock name gives the number of rocks averaged.

In many igneous series all the rocks from basic to acid can be plotted on curves that require little smoothing and this fact, taken in conjunction with the evidence of the control of crystallization, points to the basic (basaltic) magma as the parental liquid. All rocks more "basic" than the parental liquid should, theoretically, tend to give points that are characterized by a scattering and therefore by departure from smooth curves because their composition must be determined by crystal accumulation. Rocks more "basic" than basalt, namely the ultrabasic, do have this character, as a class. Points giving the oxide composition of peridotite, pyroxenite and anorthosite are shown in Fig. 37 and dotted lines joining them with the points for basalt. These lines are wholly unrelated in direction with the full curves of the liquid line of descent. Some scattering of points can occur anywhere along the lines of descent as a result of crystal accumulation, and this is the principal cause of the needed smoothing, but of only one type of rocks can it be said that scattering of points is characteristic of the class and that is the ultrabasic.[1] If basaltic rocks were normally derived from intermediate magmas then basaltic rocks as a class would be characterized by a scattering of the points indicating their oxide contents on such diagrams. We thus find strong petrologic confirmation of the geologic indications that parental magma is of a basic (basaltic) character.

[1] This matter is discussed further in Chapter IX, which treats such rocks.

CHAPTER VIII

THE GLASSY ROCKS

I N THE preceding chapter a discussion has been given of the variation diagrams of rock series and of the light which they throw upon the liquid lines of descent. We have found that the smoothed curves of variation diagrams frequently give a close approximation to the changing composition of the liquid as crystallization proceeds. There is another source of information regarding the composition of liquids which gives this composition accurately but is not as widely applicable. It is a study of the composition of glassy rocks. They are the only rocks of which we can say with complete confidence that they correspond in composition with a liquid.[1] Of most aphanites the same statement could be made with nearly as great confidence, but one must be on guard against the possibility that the fine grain is a secondary character produced, say, by granulation or replacement. Porphyritic rocks and granular phanerites often correspond very closely in composition with the liquid from which they formed, but in many instances there is great departure of such rocks from the composition of any liquid. It may be very difficult or even impossible to determine the extent of this departure with the aid of any characters revealed by an individual occurrence of a certain rock type. When the rocks of any area are viewed broadly, however, it may be possible to reach some conclusions on this question, and in a later chapter the information to be obtained from Hebridean rocks is discussed. Again, if rocks as a whole are considered, some light may be shed on the general problem.

Glassy rocks are particularly important in this latter respect. They are worthy of a more extensive study, especially in the direction of ascertaining the composition of glasses of individual rock associations, but systematic information of this kind is not yet available for individual fields or provinces. We must therefore confine our attention to a general survey of the composition of glasses. It is especially to be noted that they are not necessarily lavas but may be dikes or the glassy selvages of intrusives elsewhere crystalline, so that in mode of occurrence they are not particularly restricted.

1 We are not for the moment considering such volatile constituents as may be lost on cooling.

There are no known glasses corresponding with any of the ultrabasic types, peridotites, pyroxenites or anorthosites. It is probable that peridotite and pyroxenite magmas could not be chilled rapidly enough to give a glass so that absence of glasses of these compositions can not, in itself, be regarded as of great importance. They could, however, give spherulitic and aphanitic rocks, and to this question we shall return at a later point. A magma of anorthositic composition could be readily quenched to a glass, yet glasses of such composition are quite unknown, a fact which is, we believe, to be referred to the non-existence of magmas (liquids) of such composition.

Many other chemical differences between glasses and crystalline rocks are to be seen but they are always of one nature and are an expression of the much more limited range of composition shown by glassy types than by crystalline types. Besides the absence of glassy representatives of whole groups as already mentioned (e.g., anorthosites) there is an absence of glassy types of extreme composition within individual groups, a feature which will be discussed in connection with rocks of granitic composition.

In Fig. 38 have been plotted the percentages of the principal oxides of all rocks which have been called pitchstones or perlites (with four tachylites) and are given among the superior analysis of Washington's Tables. In all, 30 analyses are plotted. It is not apparent just what fundamental character of a glassy rock determines that it shall be called pitchstone. The resinous appearance is ordinarily considered to be connected with a high content of water but this is apparently only a tendency. Some rocks termed obsidian have as much as 10 per cent H_2O and some termed pitchstone have as little as 2 per cent. In any case it appears that no rock of a very decided alkaline character has been termed pitchstone and curves can be drawn that represent the variation of the oxides fairly well. They have the typical form of the curves of sub-alkaline rock series. Such curves are drawn in the figure. It will be noted that K_2O rises at the high-silica end to such a value as 8 per cent or more and Na_2O at the same time may fall practically to 0 per cent.

In Fig. 39 have been plotted the (superior) analyses of all other glassy rocks given in Washington's Tables. Of these there are 62 and of their number 44 are rhyolitic obsidians containing from 70 to 77 per cent SiO_2. The others are mainly of medium silica content and include not only andesitic types but also alkalic types such as trachyte obsidians. Corresponding with the fact that a very wide variation of liquid line of descent is represented, no simple curves can be drawn to indicate the change of the oxides. In medio-silicic rocks there is wide variation. Thus when SiO_2 is 61-62 per cent, K_2O varies from 1 to nearly 8 per cent, Na_2O from 3 to more than 8 per cent and other oxides may vary

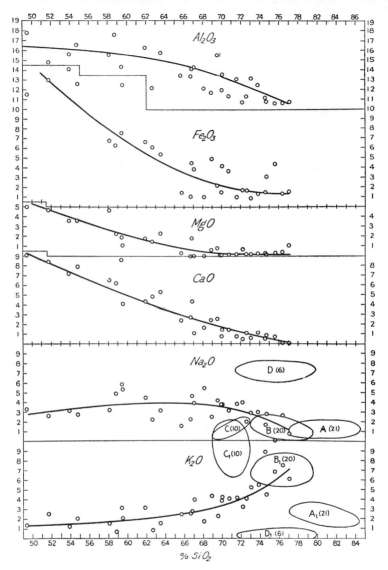

Fɪɢ. 38. Variation diagram of 27 pitchstones and perlites and 4 tachylites from Washington's Tables (1917). The areas A-D embrace the Na_2O content and A_1-D_1 the K_2O content of various types of crystalline rocks (the number of analyses given in parenthesis) of granitic composition illustrating much wider range of variation of crystalline rocks as compared with glassy rocks.

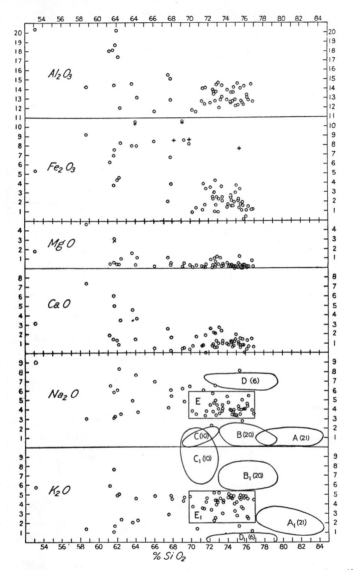

FIG. 39. Variation diagram of 62 glassy rocks (other than pitchstone and perlite) from Washington's Tables (1917). Five have less than 11 per cent Al_2O_3 and the corresponding points thus lie below the Al_2O_3 area in the Fe_2O_3 area where they have been plotted as crosses to distinguish them from Fe_2O_3 points. For explanation of areas A-D see Fig. 38.

nearly as widely. This is because the greatest divergence between the different liquid lines of descent occurs at medium silica content. In an individual field one might be able to separate different lines of descent on the basis of field association, as for example the normal and the alkaline lines of descent of the Mull authors.[1]

In the more siliceous glasses, the rhyolitic obsidians, there are, even in the presence of variation some noteworthy tendencies. There is not the same trend as in the pitchstones toward a very high K_2O value and a very low Na_2O value at the higher silica content. In only one analyzed rhyolitic obsidian is Na_2O as low as 2.3 per cent and in only one does K_2O amount to 5.2 per cent. There is indeed a general tendency for these obsidians to cluster around a K_2O content of about 4.5 per cent and an Na_2O content of about 4 per cent or a little less.

The striking feature is the much greater restriction of composition of the glassy rocks (rhyolitic obsidians) as compared with crystalline rocks of the same general class (granites, porphyries, etc.). In Fig. 39 two rectangles E and E_1 have been drawn in the Na_2O and K_2O sections of the diagram and practically all the 44 rhyolitic obsidians fall within these rectangles. Only one shows great departure and this only with respect to its Na_2O content.

The corresponding crystalline rocks are dominantly of composition analogous to the obsidians but they show in addition a great number of types of much wider variation. The principal groups with notable departure are indicated on the diagram (Fig. 39) by ovoid areas which embrace the Na_2O and K_2O contents of the granites of these groups. Thus the areas designated A and A_1 represent the Na_2O and K_2O content, respectively, of a large group of granites (the numbers in parentheses give the number of analyses) very high in SiO_2 and rather low in both alkalis. There are no corresponding rocks among the obsidians. The areas B and B_1 embrace the Na_2O and K_2O content of a group of rather siliceous granites, high in K_2O and very low in Na_2O, without corresponding rocks among the obsidians. The areas C and C_1 represent the Na_2O and K_2O content[2] of a group of granitic rocks of somewhat less silica content with extremely high K_2O and but little Na_2O. There are no corresponding glasses. The areas D and D_1 embrace the Na_2O and K_2O content of a group of siliceous granites very high in soda, with potash nearly lacking. Again there are no corresponding obsidians. One obsidian, the exceptional one already referred to, has as high Na_2O as this group, viz., about 8 per cent, but also has 4.5 per cent K_2O, the proportion being thus much more nearly balanced than in the crystalline rocks of similar tendencies.

[1] Mull Memoir, Figs. 2, 4.
[2] The area C_1 extends beyond the 10 per cent limit of the K_2O area, some of the rocks of this group containing 12 per cent K_2O.

The same general observations may be stated in the language of the norm classification. Granular rocks and rocks with porphyritic elements, such as porphyries and rhyolites, are found in all subrangs. On the other hand, of 44 obsidians 31 fall in sodipotassic subrangs and 13 are dosodic. None are persodic and none dopotassic or perpotassic.

When these ovoid areas representing the alkali compositions of extreme granites are laid on the pitchstone diagram (Fig. 38) the same avoidance of the points indicating the composition of glassy rocks is noted. There is, however, an exception. The areas B and B_1, representing the composition of high potash granites, embrace typical pitchstone compositions. This fact will be discussed more fully at a later point. Suffice it here to note the large number of crystalline rocks of granitic composition for which there are no glassy equivalents either among the pitchstones or the obsidians.

If the differentiation of rock-types is due to processes taking place while magmas are entirely liquid, say to liquid immiscibility or to transfer of material from one part of the liquid to another by gases, then all of the liquids so produced have an approximately equal chance of being represented among glassy rocks and there should be no restriction of composition of the glassy rocks as compared with crystalline rocks. On the other hand, if differentiation is due to crystallization then the glassy rocks should be restricted to possible liquids formed during the course of crystallization. Crystalline rocks will include such compositions but will also include many rocks representing crystal accumulations from these liquids, separated from their mother liquor in one way or another, but probably by filter-pressing in the case of the siliceous rocks we have just discussed. The fact that crystalline rocks do show ranges of composition unmatched in glassy rocks may be regarded as tending to confirm crystallization-differentiation.

It is not to be regarded as probable that every particular instance of departure of, say, a granitic rock from obsidian or pitchstone compositions is evidence of its derivation by crystal accumulation. No doubt many rocks of a dominantly igneous character have acquired extreme compositions through various secondary processes such as impregnation with matter transported in aqueous solutions or even in vapors. This is, nevertheless, an action distinct from igneous differentiation itself, however closely connected with and dependent upon igneous processes it may be. The restriction of composition of the glassy rocks is strongly inimical to the conception that extreme compositions can be produced in liquid magmas themselves by any process of so-called gaseous transfer. (See also pp. 295-6.)

We may now return to the fact already noted that certain high potash granites correspond in composition with pitchstones, which pitchstones may, therefore, represent the liquids from which such granites are formed. These granites are, however, fairly abundant and if they are normally formed directly from a liquid of their own composition one would expect their glassy equivalents to be of frequent occurrence. Pitchstones are quite rare and the suggestion is worth consideration that they are the glassy equivalents of pegmatites, in particular of graphic granite pegmatites. They show similar contrasts in their alkali contents but pitchstones are distinctly more siliceous, averaging about 38 per cent normative quartz whereas graphic granite averages about 25 per cent. Possibly, therefore, the pitchstones may be regarded as the glassy equivalents of pegmatites plus quartz veins. The quartz veins would then be formed from the highly aqueous residuum after the crystallization of pegmatite from a liquid with pitchstone composition. This vein-forming liquid is not to be regarded as consisting of SiO_2 and water only, but must contain considerable quantities of other substances which render silica soluble and which are themselves too soluble to be precipitated with the silica. A large amount of liquid requires to circulate through the vein in order to deposit a relatively small amount of quartz. Such liquids are in no sense magmas as the term is ordinarily used, for this usage implies a liquid which, by a single act of injection into a fissure, with subsequent congelation, is capable of filling the fissure with a solid rock.[1] Quartz veins are plainly not formed from any such liquids. If they were we should have such veins with silica glass selvages, and there is no reason why we should not have effusive rocks of silica glass. The nearest approach to this we have is the siliceous sinter deposit of hot springs which in its genetic relations, particularly in the nature of the liquid from which it is deposited, is probably to be taken as a strict analogue of quartz veins.

Glassy rocks, formed as they are by simple, rapid cooling, permit little latitude in the interpretation of the nature of the material from which they formed. Such rocks are sharply limited at a silica content of 77 per cent (see Figs. 38 and 39)[2] and probably represent the extreme of what can be appropriately designated magma in accordance with accepted usage as stated above. "Granites" of very high silica content may be the result of crystal accumulation or they may be the result of silicification of other granites or again of granitization of a highly siliceous rock such as quartzite. It may be impossible in individual instances to determine which mode of origin is to be preferred, but the absence of glassy equivalents rules out their formation by simple consolidation

1 Schaller, *Amer. Mineralogist*, 12, 1927, p. 59 ; also Brögger in Tom Barth, *Vidensk. Akad. Oslo, Skr. I, Mat.-Naturv. Kl.*, 1927, No. 8, p. 112.
2 In plotting these diagrams the analyses have not been reduced to a water-free basis, yet it so happens that all those rocks with highest amounts of SiO_2 have so little water (even among the pitchstones) that the SiO_2 content is limited at approximately 77 per cent even on a water-free basis.

of a magma. It has frequently been demonstrated that the very high SiO$_2$ content of some rhyolites is of secondary origin.[1]

There is one type of extreme granite for which some indications are forthcoming as to a choice between the above processes of origin. The absence of any glassy rocks corresponding with granite very rich in albite is important in connection with the tendency of recent investigators to refer albite rich rocks to a replacement process.[2] One would expect glassy rocks of such composition if there were magmas of corresponding composition. On the other hand, if the richness in albite is the result of a secondary process of impregnation of a solid rock with albite by circulating aqueous solutions, we should not expect glassy rocks rich in the albite molecule. Such solutions would be incapable of consolidation in toto, a process which is necessary for the formation of a glassy rock. The absence of glassy rocks rich in the albite molecule is thus to be taken as confirmatory of the concept that albite-rich rocks are formed by a process of replacement, probably in all cases. Rocks consisting of albite and calcite belong here. There is nothing in the manner of their occurrence to warrant belief in their formation by simple consolidation of a magma, nor is there in "carbonate dikes."

Summing up the evidence of the glassy rocks, we find that the notable restriction of composition of these rocks as compared with that of primary crystalline rocks is to be referred to the natural restriction of the possible compositions of the changing "mother-liquors" in a crystallization sequence as compared with the wide range of possible composition of the crystalline deposits from these liquids. The facts are destructive of any hypothesis which assumes the formation of the more extreme rock types by any process taking place in a wholly liquid magma, such as unmixing of liquids or transfer of materials from one part of the liquid to another by gases.

1 H. C. Richards, *Proc. Roy. Soc. Queensland*, 34, 1922, p. 195.
2 Larsen, *Economic Geology*, 23, 1928, pp. 398-433; Schaller, *Amer. Jour. Sci.*, 10, 1925, p. 279; Hess, *Eng. and Min. Jour.*, 120, 1925, pp. 289-98; Landes, *Amer. Mineralogist*, 10, 1925, p. 411.

CHAPTER IX

ROCKS WHOSE COMPOSITION IS DETERMINED BY CRYSTAL SORTING

INTRODUCTION

IN THE two preceding chapters attention has been directed particularly to rocks belonging to or approaching in composition the liquid lines of descent. It was intimated that certain rocks, as originally constituted and quite apart from any secondary process, may depart from such compositions, in some instances moderately, in others very markedly. This departure is believed to be due to crystal accumulation accompanied by greater or less sorting of various kinds of crystals. Attention will now be turned to these rocks. In adducing the evidence of their character and origin we shall stress certain examples because it is believed that they present the evidence in clearer form. These examples are found among the rock associations of the Scottish Hebrides.

The Western Isles of Scotland constitute an igneous province that has probably contributed as much to the growth of present-day concepts of igneous geology as any other single area upon the earth. In the days when igneous rocks were struggling for recognition as such this area was a hard-fought and hard-won battlefield and, from that time to the present, controversy has raged over various aspects of igneous geology there displayed. The reason for the great interest shown in the region lies mainly in its natural advantages. The igneous rocks are dominantly of Tertiary age and for the most part unmetamorphosed. Every expression of igneous activity from surface flows to deep-seated masses is to be found and, accompanying this, a wide range of chemical composition. The region has suffered glaciation, which makes for good exposures of fresh rock, and is at the same time one of considerable relief so that structural relations may be examined in three dimensions. It is not surprising that such a region should have attracted investigators of the first rank during a century or more.

Out of the numerous controversies upon this area there has come a large measure of agreement upon the outstanding facts and but little less agreement upon the broader interpretation of these facts. Volcanic activity has been a dominant feature and surface flows are the most widespread rocks, constituting lava plateaux, once of enormous extent but now broken up by faults and partly submerged beneath the sea. These lavas are, to a great extent, the product of fissure eruptions and in consonance with this they are principally of basaltic composition. Equally widespread are regional dikes, in part the feeders of these flows. Supplementing this general activity local centers of eruption came into being and in these, besides surface lavas, there have been formed deeper-seated masses now partly exposed by erosion. Here we find the principal variation of composition in the rocks.

The most important and the most-studied of the eruptive centers are located in Skye and in Mull. The principal study of Skye has been made by Harker,[1] though he has, of course, in some measure built upon the work of others, notably Geikie and Teall. The principal study of Mull is the result of the collaboration of a large group of workers.[2] The igneous geology is principally by E. B. Bailey, C. T. Clough, and H. H. Thomas and, of earlier workers in Mull, Judd is particularly to be mentioned. The Skye and Mull Memoirs will ever be among the classics of geology.

The authors of these memoirs are thoroughly agreed upon the dominance of crystallization as a factor controlling differentiation in their areas. On only one major point were the Mull authors in doubt and that was on the possibility of the derivation of their Non-Porphyritic Central magma-type from their Plateau magma-type by crystallization-differentiation. This question we have already discussed and have shown that a very satisfactory solution is afforded by crystallization-differentiation, whereas the suggested alternative, assimilation, is apparently out of the question.[3] There is thus notable agreement of the findings in these fields with the thesis of the present study. If, therefore, in the following, we shall raise questions not raised by the investigators of these rocks or, to certain questions, suggest answers differing from those given by them, the reader should always bear in mind the essential agreement on the central fact of the control of crystallization.

THE PORPHYRITIC CENTRAL MAGMA-TYPE OF THE MULL AUTHORS

To introduce some of these questions we shall turn first to Mull. The rocks of Mull are lavas, dikes, sheets and major intrusions in each of which intrusion form there is much variety of composition. It has already been pointed out that the term, "magma-type," has been intro-

---

1 "The Tertiary Igneous Rocks of Skye," *Mem. Geol. Surv. United Kingdom*, 1904.
2 "Tertiary and Post-Tertiary Geology of Mull, Loch Aline and Oban," *Mem. Geol. Surv. Scot.*, 1924.
3 See pp. 75-8.

duced by the Mull workers. "The conception of magma-type is based upon composition alone. In this, it differs from the conception of rock-type which takes into account texture as well as composition. Thus a basalt and a gabbro may belong to one magma-type though admittedly representatives of different rock-types."[1] In introducing the magma-type the Mull authors have performed an important service, for one can take each type and examine its range of textural variety. In this way, as we hope to show, a good deal of information is to be obtained as to the genesis of types. The probable parental magma is the Plateau magma-type which "is known with either a basaltic or doleritic crystallization attaining to gabbroic in certain massive cone-sheets."[2] Of other basic magma-types there are two, the Non-Porphyritic Central type and the Porphyritic Central type. In making this distinction a fact of fundamental importance in Hebridean igneous geology has been isolated. The Non-Porphyritic Central type "is known as a partially devitrified glass and with every grade of crystallization from this onwards to gabbroic."[3] On the other hand, the Porphyritic Central type is "always porphyritic in its finer crystallizations, with small or large phenocrysts of basic plagioclase feldspar. The base may be partially vitreous and often variolitic in chilled portions of pillow-lavas and minor intrusions; elsewhere it is anything from basaltic to gabbroic in texture. In the gabbros the porphyritic tendencies of the feldspar may be lost sight of, or at any rate much obscured."[4] Remembering now that the distinction between these two types is purely chemical it is a fact of fundamental significance that the types can be described and distinguished by textural terms. It is evident that there were liquids corresponding in composition with the Non-Porphyritic magma-type for it is "*known as partially devitrified glass.*" It is equally evident that there were no liquids corresponding in composition with the Porphyritic magma-type for it is "*always porphyritic in its finer crystallizations.*" No matter how rapidly chilled the magma may have been, it was impossible to cause all of the feldspar to enter into the glassy to aphanitic groundmass. Always some of the feldspar occurs as phenocrysts. There is only one conclusion to be reached and that is that the porphyritic feldspar never was in solution in the liquid now quenched to an aphanitic or partly glassy groundmass. The phenocrysts represent a certain excess over liquid. They were, of course, once in solution in some liquid but the liquid did not correspond in composition with the bulk composition of the Porphyritic magma-type for, in that case, there would be no reason why the Porphyritic magma-type should be "*always* porphyritic in its finer crystallizations." The

1 Mull Memoir, p. 13.
2 *ibid.*, p. 14.
3 *ibid.*, p. 18.
4 *ibid.*, pp. 24 and 25.

phenocrysts of plagioclase are usually about Ab_1An_4 and they must have accumulated from a considerable mass of liquid. The manner of their accumulation we shall not attempt to decide. Presumably at such an early stage, movement of crystals under the action of gravity is the only process available, but whether the movement would be upward or downward is not determinable with certainty. No doubt other crystals were separating at the same time but their motion differed either in magnitude or in direction. While the Porphyritic magma-type thus represents a certain liquid plus plagioclase phenocrysts it also represents an earlier liquid plus plagioclase minus olivine and/or augite. If the mass is injected into surrounding rocks at this stage, to form an intrusive of moderate dimensions, or is poured out on the surface it gives a dolerite or a basalt with feldspar phenocrysts. If, on the other hand, it cools where it had its origin or is moved elsewhere as a large mass, the continued outgrowth of crystal boundaries gives a gabbro in which "the porphyritic tendencies of the feldspar may be lost sight of, or at any rate much obscured." It is clear that the plagioclase phenocrysts are excess material. The unfailing porphyritic character not only proves this fact but, in addition, it proves that the phenocrysts of plagioclase, accumulating by whatever method, were never redissolved. Otherwise there would be available somewhere a rapidly cooled mass having the total composition of the Porphyritic magma-type and containing all of its plagioclase as a constituent of the aphanitic ground. There are no such rocks in the region, at least, none has been found.

In the face of the large amount of work that has been done on these rocks by organized surveys whose collections are preserved for comparison during each succeeding investigation, it is very unlikely that, if such rocks exist, they should have escaped collection and subsequent determination of their character in petrographic and chemical studies. Harker has noted the same fact that is so emphatically stated by the Mull workers. Of basic sills in the Small Isles he states, "A decided porphyritic character seems to be connected with the composition of the rocks which exhibit it, these having on the whole a greater preponderance of felspar than the others."[1] He thus states it rather as a tendency, yet no specific exceptions are noted. In view of these facts the emphasis of the Mull authors seems to be justified. All the more is this true when it is found in a broad survey of the composition of basaltic rocks that a porphyritic character is a perfectly general one in basaltic rocks notably rich in basic plagioclase. Such a survey is given on a later page.

It is especially to be borne in mind that we are not contending that the phenocrysts of all porphyritic rocks represent an excess of those crystals and that no liquid having the bulk composition of a porphyritic

---

1 "Geology of the Small Isles," *Mem. Geol. Surv. Scot.*, 1908, p. 122.

TABLE IX
PORPHYRITIC CENTRAL MAGMA-TYPE

	Dolerite	Gabbro			Basalt		
	I	A	B	II	III	IV	V
SiO₂	45.54	46.39	47.28	48.34	47.24	47.49	48.51
TiO₂	1.06	0.26	0.28	0.95	1.46	0.93	1.46
Al₂O₃	23.39	26.34	21.11	20.10	18.55	21.46	19.44
Cr₂O₃	....	tr.	....	....	....	....	....
Fe₂O₃	1.98	2.02	3.52	1.97	6.02	1.72	5.66
FeO	6.98	3.15	3.91	6.62	4.06	4.80	4.00
MnO	0.27	0.14	0.15	0.32	0.31	0.15	0.23
(Co, Ni)O	....	....	....	nt.fd.	0.05	0.04	0.04
MgO	4.60	4.82	8.06	5.49	5.24	4.59	5.12
CaO	11.82	15.29	13.42	13.16	11.72	13.24	12.03
BaO	....	....	....	0.10	nt.fd.	nt.fd.	nt.fd.
Na₂O	2.50	1.63	1.52	1.66	2.42	2.17	2.53
K₂O	0.44	0.20	0.29	0.98	0.15	0.42	0.25
Li₂O	....	....	....	nt.fd.	nt.fd.	nt.fd.	nt.fd.
H₂O + 105°	0.72	0.48	0.53	0.44	2.24	2.54	0.48
H₂O at 105°	0.62	0.10	0.13	0.02	0.21	0.17	0.04
P₂O₅	0.13	tr.	tr.	0.04	0.26	0.43	0.16
CO₂	....	....	....	0.11	0.19	0.08	0.09
FeS₂	....	....	....	nt.fd.	nt.fd.	nt.fd.	nt.fd.
	100.05	100.82	100.20	100.30	100.12	100.23	100.04
Spec. grav.	2.85	2.85	2.90	2.93	2.85	2.82	2.93

NORMS

	Dolerite	Gabbro			Basalt		
	I	A	B	II	III	IV	V
Q	....	....	....	....	4.02	....	4.32
Or	2.78	1.11	1.67	5.56	1.11	2.22	1.67
Ab	17.82	13.62	12.58	14.15	20.44	18.35	20.96
An	51.15	63.94	50.04	44.48	39.20	47.82	40.87
Ne	1.70	....	....	....	....	....	....
Di	7.76	9.58	13.00	17.03	13.31	13.39	13.92
Hy	....	1.06	12.81	10.32	7.00	11.69	6.70
Ol	12.90	7.47	3.69	4.25	....	....	....
Mt	3.02	3.02	5.10	2.78	8.82	2.55	8.35
Il	1.98	0.46	0.61	1.82	2.89	1.67	2.89
Ap	0.34	....	....	....	0.67	1.01	0.34
Total Feldspar	71.75	78.67	64.29	64.19	60.75	68.39	63.50
% An in Norm. Plag.	74	83	80	76	66	72	66

I—(15994, Lab. No. 389.) Sill. Hillside between two streams S. of Coire Buidhe; Mull. Small-Feldspar Dolerite, p. 285. Anal. E. G. Radley.

A—(8043, Lab. No. 18.) Major Intrusion, Cuillins. Sligachan River; Skye. Olivine-Gabbro, quoted from Harker, *Tertiary Igneous Rocks of Skye*, 1904, p. 103. Anal. W. Pollard.

B—(8194, Lab. No. 19.) Major Intrusion, Cuillins. Coir 'a 'Mhadaidh; Skye. Olivine-Gabbro, quoted from Harker, *ibid.*, p. 103. Anal. W. Pollard.

II—(14846, Lab. No. 373.) Major Intrusion, Beinn na Duatharach. Five-eighths mile N.N.W. of summit of B. na Duatharach; Mull. Olivine-Gabbro. Anal. E. G. Radley.

III—(18469, Lab. No. 445.) Lava. One-half mile S.S.W. of Derrynaculen; Mull. Basalt, Porphyritic Central Type, p. 148. Anal. E. G. Radley.

IV—(18472, Lab. No. 447.) Pillow-lava. One-quarter mile slightly E. of S. of cairn on Cruach Choireadail; Mull. Basalt, Porphyritic Central Type, p. 150. Anal. E. G. Radley.

V—(18471, Lab. No. 446.) Lava. Three-eighths mile N.E. of cairn on Cruach Doire nan Guilean, west side of a little lochan; Mull. Basalt, Porphyritic Central Type, p. 148. Anal. E. G. Radley.

rock is ever to be expected. This is plainly at variance with the facts. The augite andesites of the Mull area itself are, for example, frequently porphyritic but material of the same composition has at other times been cooled rapidly to a glassy or wholly aphanitic rock thus showing that all of its constituents entered into the composition of a liquid. They are not "always porphyritic in their finer crystallizations."

We may now examine somewhat more closely the chemical and mineral composition of the various crystalline manifestations of this Porphyritic Central magma-type. To this end we reproduce, in Table IX, the table of chemical composition as given in the Mull Memoir.

It will be noted that two of the typical representatives of this magma-type, those lettered A and B, are gabbros of the Cuillin laccolith of Skye. Of these Harker gives the mineral composition as determined by actual separation of the minerals.[1] The result is stated below in Table X.

TABLE X

MECHANICAL ANALYSIS OF SKYE GABBROS

	A	B
"Labradorite"[2]	79.50 (G = 2.735 — 2.74)	65.96 (G = 2.737)
Augite ⎱ Olivine ⎰	16.18	32.43
Enstatite	2.10	
Magnetite	2.40	1.61

When the actual mineral composition of these two gabbros is compared with their calculated normative composition as given in Table IX a remarkable degree of correspondence is shown. We may therefore take it that in rocks falling under this magma-type the "norm" represents the actual mineral composition fairly closely and the table of norms gives a satisfactory survey of mineral composition. It has already been pointed out that this magma-type occurs in the form of flows, minor intrusions and major intrusions. There is, however, a distinct trend of mineral composition apparent in the table of norms. The basalts show, on the average, the least enrichment in plagioclase, the dolerite is intermediate and among the gabbros the extreme of this condition is in evidence, one of the gabbros containing nearly 80 per cent of plagioclase. There is a very plain tendency, too, for the proportion of anorthite in the plagioclase to increase as the total amount of plagioclase in the rock increases. Thus among the gabbros are found the rocks richest in plagioclase and also the most calcic plagioclase. All of this is in accord with the deduction that these rocks represent types in which basic plagioclase has accumulated as crystals without any significant remelting or

1 Skye Memoir, p. 104.
2 Both the chemical composition and the density of the plagioclase show that the composition approaches Ab_1An_4 which is ordinarily termed bytownite.

re-solution. As the amount of accumulated crystals increases the mass becomes less capable of being poured out as a lava and the extreme types occur only as a major intrusion. It is, of course, not necessary that all major intrusions should show this enrichment in plagioclase and in anorthite. It is true, however, that only a major intrusion can show it or, to be more accurate, usually only a part of a major intrusion. This part of the intrusion is the locus of accumulation of plagioclase crystals. The mass has neither arrived there as such, nor, in the case of the extreme types, does it move thence to form other intrusive masses except under special circumstances which we shall discuss on another page.

Considering the manner of their origin it is only natural that the compositions of rocks of the Porphyritic Central magma-type do not lie on the smooth curves of variation of the Normal Mull Magma series (see Fig. 27). The rocks are unquestionably, in their mineralogy and all their features, derivatives of this series but their formation involves crystal sorting. We have already shown that only rocks which have the composition of possible liquids can give points which lie on smooth curves. Where crystal sorting is involved there must be a scattering of the points (see Fig. 37). Thus we find the Mull authors plotting these rocks on a separate diagram which is not, however, a variation diagram in the same sense as the diagram of the Normal Mull Magma series, in that it does not represent the compositions of successive mother-liquors. Indeed, if the term magma-type involves the assumption that a magma is completely molten rock material then the Porphyritic magma-type is not, strictly speaking, a magma-type at all. The very useful concept of the magma-type may be retained for these rocks, as well as the designation which has been applied to them, only if the term magma be regarded as including liquid with some crystals.

ANALOGOUS LAVAS FROM OTHER REGIONS

There is a considerable body of evidence to show that lavas analogous to those belonging to the Porphyritic Central magma-type are of widespread occurrence, though perhaps not abundant, and that they have an origin identical with that deduced for the Mull representatives. Emphasis is to be placed upon lavas in such a search because the quenching to which they have been subjected upon extrusion reveals their nature prior to extrusion, whereas in granular rocks the continued outgrowth of crystal boundaries obscures the stages of their development.

In Washington's Tables of 1917 there are 335 superior analyses of rocks termed basalt by the authors describing them. They are for the most part lavas but a few are dikes of fine texture. This number is exclusive of special types such as those termed nephelite basalt, leucite basalt, etc., and also of those containing more than a small amount (usually about 3 or 4 per cent) of normative nephelite, leucite, etc. It includes, however, such equivalents of basalt as the French "labradorite." In other words there are 335 ordinary basalts. A general survey of these shows that they tend to contain normative plagioclase of the composition Ab_1An_1. In Fig. 40 has been plotted on the abscissae the percentage of

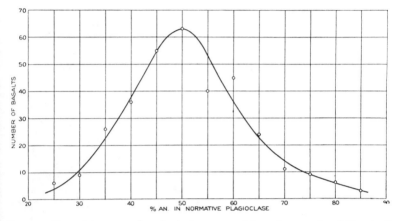

FIG. 40. Plot of the normative plagioclase composition of 335 analyzed "basalts." The number of basalts falling in each 5 per cent subdivision is plotted at the center of the subdivision.

anorthite in the normative plagioclase and on the ordinates the number of basalts having normative plagioclase of each type. In order to do this it has been necessary to divide the abscissae into parts representing equal increments of anorthite percentage and to plot at the center of each division the number of basalts falling in that division. In the actual diagram the plagioclase series was divided into parts representing a 5 per cent increase of anorthite and the number of basalts lying, for example, between 42.5 per cent and 47.5 per cent anorthite in the normative plagioclase was plotted at 45 per cent anorthite. In this manner we get a diagram showing the general trend of basalts in the matter of their normative plagioclase. The curve shows a strong maximum at 50 per cent anorthite and the number of basalts which depart from this maximum is decidedly small. Thus less than 9 per cent of the total have normative plagioclase with more than 67.5 per cent anorthite (about Ab_1An_2) and less than 3 per cent of the total have normative plagioclase with more than 77.5 per cent anorthite (about Ab_1An_3).

The reason for the marked falling off in the number of basalts on the low anorthite side is, of course, the fact that, as the plagioclase changes in that direction, the rock ordinarily ceases to be called basalt. On the anorthite-rich side, however, the cause of the falling off is quite different. It is, in fact, due to the occurrence of only a very limited number of rocks of such compositions among lavas. The cause of this limitation, in turn, is to be found in the restricted composition of possible liquids the evidence of which is obtained from the petrographic description of the more extreme types. Apparently no basalt has even been encountered which has such a composition and in which all the plagioclase enters into a fine-grained fabric. If we may assume that the canvass of rocks is adequate then such rocks are "always porphyritic in their finer crystallization" not merely in Mull but elsewhere, and the inevitable conclusion is that liquids of that nature do not exist. Mention may be made of some of the better known examples of basalts of the chemical type under discussion. Curiously enough, basalts from the New Hebrides are among the best (An = 78 per cent.)[1] They contain glomeroporphyritic crystals of basic plagioclase. The type is well represented in basalts of Madagascar and Réunion (An as high as 76 per cent) where phenocrysts of bytownite are present.[2] Many Japanese lavas belong here (An as high as 85 per cent) and they show phenocrysts of bytownite-anorthite.[3] The lavas of Pelée exhibit the type (An = 66 per cent) and again they contain phenocrysts of bytownite.[4]

Among coarse granular rocks the type is rather common and extremes are attained in these rocks that are never attained in lavas. As a striking example we may note that under the per calcic (anorthite-rich) rang, corsase, there are 13 deep-seated rocks, and under the similar kedabekase 20 such rocks,[5] without a single lava in either. No evidence is forthcoming of the existence of liquids corresponding in composition with these coarse granular rocks; indeed, all the evidence is decidedly against it.

THE LIMITATION OF THE PLAGIOCLASE COMPOSITION OF MAGMATIC LIQUIDS

The most calcic composition of the total plagioclase of a uniformly fine-grained or aphanitic, basaltic rock appears to be about Ab_1An_2. This is determined by examining the normative plagioclase of a large number of non-porphyritic basalts. To some investigators this method of determination of the nature of the plagioclase may appear to be unsatisfactory, but a little consideration will show that the norm can err only in

1 D. Mawson, "The Geology of the New Hebrides," *Proc. Linn. Soc. N.S. Wales*, 1905, pt. 3, p. 470. An = 78 per cent means that the normative plagioclase is 78 per cent anorthite.
2 A. Lacroix, *Geology of Madagascar*, III, pp. 44, 48.
3 Tsuboi, *Jour. Coll. Sci. Tokyo Imp. Univ.*, 48, p. 82.
4 A. Lacroix, *Mont. Pelée*, 1904, p. 588, analysis (f).
5 Washington's Tables, pp. 551, 667.

the direction of indicating plagioclase more calcic than the actual modal mineral. The only significant factor introducing a discrepancy between the norm and the mode in the matter of the nature and quantity of feldspar is the Al_2O_3 occurring in augite. The amount of this oxide so occurring is calculated as anorthite in the norm and the normative plagioclase is thus more calcic and more abundant than the modal plagioclase.

If the petrographic descriptions of aphyric basaltic rocks are examined with this fact in mind there is revealed a very general tendency to overestimate the anorthite content of the plagioclase as determined by ordinary petrographic methods. Thus Washington describes the Deccan traps as containing labradorite of the composition Ab_1An_2 (determined microscopically), yet eleven analyses indicate a normative plagioclase about Ab_1An_1[1] and this can not be less calcic than the modal plagioclase. Bowen[2] describes a diabase from Cobalt, Ontario, as containing $Ab_{35}An_{65}$ but analysis shows that the normative plagioclase is notably more sodic. Two rocks of basaltic composition from the Brito-Arctic region, the one a dolerite sill described by Harker,[3] the other a flow described by Holmes,[4] are uniformly fine-grained and their feldspar is said to be bytownite. Both rocks have been analyzed and their normative feldspar is as given under I and II below.

	I	II	III
	Sill, Ben Lee, Skye	Flow, Iceland	Major Intrusion, Eucrite, Rum
Or	3.89	1.11	1.11
Ab	16.77	15.72	10.48
An	31.69	30.58	35.58

It is plain that the modal plagioclase of the two rocks referred to (I and II) can not have been more calcic than Ab_1An_2. Under III is given for comparison the normative feldspar of a eucrite from Rum.[5] It illustrates the much greater extent to which enrichment of anorthite in the plagioclase may be carried in a coarse-granular major intrusion, the normative plagioclase being about Ab_2An_7.

The cause of this apparently common tendency to overestimate the anorthite content of plagioclase as determined microscopically is not far to seek. If the plagioclase is determined by the method of maximum extinction in the symmetrical zone there is unquestionably a tendency to emphasize the most basic plagioclase. Even if determined on sections

1 H. S. Washington, *Bull. Geol. Soc. Amer.*, 33, 1922, pp. 769, 775.
2 N. L. Bowen, *Jour. Geol.*, 18, 1910, pp. 659-61.
3 A. Harker, Skye Memoir, 1904, p. 248, Analysis II, and p. 249.
4 A. Holmes, "The Basaltic Rocks of the Arctic Region," *Min. Mag.*, 1918, p. 196.
5 A. Harker, "Geology of the Small Isles," *Mem. Geol. Surv, Scot.*, 1908, p. 98.

showing a combination of Carlsbad and albite twinning there is a tendency, especially in fine-grained rocks, to give particular attention to the sections of largest area. Should the plagioclase be zoned the outer zone will not be adequately weighted on this account for only the sections of small area emphasize the outer zones. To be sure the larger sections must be central sections and should afford a correct estimate of the relative proportions of inner and outer zones, but it is more than probable that the greater number of petrologists do not realize that an outer shell of a crystal requires to have only about one-tenth the thickness of the whole crystal in order to constitute one-half the volume. A drawing to scale is very illuminating on this point and renders it practically certain that in ordinary petrographic work there would be a tendency to underestimate the quantitative importance of the outer shells of zoned crystals.

These various factors working together are sufficient to account for a common tendency to overestimate the anorthite content in the plagioclase of fine-grained basaltic rocks, a tendency which is demonstrated by the comparison of the norm and the mode of the few examples just given.[1]

We may now repeat the opening statement of this section. The most calcic composition of the total plagioclase of a uniformly fine-grained or aphanitic, basaltic rock appears to be about Ab_1An_2. The two examples given of such rocks in which the plagioclase has been stated to be bytownite constitute no exception to this rule. Indeed, no exception has been found. A few uniform (aphyric) basalts such as those of New Jersey match these two in this respect but most rocks of this character fall far short of them. Uniformly coarse-grained rocks and porphyritic rocks may be much more notably enriched in anorthite. There is thus no escape from the conclusion that the process of accumulation of crystals is not supplemented by significant remelting or re-solution of the accumulated crystals. Rocks may show almost unlimited enrichment in anorthite but not so liquids. There are indeed no liquids corresponding with the extreme rocks.

Two basaltic magma-types other than the Porphyritic Central magma-type have been recognized by the Mull authors in their area and the Hebridean region in general. They are the Non-Porphyritic Central magma-type already mentioned and the Plateau magma-type. (Mull Memoir, p. 15.) From the fact that these have both been cooled in some instances to uniformly aphanitic or glassy facies they are to be recognized as having the composition of actual liquids. It is of interest, there-

---

1 A striking example has appeared since the above was written. A basalt from Saint Helena is described by Daly as containing plagioclase of the composition Ab_3An_7 though the norm shows that there must be more albite than anorthite. R. A. Daly, "The Geology of Saint Helena Island," *Proc. Amer. Acad. Arts Sci.*, 62, 1927, pp. 63-5.

fore, to compare their composition in the matter of the nature and quantity of feldspar with that of the Porphyritic Central magma-type. This may be done conveniently by calculating the normative feldspar. The results are given in Table XI below and the corresponding figures for the Porphyritic Central type are repeated for comparison. It will be noted that in the Plateau magma-type the normative feldspar tends to constitute but little more than one-half the normative minerals and anorthite is a little more than one-half the normative plagioclase. The Non-Porphyritic Central magma-type likewise has about 50 per cent normative feldspar but a more important proportion of this is now orthoclase. In addition the anorthite in the normative plagioclase tends to be distinctly less than 50 per cent. This character is quite that to be expected in *liquids* formed by the subtraction of crystals from the more basic Plateau magma-type. Table XI as a whole offers an excellent means of contrasting the composition of the usual basaltic *liquids* of the Hebridean field with that of the basaltic *rocks* (Porphyritic Central) of the same field which have been enriched in accumulated feldspar crystals.

TABLE XI

NORMATIVE FELDSPAR IN PLATEAU MAGMA-TYPE[1]

	I	II	III	C	D	E
Total Feldspar	54.38	55.03	48.64	56.01	52.69	50.47
% An in Norm. Plag.	52	56	60	52	58	56

NORMATIVE FELDSPAR IN NON-PORPHYRITIC CENTRAL MAGMA-TYPE[2]

	I	III	IV	V
Total Feldspar	50.31	55.63	50.76	51.97
% An in Norm. Plag.	51	41	42	40

NORMATIVE FELDSPAR IN PORPHYRITIC CENTRAL MAGMA-TYPE[3]

	I	A	B	II	III	IV	V
Total Feldspar	71.75	78.67	64.29	64.19	60.75	68.39	63.50
% An in Norm. Plag.	74	83	80	76	66	72	66

The two basaltic magma-types with nearly equal amounts of feldspar and femic mineral thus belong to the liquid line of descent. They are the Plateau type and the Non-Porphyritic Central type, the former being parental. In the character noted they correspond with liquids lying on the boundary curve of the investigated system, plagioclase-diopside (Fig. 16) at points where the plagioclase is of intermediate composition. Like such liquids they become saturated with both plagioclase and a femic mineral approximately simultaneously. There is thus further reason for believing that these basalts correspond with liquids lying

1 The Roman numerals and letters distinguishing the rocks correspond with those used in Table I, p. 15, Mull Memoir.
2 The Roman numerals and letters distinguishing the rocks correspond with those used in Table II, p. 17, Mull Memoir.
3 The Roman numerals and letters distinguishing the rocks correspond with those used in Table VI, p. 24, Mull Memoir.

along a boundary surface in the natural system of more components, especially when it is found that rocks which depart from this balanced proportion do so in virtue of the presence of an excess of one phase as phenocrysts (Porphyritic magma-type). We shall find that when basaltic types depart from this balanced proportion in other directions they do so in virtue of the presence of an excess of other phases as phenocrysts.

There is a very distinct tendency for plateau basalts in general to show the same character as the plateau basalts of the Hebrides. The total normative feldspar is approximately one-half the rock and this feldspar is approximately one-half anorthite.[1] The Hebridean plateau basalts seem to be somewhat more basic than the general run of such basalts but of this fact we can not be altogether assured, for in no place have detailed studies of basaltic complexes been prosecuted as they have in the Hebrides. It is these detailed studies that have there permitted the separation of the lavas of central eruptions from the (usually more basic) plateau types.

Corresponding with its basic character the Hebridean Plateau magma-type precipitated olivine among its early minerals. On page 77 it has been shown that the best means of derivation of the Non-Porphyritic magma-type from the Plateau magma-type is the separation of olivine and basic plagioclase. The derived magma and the crystal assemblage (olivine + plagioclase) are complementary products of the process of crystal *accumulation* at this stage. In some instances the accumulation has taken place under conditions permitting a sorting of the crystals and rocks rich in basic plagioclase have arisen, whose origin and character have just been discussed. Their existence suggests the presence of rocks enriched in olivine, rocks which are complementary to the plagioclase-enriched rocks in the process of crystal *sorting* at this stage. Such rocks are abundantly represented in the Hebrides and to them attention is now directed.

ULTRABASIC TYPES OF THE HEBRIDES

Rocks of the varieties known as ultrabasic are of much importance in the intrusive centers of the Hebrides. In Skye and in Rum they are found most abundantly developed and there they have received detailed study especially by Harker. In Mull ultrabasic rocks are not prominent and, although representative specimens of a wide range of ultrabasic types can be obtained, the masses from which they come are local variants of gabbro rather than independent geologic units. In Skye, on the other hand, such rocks occur both as great masses constituting parts of the composite laccolith of the Cuillins and also as dikes. We shall

1 H. S. Washington, "Deccan Traps and Other Plateau Basalts," *Bull. Geol. Soc. Amer.*, 33, 1922, pp. 765-804.

consider these rocks with special reference to their occurrence in Skye.

In some measure ultrabasic rocks have already been discussed. The analyzed gabbro of the Cuillin laccolith, containing nearly 80 per cent of plagioclase of the composition Ab_1An_4, is preferably to be regarded as an ultrabasic rock. In the magma-type classification proposed by the Mull authors it falls under the Porphyritic Central. Rocks of this magma-type have been formed, according to our deductions, by accumulation of basic plagioclase crystals which have not suffered significant remelting. Extreme examples, such as the Skye gabbro just mentioned, have certainly no effusive equivalents and possibly no dike equivalents.

It has been pointed out that during the separation of basic plagioclase other crystals, principally olivine, were separating as well. There is thus a possibility of the formation of olivine-rich rocks by a similar process of accumulation. With the growing appreciation of the strength of the hypothesis of crystallization-differentiation an increasing number of petrologists have signified their belief in the probability that rocks rich in early-crystallizing minerals have been formed by local accumulation of the substance of these minerals. There has also been a greater leaning toward the concept that the material of these rocks has accumulated as actual crystals. The great majority of those who have gone thus far have made the additional assumption that the accumulated crystals have been redissolved or remelted and that *liquids* of extreme composition have thus come into being. The evidence appears to be decidedly against this latter assumption for rocks rich in basic plagioclase. We shall now give some consideration to the evidence in the case of olivine-rich rocks.

The larger laccolithic masses of peridotite in Skye are made up of typical, coarse-granular rocks including dunite and various other types less rich in olivine. These laccolithic masses are usually so involved with later gabbro that at only two or three places do they have intrusive contacts with older rocks and these are such that they do not give clear evidence of the nature of their chilled facies. In addition to the larger masses there occur dikes and, much less abundantly, sills of peridotite and to these we may turn with some hope of finding evidence as to the nature of the peridotitic material at the time of its intrusion. A picrite sill on the adjacent small Island of Soay has, locally at least, suffered quick chilling. Harker describes and illustrates this rock and his illustration is here reproduced in Fig. 41. In the description he states "The porphyritic elements are olivine and picotite. These recur also in a second generation, but the bulk of the groundmass is of slender rods of felspar with sub-parallel arrangement and interstitial augite, the structure recalling that of some variolitic basalts."[1] It is not reasonably

1 Skye Memoir, p. 470, plate XXVI, Fig. 1.

to be supposed that the large olivine crystals of this rock were formed under the same conditions as a groundmass which is but a step removed from the spherulitic state of crystallization. Mere priority of crystallization, with the attendant greater freedom of growth, might perhaps be

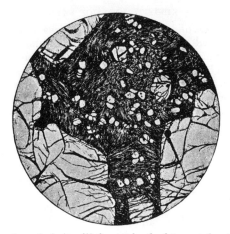

FIG. 41.   Thin section of picrite sill from Island of Soay (after Harker) showing large olivines in variolite-like groundmass of feldspar and augite.

assumed to give the olivines some slight advantage in size, but no such contrast as is here displayed can be referred to such a cause. Plainly the groundmass was the only portion of the rock that was liquid when it was finally chilled on intrusion. This fact does not in itself prove that the olivine was not, at some earlier time, entirely in solution in that liquid, but if we examine a large number of similar chilled types, and if we find that invariably the same relation obtains, we may have much confidence in the conclusion that olivine, in the quantity there shown, was never in solution in the liquid now represented by groundmass. In other words if we find that, under no circumstances, has a rock of peridotitic composition been chilled in such a way that all of its olivine enters into a fine-grained groundmass but, on the contrary, do find that, when such rocks are quickly chilled, most of the olivine is in large crystals strongly contrasted with the groundmass in grain, and of course in composition, then we may safely conclude that there was no liquid corresponding with the peridotite in bulk composition.

With these questions in mind, a study of the peridotite dikes of Skye has been made in order to supplement the information set forth by Harker. Dr. Harker has facilitated the study in every way, especially by acting as guide to the promising localities during two visits to Skye and by much helpful discussion. The description of these dikes now to be given is based largely on Harker's original work but partly on these recent examinations.

PERIDOTITE DIKES OF SKYE

The peridotite dikes of Syke are very conspicuous in the field because of their distinctive, red-weathered surface. This color arises from the weathering of olivine and is notable even in those examples that are too poor in olivine to be called peridotite in the stricter sense. As is in some measure implied in the foregoing statement the dikes show a considerable range of composition. There is not only a variation of composition from one dike to another but also a variation from place to place in an individual dike. Specimens of dunite are to be had and some of the dikes are olivine gabbro rather than peridotite, but the great bulk is picrite, to use the term in the sense adhered to by Harker in his memoir. They tend to contain 40 to 50 per cent of olivine, the rest of the rock being made up of basic plagioclase and augite with minor quantities of accessories, principally picotite.

A feature of the dikes to which attention is particularly directed is their width. Although dikes and sheets of basaltic composition in the same surroundings may vary in width from many feet down to an inch or less the typically peridotitic dikes are nearly always wide, averaging perhaps 15 to 20 feet and often attaining 40 feet. One dike, only a foot wide, on the south slope of Sgurr nan Gobhar contains nearly 50 per cent olivine, if the single specimen taken can be regarded as representative, but it is seen to outcrop for only a few yards. It is probably an offshoot of one of the adjacent larger dikes and of no great extension. The larger peridotite dikes are very persistent. A single dike may sometime be traced, outcropping continuously through a vertical range of 1500 feet or more and a horizontal distance of a mile. When small dikes of this suite are similarly persistent they contain distinctly less olivine than the larger dikes. Examples of such dikes, a foot wide or even considerably less, are to be seen in Coire Labain, crossing the floor of the corrie and mounting the walls on either side. Petrographic examination shows that their content of olivine is not more than about 25 per cent.

The tendencies displayed by these dikes strongly suggest that peridotite freely forms only wide dikes, the more so as the proportion of olivine in it increases. In this we may see some indication of the condition of the material as intruded.

Another feature of the peridotite dikes as displayed in the Cuillins is the radiate arrangement to which Harker calls attention. The center is in the heart of the mountains, about where the laccolithic mass of

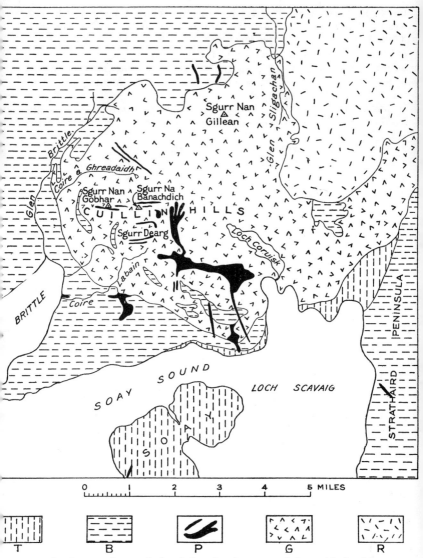

FIG. 42. Sketch map of part of Isle of Skye showing some of the peridotite dykes with their general radiate arrangement (from Harker's maps). T = Torridon sandstone; B = Tertiary basaltic complex; P = peridotite; G = gabbro; R = granite.

ultrabasic rocks occurs, and from this the dikes diverge northward, westward and southward with all intermediate directions (see Fig. 42). There is a very distinct tendency for the more extreme types to occur only close in toward the center and for those dikes which are found far out to show only moderate enrichment in olivine. This, like the contrast between large and small dikes, may be connected with the condition of the dike-substance as intruded.

Another feature of the dikes of importance in this respect is a very notable general tendency to a xenolithic character.

MARGINS OF THE PERIDOTITE DIKES

It is, however, to the margins of the dikes that we must turn for more decisive evidence of the condition of peridotite as intruded. Several samples of marginal portions were taken, some of them such that a single slice could be cut showing both dike and wall rock. Three different types of selvage were observed.

1. The selvage has a fine-grained matrix of augite and plagioclase, with perhaps a little olivine, in which lie more or less abundant olivine phenocrysts and sometimes phenocrysts of basic plagioclase.
2. The selvage is a fine-grained rock made up of augite and plagioclase without olivine and thus very different from the main body of the dike in composition.
3. In one example in which the dike had a porphyritic character even in the central portions, containing large crystals of olivine and plagioclase in a dolerite-like groundmass, the texture of the dike did not change materially as the margin was approached but, at the immediate contact, crystals of both the peridotite (picrite) and the adjacent gabbro were bent, fractured, or even granulated. This condition probably obtains only locally in any dike. It is probable that other portions of the margin of the same dike would show an effect like 1 or 2, but this has not been proved.

The marginal changes described under 1 and 2 are really variants of the same thing, for the amount of olivine in the marginal phase is ordinarily distinctly less than that in the central part of the dike. An additional change, occurring in both cases, is in the nature of the plagioclase. Harker has found that the plagioclase of the typical, coarse-granular, wider peridotite dikes is usually a bytownite-anorthite, a fact which has been confirmed by measuring refractive indices in immersion liquids. Similar measurements on the plagioclase of the fine-grained marginal material of the larger dikes, when fresh, and of the general material of the smaller dikes of peridotitic affinities show that it is much less calcic. The cores of the crystals may attain such high anorthite content but the outer zones are always much higher in albite and actually greater in amount. Some of the observations on which these statements are based may now be detailed.

The peridotite dike which is seen from Glen Brittle as a red gash on the west spur of Sgurr nan Gobhar is apparently typical and will therefore be described somewhat fully. Its wall rock is principally gabbro and where it first appears above the scree on the lower slopes it has a width of nearly 40 feet. Here, as elsewhere, it is made up of olivine, bytownite-anorthite and augite, with a little picotite. All the principal minerals are of roughly equal dimensions, average diameters being about 2 mm. Plagioclase and augite tend to occur as a groundmass for the more nearly idiomorphic crystals of olivine. The plagioclase has the composition about $Ab_{15}An_{85}$. A slight zoning is sometimes observed but it is apparently not of significant magnitude. At the lowest outcrop olivine makes up approximately 70 per cent of the dike with the remainder divided about equally between plagioclase and augite. The rock is not, however, uniformly so rich in olivine and, as the hill is ascended, specimens are obtained with decidedly less olivine. There is, in fact, a more or less regular decrease in the proportion of olivine with increasing elevation, accompanied by a like increase in plagioclase, with augite remaining approximately constant. Thus at the summit of Gobhar the proportions are 20-25 per cent olivine, 15-20 per cent augite and 55-65 per cent plagioclase, which is still the same very calcic member. Though the change in the dike is in the direction that would be expected from a gravitative effect during crystallization it probably only happens to be so.

Only one specimen of the contact of this dike was obtained. This is much altered but still reveals the original character fairly well. It formerly had about 35 per cent of olivine as phenocrysts with approximately the same dimensions as the olivine of the center of the dike. These are now largely changed to serpentine or sometimes to talc, but remnants of unchanged olivine are found. The change to talc is noteworthy as being a rather general tendency in the contact facies of these dikes. The remainder of the contact rock is altered considerably but it was made up of plagioclase and augite in about equal amounts. The plagioclase is now entirely altered but the augite is for the most part unchanged. This conspicuous dike on the face of Gobhar is much narrower above than below and where last seen is only about 6 feet wide. Here it appears to end abruptly but shortly before its termination another parallel dike comes in a few feet to the south and this continues eastward on the same general line of strike. Where this parallel dike first appears it is much like the olivine-rich portions of its fellow just described. The olivine makes up about 60 per cent, with plagioclase and augite each about 20 per cent. The rock is of even fabric, each constituent attaining dimensions of about 2 mm. There is as usual some brown translucent picotite but there is also a black, opaque ore mineral.

The contact of this dike, at least at the place where specimens were taken, is of an entirely different nature. At about six inches from the contact it has only 20 per cent of olivine, mostly in large grains averaging 1.5 mm but often attaining 3 mm in diameter. There are also some much smaller olivines but even these are somewhat set off from the groundmass. Picotite grains are again fairly common. Of somewhat lesser dimensions than the olivine are some large crystals of basic plagioclase which have broad interior areas of the uniform composition Ab_1An_5 but also outer zones of the composition Ab_1An_2 or even somewhat more albitic. These porphyritic elements are set in a doleritic mass of plagioclase and augite of distinctly contrasted grain.

Closer to the contact both the olivine and the large basic plagioclases become distinctly less abundant but not notably smaller and the groundmass acquires a finer grain. At the contact there are no porphyritic elements at all. The rock is made up entirely of very fine-grained dolerite without olivine. There is thus in this dike a progressive though rapid change close to the contact. The central part, which is rich in olivine and has very basic plagioclase, changes to a rock in which these constituents appear as porphyritic elements in a doleritic groundmass. The proportion of groundmass increases and its grain becomes increasingly finer until at the contact, and within $\frac{1}{2}$ inch of it, the rock is a plain dolerite lacking olivine and very basic plagioclase.

In Coir' a' Ghreadaidh there is a large dike of peridotite which has been split by another smaller dike. Their character and relations have been described in detail by Harker (p. 379). The smaller dike is, in accordance with the general tendency of small dikes, not notably rich in olivine. The larger dike is rather typical of the average olivine-rich wider dikes. However, it differs somewhat in texture from many of the wide dikes, among them the Sgurr nan Gobhar dike already described. There is a tendency for augite and plagioclase, ophitically intergrown, to be of appreciably though not markedly finer grain than the olivines, whereas in the Gobhar dike all constituents are of approximately equal grain, or at times, indeed, large plagioclase crystals may poikilitically enclose two or three olivines. These plagioclase crystals are, as we have noted, a bytownite-anorthite and in the matter of being thoroughly ultrabasic both in the quantity of olivine and the nature of the plagioclase the Gobhar dike is to be regarded as the more typical. An analysis of the Gobhar dike or another of similar character would thus be desirable in order to elucidate the chemical character of the more typically ultrabasic dikes. Unfortunately, however, both the olivine and the plagioclase, at least in the specimens collected, are somewhat altered and it was doubted whether analysis would give an adequate idea of the original character.

The Coir' a' Ghreadaidh dike is, however, relatively fresh and for this reason was chosen for analysis though the petrographic examination indicated that the plagioclase was not the very basic variety of the more thoroughly ultrabasic types, a fact which the analysis confirms. The analysis given under I, Table XII, was made by Miss Mary Keyes, to whom thanks are here expressed. The mineral composition is so simple that the norm unquestionably represents very closely the actual modal character. There is about 55 per cent normative olivine and 25 per cent normative feldspar which is approximately Ab_1An_2. In quantity of olivine it may be taken as representative of the general average of the peridotites. Many of them and possibly most of them have notably more basic plagioclase.

A peridotite dike near the summit of Sgurr na Banachdich is, at the point where it was examined, made up of about 80 per cent of olivine and a little picotite with a groundmass of plagioclase and augite. The olivines average from 2 to 3 mm in diameter and, in spite of their very large proportion, have uniformly a remarkable tendency to idiomorphism. The groundmass of plagioclase and augite is of much finer grain and of a diabasic texture. The dike is notably variable in composition and is almost certainly the same dike sampled by Harker, to whom it yielded a specimen consisting almost entirely of olivine mainly in large crystals of parallel orientation which gave it a platy fracture (p. 378). It was thus like a specimen obtained from a dike in Coire Labain, shortly to be described.

The striking tendency towards idiomorphism of all the olivines in a specimen with as much as 80 per cent of that mineral is a matter of some importance.[1] If these crystals experienced their full growth in a mixture containing 80 per cent of olivine there would have been so much mutual interference of outline long before they attained full size that they could have retained but little suggestion of their idiomorphic outlines. On the other hand, if they grew in a mass where they were more widely spaced in a liquid medium and were afterwards brought close together as a result of the draining-off or squeezing-out of the liquid, the observed structure might result. The structure may be taken as evidence that such action has probably occurred.

The Banachdich dike has a chilled selvage against the adjacent gabbro and one fine-grained stringer about 1 inch in width was found running off into the gabbro. The same fine-grained material occurs also

[1] This feature of the rock was called to my attention by my colleague, Dr. H. E. Merwin, who was uninfluenced by any theory of the origin,

as the matrix of a narrow zone of breccia locally along the contact. The fine-grained facies, while somewhat variable, is of the same general character throughout. It never contains more than about 35 per cent of olivine, which amount is found in the general chilled border, and here the olivines are from 2 to 3 mm in diameter as they are in the general mass of the dike. They are accompanied by a few picotites, also relatively large, and are set in a fine dolerite-like groundmass of plagioclase and augite which is increasingly finer as the contact is approached whereas olivines or picotites of full dimensions may lie right against a crystal of the adjacent gabbro.

Little tongues from this material form the matrix of the contact breccia. Small picotites have been carried into the smallest of these but only occasional olivines have been carried in and then only into the larger seams.

TABLE XII

ANALYSES OF PERIDOTITES OF SKYE

	I	II	III
SiO_2	40.90	43.30	44.61
Al_2O_3	7.56	12.71	10.86
Fe_2O_3	3.01	2.35	2.31
FeO	7.31	7.60	7.46
MgO	29.63	14.65	21.06
CaO	5.40	10.50	9.01
Na_2O	0.98	0.96	1.15
K_2O	0.37	0.22	0.19
H_2O +	2.98	4.27	1.17
H_2O −	0.13	0.33	0.05
TiO_2	1.70	2.44	2.25
P_2O_5	0.10	0.11	0.10
Cr_2O_3	0.11	0.08	sl.tr.
MnO	0.34	0.19	0.16
	100.52	99.71	100.38

I—Peridotite dike, Coir' a' Ghreadaidh, Isle of Skye. M. G. Keyes anal.
II—Olivine dolerite seam, Apophysis from peridotite dike, Sgurr na Banachdich, Isle of Skye. M. G. Keyes anal.
III—Picrite-dolerite dike, Coire Labain, Isle of Skye. M. G. Keyes anal.

The seam 1 inch wide which traverses the gabbro for several feet has only about 15 per cent of olivine, which occurs as phenocrysts now largely serpentinized, and a little picotite, also relatively large. These are unevenly distributed, having a tendency to be concentrated in bands. The groundmass is again doleritic, but like the phenocrysts has suffered alteration. As illustrative of the composition of this seam (and the fine-grained selvage of the main dike) a specimen of it was analyzed. Unfortunately this fine-grained material is always much altered so that the analyses, while adequately reflecting the general character can not be relied upon to give the accurate mineral composition. The analysis is given under II in Table XII. It shows about 4.5 per cent H_2O. Calculation of the norm gives about 16 per cent olivine, which corresponds with the amount actually observed as phenocrysts. The normative feldspar is somewhat less than 40 per cent and of the composition Ab_1An_3. It may be doubted whether the feldspar was as basic as is indicated by calculation from this altered material, especially in the light of what has already been found concerning the limitation of the feldspar composition in aphanites. Examination of the feldspar in the powdered rock shows that the highest value of γ in fresh grains is 1.572. This approaches the value for Ab_1An_3, to be sure, but it refers to the most basic plagioclase present. Other grains considerably less basic are present but it was impossible to estimate proportions. An accurate determination of the nature of the feldspar of this aphanitic groundmass thus appears to be impossible but, if the feldspar really was as basic as is indicated by calculation from the analysis, it is the most basic feldspar yet found in an aphanite.

Waters has described interesting examples of lamprophyres with small apophyses whose character is such as to indicate a strong filtration effect during intrusion into the narrow seams. He says: "In the field several apophyses about 2 inches wide can be noticed to vary in a few feet from a very black basic rock to stringers of a light-gray shade containing hardly any mafics."[1]

A dike which is traversed at about 2500 feet elevation in ascending the west spur of Sgurr Dearg has the general average character of the peridotites. It shows a fine-grained contact facies. This selvage has phenocrysts of olivine averaging 2 mm and attaining even 5 mm diameters in direct contact with the adjacent gabbro and making up 35 to 40 per cent of the whole rock. These are embedded in a fine-grained groundmass of a doleritic character which is still resolvable at 1 cm from the contact but at the contact is too fine for resolution by the microscope. Where coarse enough to be determinable it is seen that there is no olivine in the groundmass. It is made up of plagioclase and augite. Another specimen of the contact of this dike contains basic plagioclase as well as olivine among the porphyritic elements. The plagioclase is in crystals of about the same dimensions as olivine but is distinctly less abundant. It occasionally occurs in glomero-porphyritic groupings with olivine.

Other dikes of this group on the same spur of Sgurr Dearg at lower elevations tend to run rather low in olivine. About 25 per cent appears to be the prevalent amount and this is divided into two approximately equal amounts, one of which occurs as large crystals (2 to 3 mm) and

1 Aaron Waters, *Jour. Geol.*, 35, 1927, p. 167.

the other as smaller crystals (0.2 mm). The main mass of the rock is of plagioclase and augite in doleritic structure the general grain of which approximates that of the smaller olivines. The plagioclase is very strongly zoned. There is a very notable tendency in these dikes for the large olivines to be shattered and for many of the plagioclase crystals to be bent or even broken across.

A dike 20 feet wide, high up on the north wall of Coire Labain, afforded the only specimen of dunite that was obtained, though, as already noted, Harker obtained dunite from the Banachdich dike already described. A laminated structure is seen even in the hand specimen from Coire Labain. Olivine to the extent of 97 to 98 per cent, a black opaque ore mineral, and excessively rare interstitial grains of plagioclase and augite make up the rock. The laminar effect is found under the microscope to be produced by a parallel arrangement of elongated blades of olivine up to 5 mm long and ½ to 1 mm wide. The elongation of these is always γ and the long edges, though very irregular, have the general direction of the trace of the best cleavage surface of olivine (010). There are many shorter olivines similarly oriented and the rest of the rock is made up of rounded grains which interlock with each other and with the larger elongated grains. The structure is hardly to be regarded as porphyritic. The rock is rather a granulated one. The elongated blades of olivine have been produced from larger crystals by a shearing break along the plane of the best cleavage; indeed, occasionally, two or three adjacent blades can be fitted together, the protuberances of one with the embayments of its neighbor. The whole structure is such as might have been produced by the enforced flowage of a completely crystalline portion carried along during the act of intrusion of one of these peridotite dikes. It was not suspected in the field that this specimen would prove so unusually rich in olivine, so that the question of the variability of the mass laterally and vertically was not investigated. From the general prevalence of variability in these dikes it must be regarded as probable that such extreme richness in olivine obtains only locally. It is a noteworthy fact that fissility is a very common character of Hebridean peridotites where they approach dunite in composition.[1]

The smaller dikes of peridotitic affinities have the same general characters as the chilled selvages of the larger dikes. They contain phenocrysts of olivine averaging 2 or 3 mm in diameter, but sometimes attaining 5 mm, in a fine-grained dolerite-like groundmass which may be variolitic or even unresolvable close to a contact. The olivine never amounts to much more than 25 per cent but it is sufficient to give the dikes the characteristic red weathering. Sometimes, as in one well-

1 Harker, *Geology of the Small Isles*, p. 80, gives an analysis of "foliated olivine rock," and also discussion, p. 87.

exposed dike intersecting the gabbro of a huge boulder opposite the mouth of Coire Labain, there may be some 10 to 12 per cent olivine in the fine groundmass. Several small dikes found in the floor and walls of Coire Labain have the same general character, indeed, a moderate amount of olivine in the groundmass is a rather general character of these dikes. Occasionally, but by no means frequently, these small dikes contain phenocrysts of basic plagioclase (basic bytownite) and sometimes these occur in glomero-porphyritic groups. Most of the plagioclase occurs in the doleritic groundmass and is of a much more sodic character. When strongly zoned the inner zones may be as basic as Ab_1An_4 but they are abundantly compensated by outer zones richer in albite. In several cases where the refractive indices were determined in powdered rock there was no plagioclase more calcic than Ab_1An_2, and it was accompanied by more sodic varieties in at least equal quantity. The dike at the mouth of Alt Coire Labain, where it empties into Loch Brittle, is of considerable width but belongs among those which occur far out from the focus of intrusion and, in accordance with the general tendency, is not truly peridotitic. It contains large olivines about 2 mm in diameter to the extent of some 20 per cent, and some picotites in a doleritic mass of plagioclase and augite with a little olivine. The plagioclase is strongly zoned showing compositions varying from Ab_1An_4 to Ab_1An_1.

As representative of the composition of the narrower dikes of peridotitic affinities, a specimen of an 8-inch dike from Coire Labain was chosen for analysis. It is one of those just described as containing olivine in two generations. The result of the analysis is given under III, Table XII. The normative olivine, which must correspond very closely with the actual amount, is about 32 per cent. The microscope shows that about 25 per cent of the rock is made up of large olivine phenocrysts. The rest of the olivine occurs in the groundmass, of which it makes up some 10 per cent. The normative plagioclase is about 30 per cent of the rock and has the composition Ab_1An_2. It is thus labradorite and not the very basic feldspar of the more typically ultrabasic, wider dikes.

Dikes which are not true peridotite but which have decided ultrabasic affinities occur far out from the focus of intrusion in the Cuillins. Some ten miles to the east these dikes have been noted by Clough and by Harker.[1] They have a general character which may be described as intermediate between picrite and olivine dolerite.

SUGGESTED EXPLANATION OF CONTACT FACIES OF THE DIKES

It is not easy to explain the change of nature of the substance of the peridotite dikes at the contact, but it is probably to be referred to composite intrusion. Composite dikes are common in the Hebrides, the

1 Skye Memoir, pp. 383, 384.

prevalent type having doleritic borders and an acid center. Harker has explained them as the result of successive intrusion, the first filling of the fissure being doleritic. When this filling had consolidated at the margins only, the dike was then eviscerated by the inflow of salic magma.[1] The Mull authors accept this interpretation for similar dikes and in addition express the opinion that prior intrusion of fluent basic lava has often been a necessary preliminary to the inflow of viscous salic magma under hypabyssal conditions.[2] Harker considers it probable that the doleritic margins of the peridotite dikes are analogous to such margins on acid dikes. He believes that a peridotitic liquid followed after the basaltic liquid, sweeping out the central portion of the partly consolidated doleritic dike.[3] The concept advanced by the Mull authors suggests the possibility of a modification of Harker's hypothesis. The intrusion of a fluent basic magma may have been a necessary preliminary to the inflow of peridotitic material. In this case, however, preparation of the way was necessary, not because the peridotite was in the condition of a highly viscous liquid, but because it was already in an advanced state of crystallization. We thus arrive at the concept of the intrusion of the materials from a basaltic magma mass which had suffered partial crystallization, with local accumulation of the early crystals, principally olivine and basic plagioclase. The supernatant liquid, with only a limited amount of these crystals, was capable of flowing freely into any dike fissure that may have intersected the magma mass. We therefore find quite narrow dikes of peridotitic affinities. Into the wider fissures, or possibly into widening fissures, the largely crystalline, peridotitic portion of the magma mass was capable of following after the first inrush of the more liquid part. In addition the action probably involved the squeezing onward of the interstitial liquid of the peridotitic substance into the more outlying, narrower portions of these same wide fissures, the wider portions thus being left more nearly peridotitic or allivalitic. This effect is perhaps the best explanation of the fact that when a small stringer runs out from a peridotite dike into the adjacent rock it is always basalt-like in that it is nearly all labradorite and augite but a few olivines and especially a few picotites have gone with it and thus betray its origin as the interstitial liquor of the peridotite. The contrast in the nature of the peridotite dikes according to their distance from the center of divergence and again according to their size would appear to be satisfactorily accounted for by intrusion involving some filtration in the general manner described.

But whatever may be the truth it is certain that one will look in vain in Skye for evidence of the existence of peridotite liquid. Basalt, andesite

1 Skye Memoir, p. 207.
2 Mull Memoir, p. 33.
3 Oral communication from Dr. Harker.

and granophyre all have locally suffered chilling of such degree that they have become uniformly aphanitic, spherulitic and even glassy. Peridotite has never behaved similarly. On no occasion has peridotite been chilled in such a way that all of its constituents enter into an aphanitic fabric. If there was liquid consisting principally of olivine, or even very rich in olivine, it is incredible that this liquid should never have formed small seams in adjacent cold rocks or have been locally chilled to a uniformly aphanitic mass. Though the evidence is of the negative kind it is at least strongly indicative of the fact that the peridotites of Skye were never entirely liquid as such. A general survey of lavas, regarding them as quenched rocks, points in the same direction for world peridotites as a whole.

THE OLIVINE BASALTS

As is well known, there are no truly peridotitic lavas and the reason for this fact becomes apparent when we examine their nearest relatives, the olivine basalts. We find that no olivine basalt can contain more than a moderate amount of olivine without having the excess amount present as large crystals (phenocrysts) in a ground of a totally different nature as to size of crystals and, of course, composition. If a basalt must have these olivine crystals present as such, in order to acquire a composition rich in olivine, it is plain that, in order to have the content of olivine so increased that the composition would be peridotitic, the amount of olivine present as crystals would require to be so great that the mass could not be poured out on the surface. We shall now examine basalts for the evidence of this fact.

There are, as already stated, 335 superior analyses of ordinary basalts in Washington's Tables of 1917. Of this number 145 have normative quartz and 190 have normative olivine. The normative amounts of these minerals will here be used merely as a convenient means of isolating one important feature of their chemical composition. In Fig. 43 the distribution of basalts with respect to their normative quartz and normative olivine has been plotted, the method of plotting being analogous to that used in discussing the normative plagioclase of basalts. The partitioning is made at each 2 per cent increase of quartz and olivine respectively. Quartz and olivine are plotted in opposite directions on the abscissae in order to bring out a relation shortly to be discussed. It will be noted that the number of basalts with normative quartz falls off rapidly from its highest value at 0 per cent to a vanishingly small number at somewhat more than 20 per cent normative quartz. On the other hand the number of basalts having normative olivine rises from the value at 0 per cent olivine to a maximum near 10 per cent olivine and then falls off to 0 somewhat beyond 50 per cent normative olivine.

It will be obvious to anyone familiar with the chemistry of rocks that these two curves represent a continuous variation through a common point at 0 per cent normative quartz and olivine and that it is only necessary to find a common basis for plotting them in order to demonstrate this fact. This can not be done in a strictly quantitative manner on account of the variable chemical nature of the normative olivine, but an approximation can be made which is adequate for the purpose. A given quantity of normative quartz is chemically equivalent to some 2-1/3 times as much Mg_2SiO_4 or some 3-1/3 times as much Fe_2SiO_4. The number of basalts with normative olivine, falling in a certain subdivision of the continuous series of rocks, can be made comparable with the number in a certain subdivision on the other side of the zero point (the side with normative quartz) by making the subdivisions on the olivine side somewhere between 2-1/3 and 3-1/3 times as big. In addition we must then plot them in comparable subdivisions. This can be done by regarding one of these larger subdivisions on the olivine side as equivalent to the smaller one on the quartz side, thus plotting what might be called positive and negative normative quartz. Arbitrarily intermediate divisions 2-3/4 times as large have been chosen on the olivine side and the number of basalts in these divisions have been plotted on a diagram of positive and negative normative quartz. When this is done we obtain the curve given in Fig. 44. This curve combines the information of the other two into a single curve and shows plainly that basalts in general group themselves most abundantly around a composition which has what may be termed a small deficiency of silica. This small deficiency corresponds, however, with some 10 per cent normative olivine.

The number of basalts falls off rapidly on the high silica side, not because there are no such rocks, but because such rocks cease to be termed basalt. The number falls off on the high olivine side because few rocks having even moderately high amounts of normative olivine and none having very high amounts are known among effusives.

An examination of the basalts having the higher amounts of normative olivine shows, as we have stated, that they always contain large amounts of olivine as phenocrysts. A few typical examples may be briefly referred to. The olivine basalts of Hawaii have now received important detailed petrographic and chemical studies, among the more recent being those of Daly, of Cross, and of Washington. Daly describes an ultra-femic olivine basalt which has its olivine in large crystals "often more than one centimeter in diameter." There is "a rather surprising contrast in the grain of the groundmass, which is of diabasic structure, with thin tables of plagioclase, seldom over 0.1 mm in length, separated by augite granules of even smaller diameters." The normative olivine of the rock is 18.5 per cent, but actually about 32 per cent has crystallized as olivine.[1] Unquestionably these olivine crystals were present as such before extrusion. Cross notes the similarity in composition of this ultra-femic basalt with the olivine diabase ledge of the Palisades of New Jersey.[2] The similarity amounts, in fact, almost to identity. Not often could two specimens be taken of a single rock exposure which would give analyses so much alike. The Hawaiian basalt undoubtedly acquired its composition in a manner analogous to the olivine-diabase of New Jersey. The basalt came from that part of a magma mass where olivine crystals had accumulated and had not suffered re-solution. There was no ultra-femic liquid. Upon extrusion the liquid portion crystallized under altogether different conditions and gave the fine-grained groundmass. It is noteworthy that this lava did not flow freely but formed block-lava, which is not surprising considering the proportion of crystals.

Any number of rocks from Hawaii illustrate the same process as that plainly written in this ultra-femic basalt. A picrite-basalt from Kaula Gulch has such a composition as to endow it with 34 per cent normative olivine. This basalt is "very highly phyric" with "phenocrysts of olivine up to nearly one centimeter long" constituting "about one-fifth to one-fourth of the rock."[3] Though regarded by Daly as intrusive the "porphyritic gabbro of the Uwekahuna laccolith" has been chilled with sufficient rapidity that the olivine and the other constituents are strongly contrasted in grain, the olivine being already crystallized before intrusion. It has about 31 per cent normative olivine but about 40 per cent olivine as actual crystals.[4] Look where one will in the literature of Hawaiian rocks, no basalt will be found, containing more than a small amount of olivine, in which all of that olivine enters into the aphanitic ground on an equal basis with augite and plagioclase. There is but one conclusion to be drawn. Dana's old designation "chrysophyric" is eloquent both of their character and origin. No liquid with more than a small amount of olivine was concerned in the formation of any of these rocks.

The same facts are brought to light in any igneous field to which one may turn. A Linosa basalt with 28 per cent normative olivine contains phenocrysts of olivine.[5] So it is with a Japanese basalt containing 26 per cent normative olivine.[6] Again in New South Wales a basalt with 26 per cent normative olivine contains phenocrysts of that mineral.[7] In

1 R. A. Daly, "Magmatic differentiation in Hawaii," *Jour. Geol.*, 19, 1911, pp. 294-6.
2 Whitman Cross, "Lavas of Hawaii and their Relations," *U.S. Geol. Surv. Prof. Paper* 88, 1915, pp. 39, 76.
3 H. S. Washington, *Amer. Jour. Sci.*, 5, 1923, pp. 500, 501.
4 Daly, *op. cit.*, pp. 291-3.
5 H. S. Washington, *Jour. Geol.*, 16, 1908, p. 23.
6 Kozu, *Sci. Rept. Tohoku Univ.* (2), 1, 1913, p. 51.
7 D. Mawson, *Proc. Roy. Soc.*, *N.S. Wales*, 37, 1903, p. 341.

Madagascar and Réunion, which have received such careful study by Lacroix, the facts of the origin of olivine-rich rocks are well displayed. Here is found the effusive basalt with the highest amount of normative olivine (40 per cent) of any such rock listed in Washington's Tables. It has enormous crystals of olivine in a microlitic groundmass of plagioclase and augite with a little olivine. Lacroix is not sure that this material formed a separate flow. It may be merely streaks or layers carried along in a flow having in general much fewer crystals of olivine.[1] Ranging from this extreme example, a number of basalts are described and analyzed which have more moderate amounts of normative olivine

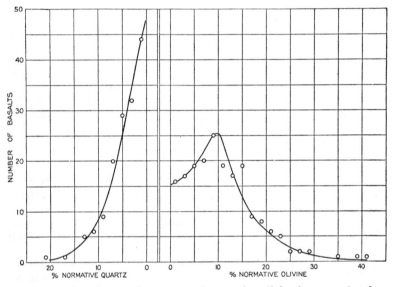

FIG. 43.   Plot of normative quartz and normative olivine in 335 analyzed "basalts."

but always with the same large olivines showing outstanding contrast in grain with the rest of the constituents. Never does more than a small amount of olivine enter into the aphanitic groundmass. Droplets and Pele's Hair, which have been quenched almost entirely to glass, may show as much as 10 per cent normative olivine,[2] corresponding in amount with the composition around which basalts tend to cluster (see Figs. 43 and 44), and this seems to be the most "basic" liquid concerned in the

1 A. Lacroix, Compt. rend., 154, 1912, p. 251.
2 Lacroix, op. cit., p. 253, Analyses h and i.

formation of the Madagascar rocks. When they became rich in olivine they did so through the accumulation of crystals of that mineral and without any significant remelting of these crystals. Deep-seated, coarse-granular rocks have, here as elsewhere, attained extremes never reached by effusives.

One other rock may be mentioned in this connection. It is a picrite-basalt from Juan Fernandez, a dike, not a lava, but quenched so as to

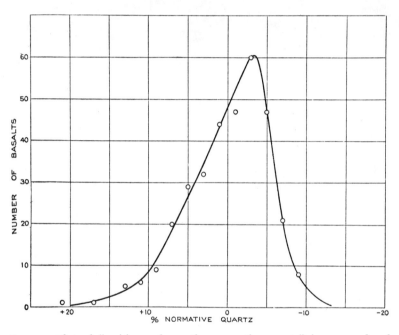

FIG. 44. Plot of "positive and negative normative quartz" in 335 analyzed "basalts." An adaptation of Fig. 43 to emphasize the continuous nature of the variation shown in the two curves of that figure.

reveal the facts of its origin. In it is shown the highest amount of normative olivine (53 per cent) of any rock termed basalt by the author describing it. Great crystals of olivine lie in an aphanitic ground composed mainly of plagioclase and augite (see Fig. 45). Some of the olivine basalts of this island group are, locally at least, about as rich in olivine as this dike, but they have not been analyzed. Their high olivine content is invariably due to an increased amount of phenocrysts of olivine of about 1 cm diameter.[1] Plainly these crystals were not in solution in the dike or flow material at the time of its intrusion or extrusion. As has been said before, this fact does not prove that they were not in solution in that material at an earlier time. But if one finds the condition shown by these basalts to be invariably true of all rocks rich in olivine which have suffered quenching, one must conclude that large amounts of oli-

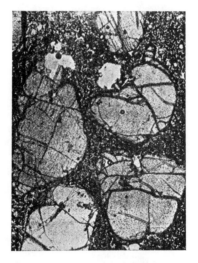

FIG. 45.   Picrite-basalt from Juan Fernandez (after Quensel) showing enormous contrast of size of olivine crystals with crystals of the other constituents.

vine never occur in solution in magmatic liquids. A survey of igneous provinces leaves no question that such rocks do have this character, that is, they always contain either all of their olivine or all in excess of a quite small amount (apparently some 12 to 15 per cent) as relatively large phenocrysts. They therefore force acceptance of the stated conclusion. Some examples encountered in the survey have been mentioned in the foregoing more for the purpose of emphasizing the nature of the inquiry than of indicating its scope.

ROCKS ENRICHED IN BOTH OLIVINE AND BASIC PLAGIOCLASE

Having found that rocks can become rich in either olivine or basic plagioclase only by accumulation of crystals of the one or the other which persist as such, we must expect that a special richness in both can come about only by accumulation of both. The allivalites of the

1 P. D. Quensel, *Bull. Geol. Inst. Upsala*, 11, 1912, pp. 286-7, 264-6.

Hebrides, which are made up of olivine and a plagioclase approaching anorthite, represent the extreme of this condition. Regarding their origin they offer no evidence of the kind we have been pursuing because they are never quenched, just as the extreme of enrichment in olivine alone (dunite) is never quenched. But rocks showing less extreme enrichment exhibit characters proving them to be formed by accumulation. Harker has described in Skye a group of dikes which he has termed the Beinn Dearg type. They contain phenocrysts of both olivine and plagioclase. The chemical analysis of a typical specimen shows it to have 21.5 per cent normative olivine and about 65 per cent anorthite in the normative plagioclase.[1] No dikes having like chemical characters are known in the area which do not have phenocrysts of olivine and plagioclase. For this particular area alone this fact might perhaps be referred to imperfection of the record but examination of a number of instances of rocks having like composition renders it very improbable that any such explanation applies. The rocks examined must be quenched rocks, in order that evidence on the question may be afforded. Rocks termed basalt are again turned to. We find that a fair number have both relatively high olivine and relatively high anorthite in the normative plagioclase. The petrographic descriptions show that they have phenocrysts of both olivine and plagioclase. Examples come principally from localities already discussed, Madagascar, the New Hebrides, Hawaii, and need not be further enlarged upon. A fine example in the form of a dike is described from the Nordingrå region of Sweden. It has phenocrysts of olivine and plagioclase which remain of substantially the same size in the chilled selvage though the groundmass is there quenched to a glass.[2]

ROCKS ENRICHED IN PYROXENE OR HORNBLENDE

The same sort of reasoning that has just been followed may be applied to rocks enriched in pyroxenes. A survey of lavas reveals the types that have been called ankaramites which are very rich in pyroxene but only in virtue of the presence of much of the pyroxene as phenocrysts. It may merely be stated that evidence of the same kind indicates that pyroxene-rich rocks are formed by the accumulation of crystals without remelting. There are no effusive equivalents of the pyroxenites, whether the characteristic mineral be ortho- or clino-pyroxene, which fact is to be referred to the non-existence of liquids of corresponding composition. Certain rocks very rich in clino-pyroxene but having a strongly alkalic base are known among effusives but they present a somewhat different problem which is treated on a later page.

Rocks rich in hornblende have probably been formed by the accumu-

1 Skye Memoir, p. 325.
2 J. M. Sobral, *Geology of the Nordingrå Region*, 1913, pp. 135-9.

lation of hornblende crystals without remelting or re-solution and there are consequently no effusives strictly equivalent to hornblendite. The question of re-solution of accumulated hornblende is, however, a more complex problem than the similar problem for olivine and basic plagioclase and can be treated only after certain principles governing it have been developed.

GENERAL CONSIDERATION OF THE ULTRABASIC ROCKS AND SUMMARY OF CONCLUSIONS

The view that ultrabasic and related rocks have been formed by the accumulation of early crystals is now rapidly gaining general acceptance, but the assumption is usually made that the accumulated crystals have been redissolved or remelted in depth. Since his recent conversion to crystallization-differentiation Vogt has been a champion of this assumption. With full realization of the requisite temperatures he boldly sets down such values as 1500-1600°C, necessary for the re-solution of olivine to give peridotite liquid, and 1500°C, necessary to give certain anorthosite liquids.[1] In view of the failure of inclusions of familiar rocks to show the remarkable effects they would undoubtedly experience upon immersion in such liquids, the temperatures mentioned can not be accepted. To these same high temperatures requisite to melt peridotite and anorthosite Vogt would refer the absence of effusive equivalents. He believes the liquid can not be maintained at this high temperature long enough to permit its pouring out at the surface from the great depth at which he supposes these magmas to originate by the remelting process. But rocks of this character are by no means always deep-seated. In the Hebrides they were formed under what was probably a moderate covering of surface lavas and if entirely liquid it is remarkable that they are never represented among these lavas. A high temperature of consolidation is in fact no bar to the occurrence of any liquid as a surface flow. In the filling of the feeding fissure the formation of a border selvage provides a hot, ill-conducting couche and it is ordinarily a long time in a mass of even moderate thickness before the central portions experience any cooling whatever. Undoubtedly such liquids, in view of their high fluidity, would be transferred rapidly and would pour out freely upon the surface. The evidence is thus in every way decidedly against the existence of these very hot liquids.

Realizing this fact Harker has turned to volatile components as an aid to the re-solution of the accumulated crystals. He believes that in some manner the mass of crystals receives an accession of volatile components adequate to bring about its liquefaction at moderate temperatures.[2]

[1] *Economic Geology*, 21, 1926, p. 232.
[2] Oral communication.

Against this view Vogt has quite definitely set his face on account of lack of evidence that the rocks concerned ever contained the requisite large amounts of volatile components, a fact to which attention was called long since.[1] There is, in addition, the difficulty, frankly admitted by Harker, of picturing the process of addition of volatile components to the crystalline mass. Absence of effusives is again unaccounted for.

These difficulties disappear entirely if the rocks are accepted as the result of crystal accumulation without remelting. The flow of the resulting material into fissures would seem to be a quite definite possibility if the proportion of liquid is not too small, but the factors influencing intrusion of this kind are too obscure for adequate discussion. No doubt the rate of application of the force is an important factor.

If the conclusions reached regarding the Skye peridotites may be accepted the intrusion of the largely crystalline mass may have the way prepared for it by the prior intrusion of a liquid material of more complex (polycomponent) constitution. Even a portion of the mass which is entirely crystalline may apparently be carried along with the rest during this kind of intrusion and we may thus obtain a dike which is locally dunite, but this is accomplished only with much granulation of the olivine and the development of a general fissility of the mass. In some cases it may be that a mass which gives rise to a dunite dike may have had a small amount of interstitial liquid (not dunitic) but this liquid is squeezed out during the passage of the mass into and along the fissure. The purely mechanical effect could, of course, never remove all the liquid but this may be supplemented by a re-solution of crystals at points of contact, where the brunt of the pressure is borne, and redeposition elsewhere. In general it is necessary to appeal to some such action to obtain a monomineralic mass of any kind. The mere accumulation of crystals can not give such a mass. In some cases it may be that no great pressure is applied in order to convert a mass of crystals into a solid cake with interstitial liquid removed. It is known that if a mass of discrete crystals of a soluble salt lie in the bottom of a beaker of saturated solution the mass will gradually be converted into a solid cake of the salt. It seems possible that in some cases the accumulation of crystals in a magmatic chamber may be sufficiently slow that as the mass grows each new layer of crystals is converted by this action into a solid cake. What the fundamental control over this action may be we do not know and need not know for our present purpose. These effects that must be assumed as supplementary to crystal accumulation in the case of monomineralic rocks offer some difficulties, it must be admitted,

1 N. L. Bowen, "The Problem of the Anorthosites," *Jour. Geol.*, 25, 1917, pp. 209-43.

but the alternative (viz., the existence of liquids of extreme composition) is very unlikely on general grounds and is altogether negatived by the characters of "quenched" rocks. By this is meant rocks which are uniformly aphanitic to glassy and rocks which have a groundmass of that character. Thoroughgoing ultrabasic rocks are not found in either category. Among the uniformly aphanitic (aphyric) rocks there are none that remotely approach ultrabasic compositions. Among porphyritic rocks with an aphanitic base there is a greater approach to such compositions but the degree of approach is dependent upon the presence of relatively large phenocrysts of the minerals characteristic of the ultrabasic rocks. These porphyritic rocks have no aphyric equivalents and must be formed by the accumulation of crystals of their porphyritic elements. This process, carried to extremes, gives rise to the thoroughly ultrabasic rocks, which are therefore not represented among lava flows and only rarely among dikes, in which case they have been intruded under special conditions.

BANDED GABBRO

Banded gabbro is appropriately considered under rocks formed by crystal accumulation with sorting. Banded gabbros are rocks with the mineral composition of gabbro in which there is a banding or layering marked by special richness of one mineral in a certain layer and special poverty of that mineral in an adjacent layer. They are abundantly developed in the Hebrides and apparently all who have studied them there, certainly Geikie, Teall and Harker,[1] have concluded that they are the result of intrusion of an inhomogeneous liquid. Harker is no advocate of liquid immiscibility of silicates but he does believe that the separate liquid portions did not succeed in mixing and is frankly agnostic as to how the inhomogeneity originates. Referring especially to the Duluth gabbro, Grout concludes that the banding was caused by convection while crystallization was going on.[2] This requires the further assumption of a rhythmic crystallization, plagioclase and augite alternating, a condition which is difficult of acceptance. It has therefore been suggested that the banding results from deformation in a crystallizing mass in which crystal accumulation is occurring. The action is of the nature of the intrusion of the more completely liquid portions into rifts in the crystal mesh of the part of the mass in which the proportion of crystals is much greater. The process is thus a sort of auto-intrusion.[3] Further crystallization-differentiation by gravity in the liquid layers thus intruded may give extreme effects, for this liquid virtually

1 Skye Memoir, p. 120.
2 F. F. Grout, *Jour. Geol.*, 26, 1918, pp. 481-99.
3 N. L. Bowen, *Jour. Geol.*, 27, 1919, pp. 417-22 ; W. J. Mead, *Jour. Geol.*, 33, 1925, p. 697.

begins its crystallization anew under conditions that are ideal for producing such effects. An interesting observation upon the banded gabbro of the Bushveldt has been made by Wagner who shows that elongated crystals are not aligned but merely lie in a common plane, assuming all azimuths in that plane. Such a structure is plainly a sedimentation structure.[1] Even differentiation by diffusion to the border might take place in these liquid layers, for the limitation placed upon the effects of diffusion would no longer be operative. This limitation, it will be recalled, is due to the fact that transfer of heat by diffusion is so much more rapid than transfer of substance by diffusion. But in the present case transfer of heat has to be effected through great distances, the cooling rate being that of an enormous mass, whereas transfer of substance must take place through the very small distance represented by the thickness of the layer. It seems possible, therefore, that drastic differentiation effects could be brought about in such sheets and that seams and layers even of pure minerals might result. These layers would have no necessary relation to the boundaries of the mass as a whole but would have to the boundaries of the particular sheet from which they formed. Layers of ilmenite, of picotite and of plagioclase in banded gabbro may perhaps be referable to such processes. That they were never liquid as such is demonstrated by the fact that they never behave independently and are never found outside their parent gabbro.

Whatever may be the details of the method whereby banding of Hebridean gabbros has resulted, a general survey of Hebridean rocks makes it plain that it is a condition brought about during crystallization and that it was not the result of intrusion of an inhomogeneous liquid. One encounters the expression "banded gabbro" time and time again in the literature of these rocks, but one will look in vain for the expression "banded dolerite" or "banded basalt." If inhomogeneous liquids of a gabbroid character were intruded in the Hebrides the evidence of their inhomogeneity would certainly be best preserved in the quickly chilled minor intrusions or in the lava flows and there would undoubtedly be banded dolerites and banded basalts. But such rocks are entirely lacking. The banding originates only upon crystallization and only when that crystallization is slow enough to permit the growth of the large crystals characterizing gabbro. The occurrence of minor intrusions which are banded only serves to emphasize this fact for they are then gabbro and not dolerite or basalt. If, therefore, the banded Hebridean gabbros were intruded in an inhomogeneous condition the inhomogeneity was due to uneven distribution of crystals already separated. The olivine basalts of

1 Percy Wagner, "Magmatic Nickel Deposits of the Bushveldt," *Geol. Surv. U.S. Africa. Mem.* 21, 1924, p. 81.

Réunion, having certain layers crowded with enormous olivine crystals, illustrate the possibility of such conditions in magmas which have even broken their way through to the surface. No uniformly aphanitic rock of gabbroid composition is ever banded. When this fact is compared with the general impression one gains from the literature that there is scarcely a mass of gabbro of notable dimensions without its banded portion there is no escape from the conclusion that banding originates during slow (coarse) crystallization. It is not merely a matter of slow cooling but plainly one of slow crystallization, with the attendant opportunity for crystal sorting, auto-intrusion, further crystal sorting and so forth.

ANORTHOSITES

The facts that have already been set down concerning the nature and origin of rocks enriched in basic plagioclase need no further amplification. It is plain that a general survey of rocks reveals no support for the concept that there are liquid magmas of the approximate composition of basic plagioclase. The same considerations may be applied to plagioclase rocks in general and the same conclusion is reached. There are no liquid magmas corresponding in composition with any plagioclase, not even albite. (See p. 132.) An origin by crystal accumulation is not, to be sure, advocated for rocks consisting entirely or almost entirely of albite or closely related plagioclase, but with increase of the lime content and probably from andesine on to anorthite, the origin by crystal accumulation becomes increasingly clear. It is to such rocks that the term anorthosite is ordinarily applied and will be here applied. A decade has passed since the probability of the origin of anorthosite by crystal accumulation was first clearly set down. Petrologists have sought diligently for evidence that anorthosites were liquid but the search has not been attended by success. No glasses, no aphanites, no lava flows have been found. Vogt has noted that the anorthosite of Ekersund is finer-grained near its borders and interprets this as due to contact chilling of a completely liquid mass. But even at the border the rock is coarse grained.[1] If due to chilling and not, say, to granulation, the finer grain of the border may be the result of restriction of the further outgrowth of crystal boundaries in a mass largely crystalline. Dikes of anorthosite have been found by diligent search but they are excessively rare, although dikes may be formed from a nearly crystalline mass, as demonstrated in the study of peridotites given on an earlier page. It is to be noted that, in considering the significance of such intrusion phenomena as anorthosite may display, careful consideration must be

1 J. H. L. Vogt, *Vidensk. Selsk. Skr. I. Mat.-Naturv. Kl.*, 1924, No. 15, p. 84.

given to the purity of the mass. The amount and nature of the foreign material may be such as to permit the existence of a moderate amount of liquid. Thus, in discussing a Canadian anorthosite, Mawdsley points to definite evidence of the intrusion of andesine anorthosite into labradorite anorthosite. He states that this seems "to imply that the andesine anorthosite as a whole did exist in a liquid state and with the composition of andesine anorthosite."[1] But while this is a possible conclusion it is by no means a necessary conclusion. The andesine anorthosite contains from 5 to 17 and exceptionally even 25 per cent alkalic feldspar.[2] It could have had 10, 20, perhaps even 30 per cent liquid at a temperature at which many syenites are still liquid and such quantities of liquid are ample to explain the intrusion phenomena displayed, especially when protoclastic structure is displayed as well. After deciding that "the andesine anorthosite, as a liquid, may have advanced to its present position relative to the surrounding rocks" Mawdsley then goes on to say, "If this did happen, there is no reason for supposing that the earlier-formed, labradorite anorthosite did not also come into place in a liquid condition." But the one does not necessarily follow from the other. The facts are that neither anorthosite can be regarded as belonging in a liquid line of descent. If either were completely liquid it must have been formed by accumulation of crystals which were subsequently remelted. The difficulties in the way of accepting this remelting are greater for the labradorite anorthosite than for the andesine anorthosite. One might thus find evidence of a completely liquid condition of andesine anorthosite without any necessity that associated labradorite anorthosite should ever have been in a similar condition. But the difficulties in the way of accepting completely liquid andesine anorthosite are themselves sufficiently great that, unless the evidence of that condition is absolutely unequivocal, it should be entertained only with the strongest reservations. It is not too much to demand the production of such evidence. Every rock which we are led to expect, from physico-chemical considerations, as a member of the liquid lines of descent promptly justifies our expectations by having aphanitic and glassy equivalents and by occurring as an effusive. The persistent failure of anorthosite to do any of these things can only be regarded as due to the fact that it does not lie in any liquid line of descent and is not the result of remelting of accumulated crystals.

The protoclastic structure of anorthosites is very important in indicating the manner and conditions of their intrusion. It is of course possible for any rock, even one belonging in a liquid line of descent, to suffer movement when it approaches complete crystallization and to

1 J. B. Mawdsley, *Geol. Surv. Can. Mem.* 152, 1927, p. 33.
2 Mawdsley, *op. cit.*, p. 24.

develop a protoclastic structure. But of no such rock can it be said that
a protoclastic structure is an almost universal character, as it can of
anorthosite. The structure is evidence of the fact that when an anorthosite mass moves it does so only with accompanying granulation, a behavior which is displayed by the dunite of Skye when it occurs as dikes.

A NOTE ON "MAGMATIC ORE DEPOSITS"

The bodies of ilmenite associated with anorthosite, of chromite associated with peridotite and similar masses of ore minerals in other
associations have led many investigators to believe in the injection of
liquid of the composition of these bodies. When one objects to such a
conclusion on the basis of the excessively high temperatures necessarily
involved, the question is often asked, "But could you not have such
liquids with enough volatile components to lower their temperatures of
consolidation to a reasonable value?" The only answer is, "Probably you
could," but the question is beside the issue. The appropriate question is
not, "Are such liquids possible?" but rather, "Have such liquids any
possible provenance in Nature?" None has ever been suggested. Certain
magmas such as the highly silicic or highly alkalic are, in virtue of
their character as residual liquids, the natural home of concentration
of volatiles. But the oxide ore bodies, like the ultrabasic rocks, are a
very unnatural place to expect concentration of volatiles. It is probable
that not all masses of the oxide ores are crystal accumulations. Some are
perhaps the result of a secondary rearrangement of the rock materials
by circulating solutions, but that any of them are the result of the
injection of a molten magma, even a "wet" magma, is exceedingly
unlikely. From this conclusion some sulphide ore-bodies may be excepted.
Sulphide mixtures have moderate melting temperatures. The liquids
are known to be but sparingly miscible with silicate liquids. Sulphide
liquids may separate from silicate magmas, carrying with them only
their appropriate share of the volatile components in their partitioning
between the two liquids, and giving thus a sulphide liquid, with moderate amount of volatiles, which is as definitely a magma as the associated
silicate magma. To the accumulation by gravity of such a sulphide
magma and occasionally to its injection into surrounding rocks some
sulphide bodies, especially those associated with norite, have been referred, probably correctly. There are evidences of the effects of solutions
in the formation of these bodies and upon these much stress has recently
been laid. It is probable that these are, in many cases, no more than the
effects of the residual solutions existing during the late stages of the
consolidation of the sulphide magma and strictly comparable with such
effects in the associated silicate masses during the late stages of their
consolidation.

ADDENDUM TO N. L. BOWEN'S "ULTRABASIC TYPES OF THE HEBRIDES"[1]

E. B. BAILEY

IN August 1927 Bowen published a very suggestive paper entitled *The Origin of Ultrabasic and Related Rocks*. In it he develops the view that all anorthosites and peridotites have originated as crystal concentrates and have never possessed more than a meagre proportion of molten matrix. Previous to complete consolidation, such solid-liquid emulsions might be fluid enough to act as intrusions under favorable conditions, and yet be too viscous to reach the surface. Thus Bowen accounts for the absence of representatives of anorthosite and peridotite amongst analyzed lavas. He might, perhaps, have emphasized his point a trifle by adding that peridotitic bombs and inclusions ("olivine nodules") are well known in certain volcanic fields.

On reading Bowen's article, it immediately occurred to me that his theory would account for a very definite feature of Carboniferous and Permian intrusion history in Central Scotland, where six familiar picrite sills are equipped with subordinate marginal layers of teschenite or allied dolerite. These picrite sills occur, from east to west, at Barnton, Inchcolm, Blackburn (two), Lugar, and Ardrossan; and exposures are full enough in every case, except that of the main Blackburn sill, to prove that the marginal teschenite occurs at the base as well as at the top of each composite intrusion. There is no doubt that a general principle is involved. Although differently constituted, this picrite-teschenite association reminds one of hundreds of composite minor intrusions of the Tertiary Hebridean suite, in which an acid interior is persistently margined by subordinate basalt. H. H. Thomas and I have offered an explanation of these acid-basic composite intrusions, suggesting that viscosity of the acid magma, when confronted with a cold environment, has been a controlling factor. Obviously a similar explanation would account for the picrite-teschenite association if one could only be sure of the viscosity of a picrite magma under "hypabyssal" conditions. I confess that I have for years expected that evidence of such viscosity would eventually be afforded. Bowen's paper of last August exactly meets the case.

I wrote to Bowen pointing out this possible application of his views. He replied: "I am much interested in your conclusion regarding the picrite-teschenite sills and agree with you entirely. In fact I have reached the same conclusion regarding the peridotite dikes of Skye. . . . This conclusion I have had written up for some time." He also kindly posted me the typescript of his *Ultrabasic Types of the Hebrides*, in which this

1 Mr. Bailey long since had this note ready for separate publication but upon learning that I was working on similar matters he was unwilling to publish it prior to the appearance of my conclusions. We have taken the present method of obtaining simultaneous publication. It is given here without change of any kind from the original form.—N.L.B.

conception is developed in a manner which speaks for itself. My contribution to the subject therefore takes the form of an additional list of examples. Presumably any other geologist who has worked among fairly coarse-grained minor intrusions of ultrabasic composition could perform a similar service.

It should be explained that the picrites which have been mentioned above are not all of such extreme composition that they cannot be chemically matched amongst lavas. In fact analyses show that some of them have approximately the composition of nepheline basalt. They are, however, in every case coarse-grained rocks for which it is easy to postulate a very considerable crystallization of the type Rosenbusch has styled "intratelluric." They are therefore likely to have been affected by crystal concentration prior to final intrusion.

It is delightful to find in Bowen's theory of peridotites a solution of the long-standing picrite-teschenite puzzle of Central Scotland—though of course this coincidence does not in itself constitute a proof of the theory. Let us hope that Bowen may return to Skye, either literally or metaphorically, and furnish us with a similarly constituted story of the banded xenolithic peridotites and banded gabbros[1] of that glorious island. Perhaps he has already some such contribution to petrology in reserve. It is obviously a field of research where experimental analogies could be devised to work under everyday conditions of temperature.

SELECTED BIBLIOGRAPHY

1897. A Geikie, *Ancient Volcanoes of Great Britain*, vol. I, especially Barnton near Edinburgh, p. 450.

1907. J. D. Falconer, "The Geology of Ardrossan," *Trans. Roy. Soc. Edin.*, vol. XLV, p. 601,—especially section of Castle Crags picrite, Fig. 1, p. 606.

1908. R. Campbell and A. Stenhouse, "The Geology of Inchcolm," *Trans. Geol. Soc. Edin.*, p. 121,—especially Pls. vi, vii, and p. 123.

1910. "The Geology of the Neighbourhood of Edinburgh," *Mem. Geol. Surv. Scot.*,—especially Blackburn and neighbourhood, J. S. Grant Wilson, pp. 280, 281; Blackburn analyses by W. Pollard, p. 300; Inchcolm analysis by J. W. Judd, p. 300. Judd's analysis of the Inchcolm picrite shows 22.9 per cent magnesia. T. C. Day allows me to say that he has got a similar high value in an analysis of fresh material which will shortly appear in a paper by A. Stenhouse.

1917. G. W. Tyrrell, "The Picrite-Teschenite Sill of Lugar (Ayrshire)," *Quart. Jour. Geol. Soc. London*, vol. LXXII, p. 84,—especially Fig. 4, p. 96, and analyses by A. Scott, pp. 104, 110, 114.

1924. "Tertiary and Post-Tertiary Geology of Mull, Loch Aline, and Oban," *Mem. Geol. Surv. Scot.*,—especially Composite intrusions, H. H. Thomas and E. B. Bailey, pp. 32, 33.

1927. N. L. Bowen, "The Origin of Ultrabasic and Related Rocks," *Amer. Jour. Sci.*, vol. XIV, p. 89.

1 An earlier section of the present chapter deals very briefly with the banded gabbros of Skye. N.L.B.

CHAPTER X

THE EFFECTS OF ASSIMILATION

AS A rule those variations of igneous rocks which have been brought about by solution of foreign material have been treated as something apart from the effects of (fractional) crystallization. The solution of solid rock is, however, so intimately related to the separation of solid crystals that, for the most part, assimilation is best treated as a sort of corollary of crystallization and as governed by the same general laws. This it is now proposed to do.

Many igneous rocks contain inclusions of foreign material and not infrequently these inclusions show evidence of having been attacked by the magma, some to a moderate extent and others to such an extent that only traces of the inclusion remain. To some petrologists these inclusions are but the remnant of a great host, most of which has been completely incorporated in the magma, and to such incorporation or assimilation of foreign matter they would assign the principal variations of igneous rocks. The variations are not usually regarded by these petrologists as the result of assimilation alone but of assimilation followed by the differentiation of the syntectic magma which is supposed to have special powers of differentiation not possessed by the original magma. Other petrologists believe that magmas cannot be expected to have the energy content necessary for the solution of a significant amount of foreign material; that the amount of solution actually observed at and near contacts is an approximate measure of the total and that the variations of igneous rocks are quite independent of these slight additions, being due to spontaneous powers of differentiation possessed by original, uncontaminated magmas.

In this chapter it is proposed to discuss the behavior of inclusions with special reference to these questions. By application of the principle of the reaction series, as developed in Chapter V, it is hoped to effect a certain amount of reconciliation of these extreme views.

HEAT EFFECTS OF SOLUTION

Since one of the important questions involved is that relating to the heat effects resulting from solution it is desirable to consider the information available on these effects. The ordinary equilibrium diagram, commonly regarded as a freezing-point diagram, is at the same time a solubility diagram. It gives the change of solubility of any phase with temperature. But the change of solubility with temperature depends mainly[1] on the heat effect involved in solution, and the equilibrium diagram contains complete information on this heat effect. Unfortunately the information may be very difficult of extraction; in the present state of knowledge, often impossible. One solubility diagram, that of the plagioclase feldspars, has proved particularly tractable in this respect. This diagram has been shown in Fig. 9, and is repeated in somewhat different form in Fig. 46, the curves being calculated on the basis of

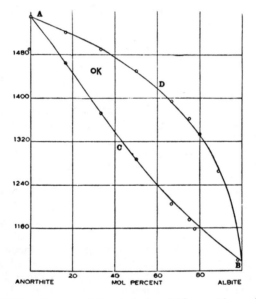

FIG. 46. Equilibrium diagram of the plagioclase feldspars. The small circles indicate determined points. ACB and ADB are calculated curves assuming no heat of mixing.

a latent heat of 104.2 cal. per gram for anorthite and 48.5 cal. per gram for albite. The determined points are given by the small circles and their correspondence with the calculated curves is very remarkable. Since the curves were calculated on the basis of constant latent heats (solution heats) this correspondence simply means that there are no mixing-heat

1 The volume change is involved also but is relatively unimportant.

effects and that the heat of solution of any plagioclase in liquid plagioclase is simply the latent heat involved in the change from solid to liquid.[1]

In Fig. 47 is plotted the equilibrium diagram for anorthite and diopside. Now we know the latent heat of anorthite from the calculated

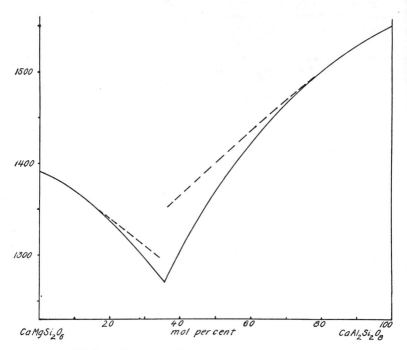

FIG. 47. Equilibrium diagram of diopside and anorthite. Determined curves in full lines. Broken curves calculated on the assumption of no mixing heats.

results of Fig. 46 and we may calculate, using this value, a curve of freezing-point depression for anorthite according to the equation for ideal concentrated solutions, viz.:

$$T = \frac{T_0}{1 - \dfrac{2T_0}{Q}\ln x}$$

[1] See N. L. Bowen, "Melting Phenomena of the Plagioclase Feldspars," *Amer. Jour. Sci.*, 35, 1913, p. 590.

where T is the melting temperature (saturation temperature) of anorthite in a solution of mol fraction x in anorthite; Q is the latent heat of melting per mol of anorthite (= 29,000 cal.) and T_0 its melting-point (= 1550°). As a result of this calculation we obtain the right-hand dotted curve. It will be noted that the determined curve corresponds with the calculated dotted curve in the upper portion (as far as about 80 per cent anorthite) and then falls below. In other words, we could have calculated the latent heat of anorthite from a point on the upper portion of the determined curve. If now we do this for the diopside curve, that is, calculate the latent heat of diopside from a point on the upper portion of its solubility curve, we find a latent heat of 23,420 cal. per mol or 108 cal. per gram.[1] Calculating the further course of the curve with the use of this value we obtain the dotted curve. This calculated curve also lies above the determined curve at points distant from pure diopside. There is one factor which can cause such a deviation of the freezing-point curve from the theoretical curve of the above equation, viz., a heat of mixing of the liquids, and Van Laar has developed an equation which enables one to calculate the differential heats of mixing involved. The equation is

$$T = T_0 \frac{1 + \dfrac{a(1-x)^2}{[1+r(1-x)]^2}}{1 - \dfrac{2T_0}{Q}\ln x}$$

where a and r are coefficients from the Van der Waals equation of state for binary mixtures. The numerator gives the number of times the solution heat, at the concentration x and temperature T, is greater than the melting heat Q, and the difference between this and Q is the differential heat of mixing.[2] From this equation we find the following values of the differential heats of mixing per mol (q).

TABLE XIII

MOLAL DIFFERENTIAL HEATS OF MIXING IN ANORTHITE-DIOPSIDE MIXTURES

For anorthite	x	0.8	0.5	0.4	0.35
	q (cals.)	100	600	1100	1250
For diopside	x	0.7	0.65	....	....
	q (cals.)	120	340	....	....

Now these differential heats of mixing are the heats of mixing of one mol of liquid with a very large amount of solution of the various con-

1 A direct determination by W. P. White gave 106 ± 15 cal. *Amer. Jour. Sci.*, 28, 1909, p. 486.
2 *Z. physik. Chem.*, 8, 1891, p. 188.

centrations referred to. These in themselves have no particular interest from the present point of view but from them the so-called integral heats can be calculated by a graphical method. Roozeboom[1] shows that if the curve of integral heats of mixing is plotted against mol fractions of the components then the intercept (on the heat axis) of the tangent to the curve gives the differential heat of mixing for the composition represented by the point. Thus in Fig. 48 if the curve ABC represents the heats of mixing of diopside liquid and anorthite liquid in various pro-

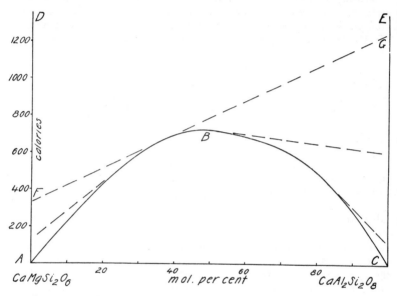

FIG. 48.   Curve of integral mixing heats of diopside and anorthite liquids (ABC). Showing graphical method of determining integral mixing heats from differential mixing heats.

portions to form one mol of mixture, then the differential heat of mixing of one mol of anorthite in a large amount of liquid of the mol fraction x is given by the intercept on CE of the tangent at x. Also the differential heat of solution of diopside in this same mixture is given by the intercept on AD of the same tangent. Now in our particular case we have determined the differential heats by calculation from the freezing-point curves and we wish to know the integral heat. This is the reverse of the above problem and while not as straightforward can nevertheless

1 *Die Heterogenen Gleichgewichte*, Zweiter Heft, Erster Teil (1904), pp. 287-90.

be solved in the same way. We have, fortunately, for the composition of the eutectic, the differential heats of mixing for both diopside and anorthite which completely fixes the tangent at the composition of the eutectic. Fig. 48, which has been referred to for purposes of illustration, is also the actual figure for diopside and anorthite determined graphically from the calculated differential heats of Table XIII. Thus FG is the tangent at the composition of the eutectic joining the value of the differential heat of mixing for diopside liquid in the eutectic with differential heat of mixing for anorthite liquid in the eutectic liquid. This fixes immediately that the integral heat of formation of one mol of the eutectic liquid from its liquid components is 650 cal. Knowing that the curve of integral heats is tangent to this line at $x = 0.65$, we may readily draw a complete curve (the envelope of the family of tangents) that satisfies the known values of the differential heats. When this is done we get the curve ABC. This curve, then, gives us directly the amount of heat evolved when liquid anorthite and liquid diopside are mixed in any proportion. Thus when ¼ mol diopside liquid is mixed with ¾ mol anorthite liquid 500 cal. are evolved. The maximum amount of heat per mol of mixture (720 cal.) is evolved when about 0.047 mol of anorthite liquid is mixed with 0.53 mol of diopside. This happens to be about equal weights, so that the maximum amount of heat is evolved when about equal weights are mixed and is equal to about 3 cal. per gram of mixture. This heat is sufficient to heat the mixture about 10°.

If now we turn to the equilibrium diagram of diopside and albite and calculate a curve of freezing-point depression for diopside, using the value of the latent heat found from the anorthite diagram, we find again that the calculated curve coincides with the observed curve in its upper portion and then deviates from it (Fig. 49), but in this case in the opposite direction to that found for the other diagram. Again, this deviation can be interpreted as due to a heat of mixing of the liquid but now of the opposite sign. If we apply the Van Laar equation we can calculate the differential heats and from these determine graphically as before the integral heats of mixing. Thus we get Fig. 50.

We find, then, that albite liquid and diopside liquid mix with absorption of heat, the maximum absorption (790 cal.) taking place when 0.48 mol albite is mixed with 0.52 mol diopside. This also is about equal to 3 cal. per gram of mixture.

Assuming that the theoretical basis of our calculations is to be relied upon we have proved that anorthite and diopside, both characteristic molecules of basic rocks, mix in the liquid state with evolution of heat, whereas albite, one of the most characteristic molecules of acid rocks, mixes with diopside in the liquid state with absorption of heat and with anorthite without significant heat effect. These results are the opposite of what is often assumed to be the case. Many statements are to be found in the literature to the effect that acid and basic rock material will mix with evolution of heat. There is nothing in the results

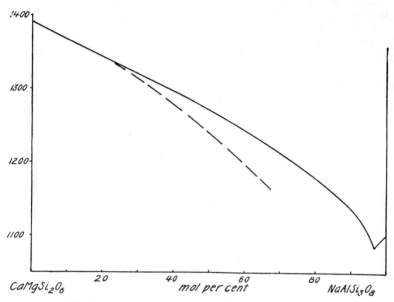

FIG. 49. Equilibrium diagram of diopside and albite. Determined curves in full lines. Broken curves calculated on the assumption of no mixing heats.

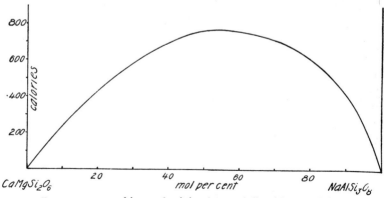

FIG. 50. Curve of integral mixing heats of diopside and albite.

of studies of phase equilibrium to warrant such a statement; indeed, there are, as we have seen above, good reasons for doubting it.

Possibly in rare cases, where more fundamental reactions are involved, greater heats of mixing are available, but the mixing of two liquids has, in itself, little importance in petrogenesis. Our particular interest lies in the heat effect on mixing solid rock-matter with liquid magma. In that connection the heat of mixing of liquids is not without significance, for the heat of solution of a solid is to be regarded as the resultant of two heats, the heat of melting of the solid and the heat of mixing of the liquids. Now the heats of mixing that we have calculated above are really very insignificant as compared with the heats of melting of the solids, and the heats of solution of the solids are, therefore, very nearly equal to the latent heats of melting. For the solution of solid anorthite in diopside a little less than the latent heat is required, for the solution of solid albite in diopside a little more than the latent heat is required. The differences are noteworthy in connection with any theory that postulates an evolution of heat when an acid rock is immersed in basic magma, but for the purposes of the present inquiry which seeks to find merely the order of magnitude of the heat effect when solid rock is added to magma, we may state that the heat of solution of solid anorthite, albite, or diopside in any liquid mixture of them is substantially equal to the latent heat of melting. Moreover it is probably true of silicates in general that the solution of the solid is attended by a large absorption of heat, though not many determinations lend themselves to interpretation in the same way as the above. Thus when one attempts similar calculations from the equilibrium of anorthite with nephelite and of anorthite with silica it is found necessary to assume molecular association in nephelite and silica, and, while this is not surprising since both occur in more than one crystal form, it nevertheless so complicates the case that interpretation becomes impossible. One thing is certain, namely, that all the solubility curves of silicates yet determined show a marked increase of solubility with temperature, which means a strong absorption of heat upon solution. No example of retrograde solubility is known. In conclusion, then, we may state that the solution of a silicate in a magma is usually accompanied by a large absorption of heat, probably of the order of magnitude of the heat of melting.

THE QUESTION OF SUPERHEAT

Having thus arrived at a general conception of the heat required for solution we may consider the question of the heat available. One aspect of this is concerned with the superheat of the magma, that is, the excess of temperature above that at which crystallization begins. Somewhere within the earth there is material (whether a rigid liquid or a crystalline solid we need not here consider) which is capable of becoming fluent liquid. If the action is a remelting of crystals by release of pressure there could be no superheat. At great depths there may be excessively hot material capable of giving excessively hot liquid, but this is hardly concerned with matter that ever comes to the light of day as an igneous rock. There must be a transition zone, representing a passage from stable crystal material incapable of becoming magma, through a zone capable of giving magma with some suspended crystalline matter, to a zone capable of giving a completely molten magma. It is from the zone giving rise to magma with some suspended crystals, rather than from the zone giving a simple liquid, that we must expect the great upwellings of magma to come, because of the advanced position of the partly crystalline material with respect to an action producing such upwelling and because it may be readily mobile when the amount of suspended crystals is small. As the magma rises in the crust several factors may combine to reduce the amount of crystals. Among these factors is the increased solubility of the crystals resulting from the lowered pressure, for the solution of silicates usually takes place with increase of volume. Besides this there may be an actual addition of heat as a result of the Joule-Thomson effect[1] and of possible exothermic reactions ensuant upon reduced pressure. It is probable that, in the usual case, the magma still retains some crystals even when it rises to shallow depths within the crust for, though the intensity of the above actions is thereby increased, it is also passing into colder and colder surroundings and must lose a corresponding amount of heat. It is possible that in some cases a magma may have lost all its crystals by the time it has risen to shallow depths and perhaps may have a temperature above the saturation temperature under the conditions there prevailing. However, even if we make the liberal assumption that it is superheated $100°$ (say its temperature is $1100°$ and the saturation temperature $1000°$) and that this condition is acquired when it has reached a level 5 km. below the surface, it should be noted that the mere immersion of blocks of country rock amounting to about 10 per cent of the mass of magma would be sufficient to wipe out this superheat, for the original temperature of blocks even at this quite considerable depth would be only $200°$.

This, too, is quite apart from the amount of heat that must be lost by the magma in heating up a large amount of wall rock other than the immersed fragments. If we consider greater depths, where the surroundings would be at a higher temperature, it must be remembered that we are thereby limiting the amount of superheat that may have been acquired as a result of those processes attendant upon the rise of the magma and proportional in amount to the extent of rise.

1 L. H. Adams, *Bull. Geol. Soc. Amer.*, 33, 1922, p. 144.

It is plain that there can seldom be more than a very small amount of superheat left after the heating up of immersed fragments. Occasionally, perhaps, such small amounts may be available for the direct solution of fragments but it would require only an excessively small amount of solution to use up this heat, a fact that becomes apparent when the enormous discrepancy between the specific heats of silicate liquids (0.2-0.3 cal.) and the solution heats of silicates (50-125 cal.) is realized.

We have thus deduced, from general considerations, that magmas cannot be expected to have much superheat and in particular that they cannot retain a significant amount after the immersion of foreign fragments in such quantity that an important effect upon the composition of the magma would ensue if solution did occur. This deduction need not be regarded as contradictory to the observed fact that on very rare occasions inclusions of foreign rock have been found converted into glass by magmas and that the measured temperatures of lavas at certain active volcanoes are sometimes high enough to indicate a condition probably significantly above that of beginning of crystallization. It is becoming increasingly apparent that in central volcanoes there are sources of heat, probably from exothermic gas reactions, that are capable of producing the temperatures observed, but these must be regarded as locally concentrated where the gases have their vent. Moreover, inclusions converted to glass are found most commonly in such extrusive rocks. Even their rare appearance in intrusive masses need not militate against the general conclusion reached, for, given the proper contrast of composition, some fusion may occur when the magma is not superheated, a fact which will be brought out in the sequel. Admitting a certain amount of superheat, the first few inclusions added would receive the full benefit of it and might be fused, but the magma would suffer a correspondingly marked loss of heat. In other words, the net result of the addition of an amount sufficient to produce a significant change of composition, say 10 per cent, would be the same even though they were added slowly and the first additions were drastically affected. Not only do magmas commonly fail to convert inclusions into liquid but they also fail to effect such changes as the transformation of quartz into tridymite and of wollastonite into pseudo-wollastonite, which fact must be regarded as indisputable evidence of the prevailing low temperatures.

There seems no reason, therefore, to doubt that direct solution of foreign material in superheated magmas cannot be a factor of importance in petrogenesis. However, the importance of superheat has been greatly exaggerated both by those who adhere to the view that magmas dissolve large quantities of foreign matter and by those who deny it. We shall find in the sequel that even saturated magmas may produce very marked effects in the way of incorporation of foreign material.

EQUILIBRIUM EFFECTS BETWEEN "INCLUSIONS" AND LIQUIDS IN INVESTIGATED SYSTEMS

In the following discussion of the effects of liquids upon inclusions, whose composition is embraced within investigated systems, attention will be confined, unless otherwise stated, to saturated liquids. This is done because we have found in the foregoing discussion that the super-heated condition has little importance in nature. But lest it be thought that this is going too far in the way of eliminating superheat we have made another assumption that should amply compensate, namely, that the inclusions are already heated to the temperature of the liquid before immersion in it.

It is proposed to discuss the magnitude of the heat effects involved when reactions go on between solid inclusions and liquid in systems that have been experimentally investigated and where the equilibrium relations at various temperatures and the approximate heat effects are known. It is not expected that the numerical results so obtained will have any direct applicability to natural magmas, but it is hoped that some principles of general significance may be thereby brought to light.

As a beginning we shall refer to equilibrium in the plagioclase feld-spars which is given in Fig. 46. In the figure we note that liquid of composition Ab_1An_1 begins to crystallize with the separation of crystals of composition about Ab_1An_4. As the cooling proceeds, if perfect equilibrium obtains, the crystals will be made over by the liquid so that the composition of the crystals changes along the curve ACB. It is plain, then, that if one had a mass of liquid Ab_1An_1 at 1450°, that is just saturated, and added to this mass some crystalline material of the composition Ab_1An_4 (already heated to 1450°) which we shall now call foreign inclusions, the liquid would, if perfect equilibrium were attainable, make over these inclusions as the temperature fell so that their composition followed the curve ACB. How much of this work the liquid will be able to accomplish in any individual case will depend on such factors as the size of the inclusions, their permeability to the liquid, and the rate of cooling, but the tendency is very plain. We can thus have a liquid exerting a marked influence upon inclusions even though these are precisely of the composition in equilibrium with the liquid, provided the solid is a member of a solid solution series. It should be noted, however, that this action is without effect on the course of the liquid. The composition of the liquid follows the curve ADB whether the inclusions are present or not. The career of the liquid may, however, be brought sooner to a close as a result of reaction with the inclusions, that is, the liquid may be entirely used up at a somewhat higher temperature.

We may now examine the case of adding to the plagioclase liquid some solid plagioclase more calcic than that with which the liquid is in equilibrium. To 50 g. liquid Ab_1An_1 at 1450° (just saturated) let us add 50 g. solid Ab_1An_9, already heated to 1450°. Equilibrium will be established, if the temperature is kept constant, only when the solid is completely changed to that with which the liquid is in equilibrium, Ab_1An_4. Since the total composition is represented by the point K we can easily determine the proportions of liquid and crystals. There will be 33 per cent liquid of composition Ab_1An_1 and 67 per cent plagioclase of composition Ab_1An_4. We have now 54 g. An and 13 g. Ab in the crystalline state. We had formerly only 45 g. anorthite and 5 g. ablite. Equilibrium will therefore be established with evolution of heat to the amount of $9 \times 104.2 + 8 \times 48.5$ cal. In order to keep the temperature constant, then, we should have to abstract 1325 cal. If, on the other hand, no heat were abstracted the temperature would rise somewhat and equilibrium would be established at a slightly higher temperature. For the particular case we have assumed we may readily calculate that equilibrium would be established at about 1458°, when the mass would consist of about 62 per cent crystals of composition Ab_1An_5. Even when the reaction takes place adiabatically, there is an increase in the proportion of crystals. The reaction is in no sense a solution of foreign material. Rather by a making over of the foreign material it becomes no longer foreign but identical with the crystalline matter with which the liquid is in equilibrium. Moreover, if the originally foreign matter is more calcic than the crystals with which the liquid is in equilibrium the reaction is an exothermic one. How much of this reaction would take place in any individual case cannot be predicted. It will depend upon rate of cooling and other factors that readily suggest themselves, but there is plainly a tendency toward such a reaction and the reaction is exothermic.

Let us now examine somewhat more minutely into the cause of this exothermic reaction and we shall find that it is due to a general principle and is not dependent upon the particular properties of the plagioclase series discussed. During the crystallization of a plagioclase mixture a small decrement of temperature results in the reaction:

plagioclase + liquid = a little more plagioclase of somewhat more sodic composition.

Since this is an *equilibrium* reaction taking place with falling temperature, it must be exothermic. When we add the plagioclase Ab_1An_9 to liquid Ab_1An_1 we merely integrate this reaction over the temperature range 1500°-1450° and the composition range Ab_1An_9-Ab_1An_4. Thus we could start with a liquid of composition Ab_1An_5, permit it to crystallize until at 1500° the crystals would be of the composition Ab_1An_9, filter off the crystals, permit the liquid to cool to 1450°, when it would attain the composition Ab_1An_1, and then add the foreign crystals Ab_1An_9 that we filtered off. It is plain that we would have available by the making over of these crystals at 1450° all the heat that would have been evolved between 1500° and 1450° by the continuous process of making over of these crystals had they been left in contact with the liquid. Though we have used the plagioclase series as an illustration, it is clear that the exothermic reaction taking place at 1450° is merely a deferred result of the principle of Le Chatelier which states that an equilibrium reaction proceeds, with falling temperature, in the direction resulting in evolution of heat. It is a perfectly general property of any solid solution series that if, at any temperature, crystals which are at equilibrium with liquid at a higher temperature are added to saturated liquid, the reaction which ensues between liquid and crystals is exothermic.

The case of the addition of an inclusion of composition nearer the low temperature (more sodic) end of the solid solution series may now be examined. A liquid of composition Ab_1An_2 is just saturated at 1490°. We cannot add to it a more sodic solid inclusion at the same temperature because any more sodic inclusion will be liquid at this temperature. But if we add inclusions of Ab_2An_1 at such a temperature that they are solid, say at 1200°, it is plain that the actual temperature of the liquid is adequate to melt these inclusions; the only question is the source of the quantity of heat. The liquid must, of course, be cooled off in supplying the heat required to heat up the inclusions, but, since the liquid is saturated, it cannot be cooled without some crystallization taking place. However, it will take a very small amount of crystallization to supply the heat necessary to heat up a considerable amount of inclusions. The enormous discrepancy between the specific heats of silicates and their solution heats is plainly of double significance in connection with these problems. Not only can the heat necessary to heat up the inclusions be supplied by crystallization of some of the liquid, but so also can the heat required to melt the inclusions. To accomplish this it will require, however, the formation of crystals approximately equal in amount to the amount of inclusions melted. If we added 20 per cent of inclusions of Ab_2An_1 at 1200° to a liquid of composition Ab_1An_2 at 1490° we would obtain (assuming that these thermal adjustments took place very rapidly compared with concentration adjustments) a mass of liquid about (Ab_2An_3) containing in it about 20 per cent of crystals of the kind in equilibrium with it (about Ab_1An_5) and containing also the inclusions converted to liquid, the whole at a temperature of about 1470°. There is no objection, therefore, to the conversion of an inclusion into liquid if the inclusion is sufficiently contrasted with the magma in composition in the proper direction; nor is such melting of an inclusion to be regarded as evidence of superheat in the magma.

This condition, in which liquid inclusions are contained in the magma, is of course a temporary one and it is rather the end result that has particular significance in petrogenesis. Final adjustment of concentration takes place by formation of a single homogeneous liquid and adjustment of the composition of the crystals to that in equilibrium with this liquid. This necessitates a further drop in temperature to about 1460° where the whole mass consists of the liquid $Ab_{45}An_{55}$ containing somewhat less than 20 per cent crystals of the composition about Ab_1An_4. Thus the net result of the addition of inclusions of composition more sodic than the liquid is that the inclusions become a part of the liquid and at the same time calcic crystals are formed in amount slightly less than the amount of inclusions added, the heat needed in order to make the inclusions part of the liquid being supplied by the formation of crystals. In the sense that they become a part of the liquid the inclusions are dissolved, but the process is not simply the formation of a liquid whose composition is the sum of that of the magma and the inclusions.

Let us now observe how these effects are carried over into more complex systems. Fig. 51 is the equilibrium diagram of the system diopside: anorthite: albite for which we have discussed the heat quantities on an earlier page and found that the solution heat of any solid phase can be regarded as substantially equal to its latent heat of melting. A liquid of composition A is, at 1250°, just saturated with plagioclase of composition Ab_1An_4 approximately. If foreign inclusions consisting of the plagioclase Ab_1An_9 were added to this liquid we would have an effect strictly analogous to that described for simple plagioclase mixtures. The liquid would tend to make the inclusions over into Ab_1An_4 which takes place with evolution of heat. If the rate of withdrawal of heat were very slow this reaction might result in an actual rise of temperature and the establishment of equilibrium at the higher temperature where the plagioclase crystals would be slightly more calcic than Ab_1An_4. If the liquid A had come into being as a result of the partial crystallization of another liquid, and if it carried plagioclase crystals suspended in it that showed zoning, the addition of the foreign inclusions mentioned might therefore result in a reversal of the zoning.

The "attack" of the liquid upon the inclusions would be facilitated at the margins and along any channels where the inclusion happened to be more readily penetrated. The replacement of a small unit of Ab_1An_9 by Ab_1An_4 means a large local increase of volume so that the action is bound to have a disintegrating effect upon an inclusion suspended in liquid. Thus the material of the inclusion, as it is gradually made over, tends to become strewn about in the surrounding liquid but there is no actual solution, no increase in the amount of liquid—indeed, there is a diminution. Such should be the behavior of an inclusion richer in calcic plagioclase than the crystals with which the liquid

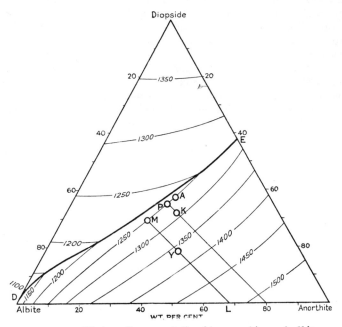

FIG. 51. Equilibrium diagram of diopside, anorthite and albite.

was in equilibrium. At a later point it will be shown that the observations of many petrographers upon actual rock inclusions strongly suggest just such an action.

If we turn now to inclusions of plagioclase less calcic than the crystals with which the liquid is in equilibrium we find that quite a different condition obtains. We may begin with the liquid A, again at 1250°, and add to it foreign inclusions of Ab_1An_1 already heated to 1250°. It will be noted that we are adding plagioclase of the same composition as that in the liquid since liquid A is a mixture of Ab_1An_1 and diopside. Suppose that the reaction takes place at first between a thin layer of liquid adjacent to the inclusion and an equal weight of the peripheral part of the inclusion. The composition of this reacting mass is then to be represented by the point Y. Suppose further that the reacting mass is a very small portion of the total so that no significant drop of temperature occurs, for in this case the reaction absorbs heat. Equilibrium will be established in this reacting layer sensibly at 1250° when the composition of the crystalline outer crust of the inclusion is Ab_3An_7 (L) and that of the adjacent thin layer of liquid is M, which is a mixture of Ab_2An_1 and diopside. If the general cooling of the liquid were proceeding rapidly so that this condition were "frozen-in" we would have a central core of unaltered inclusion of composition Ab_1An_1, an altered crust of the inclusion of composition Ab_3An_7, a reaction rim (formerly liquid) about this, consisting of a mixture of diopside and Ab_2An_1, all surrounded by the main mass consisting of a mixture of diopside and Ab_1An_1. The reaction rim is not intermediate in composition between the inclusion and the main mass, nor yet between the altered crust of the inclusion and the main mass. This is a commonly observed feature of reaction rims and has been referred to "selective diffusion," an explanation which is correct but rather misleading. The effect is really due to equilibrium effects between the inclusion and the liquid and such diffusion as occurs is selective only because equilibrium selects certain constituents to become part of the liquid and others to be fixed in the solid phase. In the case we have described, more sodic plagioclase is emphasized in the new liquid formed and more calcic plagioclase is emphasized in the solid product. We shall find it to be a general result of reactions of this kind that the liquid should be enriched in the constituents toward the low temperature end of a solid solution series and the solid in those toward the high temperature end.

We have seen, then, that even when we add inclusions of a plagioclase of the same composition as the plagioclase existing as liquid in the magma, quite marked reaction effects may be found.

Let us now examine the same example but make the cooling of the liquid very slow. In other words, we shall discuss the end result of the action described above, which gives a reaction rim only as a temporary condition. In this case we shall imagine that the liquid formed about the inclusion becomes a part of the main mass of liquid as a result of diffusion and convection, and also that the solid products of the reaction become distributed through the liquid upon disintegration of the inclusion, due to the formation of local pockets of liquid and to the volume changes taking place in the change of composition of the solid phase. Thus the inclusion completely disappears as a distinct entity. Let us imagine that the amount of the inclusions was 10 per cent of the total and determine what the effect on the mass as a whole will be. The bulk composition of the mass is represented by the point K and if the temperature were maintained at 1250° the inclusions, formerly Ab_1An_1, would be changed to crystals of a composition close to Ab_1An_4 and the liquid would acquire the composition P. The crystals would amount to only about 6 per cent of the mass and to maintain the temperature constant would require addition of heat. If the only heat available were that of the mass itself its temperature would be lowered and a slightly greater amount of crystals formed. The net result would be, however, a marked change in the composition of the inclusions, a moderate decrease in the actual amount of solid material with corresponding increase in the amount of liquid. The liquid, too, becomes more albitic, that is, is pushed onward upon its normal crystallization course.

If, instead of inclusions of Ab_1An_1, more sodic inclusions were added, say Ab_2An_1, there would be a somewhat greater increase in the amount of liquid and a markedly greater enrichment of the liquid in more albitic plagioclase. Inclusions as rich in albite as $Ab_{10}An_1$ might be completely melted by the liquid before becoming a part of the general liquid, their melting being accomplished by precipitation of more basic plagioclase.

Summing up the result of adding to a liquid various members of a solid solution series with which it is saturated we find that, if the added inclusion is nearer the high temperature end of the series than the crystals with which the liquid is saturated, the reaction is such as to decrease the amount of liquid and is exothermic. If the inclusion is nearer the low-temperature end of the series than the crystals with which the liquid is saturated, the reaction is such as to increase slightly the amount of liquid and is endothermic. The liquid, too, is enriched in this case in the constituents of the low-temperature end of the crystallization series. Even inclusions consisting of the precise crystals with which the liquid is in equilibrium must react with the liquid as the temperature falls.

Whatever the composition of the inclusions, then, the liquid may show very marked effects upon them even though it is saturated, and consequently such effects would not constitute evidence of superheat. A little consideration will show, too, that these reactions have no significant effect upon the course of the liquid. Regarding the progress of the liquid as resulting from fractional crystallization, the liquid will in all cases run along the boundary curve (ED) as the temperature falls. The point it will eventually reach on this curve will depend upon the perfection of fractionation, which in turn depends upon the rate of cooling. Inclusions of the more calcic kind tend to limit the career of the liquid by using it up, but at the same time furnish heat that tends to slow up the rate of cooling. Inclusions of the more sodic kind tend to push the liquid onward upon its course of crystallization, but hasten the cooling. The original unaffected liquid has all the differentiation potentialities that the liquid has after entering into the reactions mentioned.

In the discussion relative to liquids of the diopside : anorthite : albite system nothing has been said regarding the effect of adding solid diopside to liquids saturated with diopside for the reason that there is no effect. The solid diopside simply remains as such on account of its being a pure compound of definite composition. But rock-forming minerals are seldom so simple. Quartz is the only important example of a rock mineral of definite composition, practically all others being of a variable nature, that is, solid solutions. The rock-forming pyroxenes do not fall behind in this respect and the addition of pyroxene to a natural magma saturated with pyroxene would in general be attended by reaction phenomena similar to those we have described for the plagioclases. The precise nature of the reactions in the case of the pyroxenes cannot be stated except for the clino-enstatite-diopside solid solution series. (Figs. 18-20).

In . Chapter V we have discussed solid solution series and offered reasons for calling them *continuous reaction series*. The importance of the reaction relation between liquid and crystals was there discussed in its bearing on crystallization. Here we have seen its importance in connection with the behavior of inclusions.

There is another type of reaction relation between liquid and crystals that is exhibited in the *reaction pair* and the *discontinuous reaction series*. The existence of such series is again of great significance in connection with the behavior of inclusions. An important reaction pair are the olivine, forsterite, and the pyroxene, clino-enstatite. Their relation is exhibited in its simplest form in the binary system, forsterite-silica, of which the equilibrium diagram is shown in Fig. 52. The effect of adding crystals of the first member of a reaction pair to a liquid saturated with the second member is well illustrated by this system. A liquid of composition (M) is saturated with clino-enstatite but lies on the unsaturated side of the metastable prolongation of the forsterite liquidus. It is therefore unsaturated with forsterite and we may imagine that around each added forsterite crystal a small quantity of liquid of composition (N) may form. This condition is, however, metastable and from this liquid clino-enstatite would immediately be precipitated with formation of the liquid (M). Through constant repetition of this formation from forsterite of an infinitesimal quantity of the metastable liquid, with immediate precipitation of clino-enstatite, the forsterite is converted into clino-enstatite. Effectively, then, the liquid (M) is supersaturated with forsterite. It cannot dissolve forsterite but can only convert it into clino-enstatite, the phase with which it is saturated. This principle is capable of general application and we may state that a liquid saturated with any member of a discontinuous reaction series is *effectively* supersaturated with all higher members of the series; it cannot dissolve them but can only convert them into the phase with which it is saturated.

In connection with the specific case we have discussed, it should be

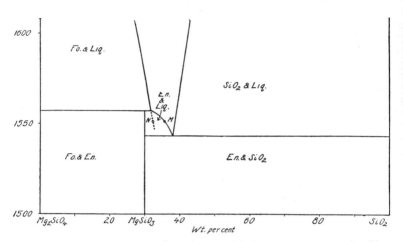

FIG. 52. Lower temperature portion of the equilibrium diagram, forsterite-silica.

noted that the amount of clino-enstatite formed is not simply the chemical equivalent of the forsterite changed. The chemical equivalent would be somewhat less than one and one-half times the forsterite, whereas the amount of clino-enstatite formed is much greater, being in fact about five times the amount of forsterite. In other words, the action is not simply an addition of silica to the forsterite, with consequent impoverishment of the liquid in silica, for the liquid cannot have silica subtracted from it without passing under the clino-enstatite saturation curve, that is, without precipitating clino-enstatite. In order to convert the forsterite inclusions into clino-enstatite, the liquid must precipitate a large amount of clino-enstatite from its own substance, and the action uses up a large amount of the liquid. The liquid left has, however, in this binary case, the same composition as the initial liquid, if the temperature is kept constant. The liquid is not at all desilicated, even though it has caused the conversion of the inclusions into a more siliceous phase. By precipitation of the phase with which it is saturated, it has adjusted its composition in such a way as to remain on the same saturation curve. This again is a principle that can be applied in general to the reaction between a liquid and inclusions belonging at an earlier stage of the reaction series than the phase with which the liquid is saturated.

Further considerations relative to this reaction pair will be developed in connection with their behavior in the more complex liquids containing anorthite as worked out by Andersen[1] and shown in Fig. 53. From this figure we may very readily predict what would happen to "inclusions" of the various solid phases immersed in liquid. Let us take first a liquid just saturated with forsterite (say M at 1450°). Ordinarily

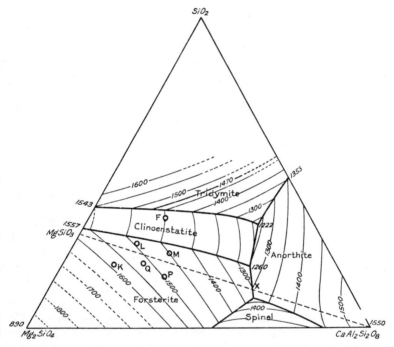

FIG. 53. Equilibrium diagram of the system, anorthite-forsterite-silica (after Andersen).

crystals of forsterite would separate first; they would then react with liquid to form pyroxene; pyroxene would continue to separate for a time and would then be joined by silica; finally, at the ternary eutectic 1222°, anorthite would join these, and the product would consist of pyroxene, silica, and anorthite. If to the original liquid some "foreign inclusions" of forsterite were added, the liquid would, on cooling, attempt to make the forsterite over into pyroxene, but now there would not

1 "The System Anorthite-Forsterite-Silica," *Amer. Jour. Sci.*, 39, 1915, p. 440.

be enough liquid to accomplish this entirely, and it would be used up at 1260° while there was still some forsterite left; so that the solidified mass would now consist of pyroxene, forsterite, and anorthite. Thus we see that even though it is precisely the substance with which the liquid is in equilibrium, the addition of forsterite has a considerable effect upon the crystalline product formed. The exact effect is a tendency to limit the scope of the products to the early members of the crystallization sequence. In this case, silica, a later member of the sequence, does not appear. It is plain, too, that the effect does not depend on the particular properties of this reaction pair and that we may conclude that it would be true of any discontinuous reaction series. We have already seen that the same effect is found in the continuous reaction series (solid solution series). We may state it as a general law therefore that a saturated liquid in any system dominated by reaction series will not remain indifferent to inclusions even of the exact composition with which it is in equilibrium, and that the effect of the addition of such inclusions is a tendency to limit the scope of the crystalline products to adjacent members of the reaction series involved.

Let us now examine what happens when inclusions of an early member of a discontinuous reaction series (or reaction pair) are added to a liquid saturated with a later member. The liquid F is just saturated with clino-enstatite at 1450°. If it is cooled, clino-enstatite will separate first, it will later be joined by silica and then by anorthite, when the whole mass will solidify at 1222°. But if inclusions of forsterite are added to this liquid it will instantly start to react with these inclusions and make them into the mineral with which it is saturated, viz., clino-enstatite. If there is perfect opportunity for reaction we may readily predict what the end result will be for varying amounts of added inclusions.

Suppose first that the amount of inclusions was such as to give a total composition represented by the point K. The liquid would then be completely used up by the reaction while some inclusions yet remained, so that the finally solidified mass would consist of forsterite, clino-enstatite, and anorthite. If, on the other hand, the amount of inclusions was such as to give a total composition represented by the point L, all the inclusions would be changed to clino-enstatite while some liquid yet remained, and this liquid would then pass onward to the deposition of silica and anorthite in the ordinary way. Thus the addition of inclusions of this kind, if in sufficient quantity, restricts the career of the liquid and confines the crystalline products to adjacent members of the reaction series.

It is perhaps not necessary to add that if the liquid did not react with the inclusions no effect on the course of the liquid would ensue.

Just as was the case in the binary system, the liquid is not desilicated by this addition of silica to the forsterite inclusion, but by precipitating an appropriate amount of the phase with which it is saturated (clino-enstatite), it maintains its position on the same (clino-enstatite) saturation surface.

Nothing has been said of the heat effect of this reaction and it has been tacitly assumed that the temperature remains constant. The reaction is, in fact, exothermic, as can be readily shown by applying the same reasoning as was applied to the similar case in a continuous reaction series. If no heat were abstracted from the system it would heat itself up and equilibrium would be established at a slightly higher temperature with a somewhat more magnesian liquid than the initial liquid. But the formation of this somewhat more magnesian liquid is not properly to be taken as an indication that the net result of the process is a direct solution of some of the inclusions. A direct solution of inclusions would mean a decrease in total solids and an increase in liquid, whereas the reaction referred to results in a diminution in the amount of liquid and a corresponding increase in the amount of solids even when this heating effect takes place. If heat is being taken from the system this process would act as a deterrent upon the rate of cooling.

All of these effects we have found to be true of analogous inclusions in the case of the continuous reaction series.

There is yet to be examined the example in which a late member of a discontinuous reaction series is added to a liquid saturated with an early member. To the liquid P at 1500°, where it is just saturated with forsterite, inclusions of clino-enstatite are added in an amount sufficient to give a total composition Q (about 20 per cent). If the temperature were kept constant equilibrium would be established when the mass consisted of 4 per cent forsterite and 96 per cent liquid, that is, the inclusions have been changed into the phase with which the liquid is saturated and there has been an increase in the amount of liquid. In order to effect this change heat would have to be added to the system. If the system is self-contained, that is, if the only heat available is the heat of the system itself, a cooling would result and this would necessitate the crystallization of a further amount of forsterite until the necessary heat was supplied by this crystallization and the cooling of the mass. The net result would depend entirely on the relative heats of solution of forsterite and clino-enstatite in the liquid. These are probably of the same order of magnitude, so that equilibrium would be established at about 1475° when the mass consisted of 10 per cent forsterite and 90 per cent liquid approximately. The net result, then, has been the conversion of the inclusions added (clino-enstatite) into the phase with which the liquid is saturated (forsterite), an enrichment of the liquid in the material added, with, at the same time, a pushing onward of the liquid along its usual course of crystallization. Or it could be stated that the inclusions pass into solution by precipitating their heat equivalent of the phase with which the liquid is saturated.[1] This, then, is the result of adding inclusions which belong to a discontinuous reaction series and are later in that series than the phase with which the liquid is saturated. It is sensibly the same result as that obtained in the corresponding case in the continuous reaction series.

EFFECTS OF MAGMA UPON INCLUSIONS OF IGNEOUS ORIGIN

In the development of the conception of the reaction series it was shown that these series are very prominent in rocks. An attempt was made to arrange the minerals of rocks as reaction series and it was found that there are, toward the basic end of rock series, two fairly distinct reaction series that finally merge into one in the more acid rocks. This was expressed diagrammatically in Table II, p. 60.

On the basis of the principles that we have found to govern the behavior of inclusions belonging to reaction series we may deduce with considerable confidence the effects of liquid upon inclusions in this more complex series.

It may be stated immediately that any magma will tend to make inclusions over into the phase or phases with which it is saturated, in so far as the composition of the inclusions will permit. It may be stated also that any magma saturated with a certain member of a reaction series is effectively supersaturated with all higher members of that reaction series. It cannot, in any sense, dissolve inclusions of such higher members but can only react with them to convert them into that member of the reaction series with which it is saturated, often by passing through other members of the series as intermediate steps. The material used to effect these changes cannot be regarded as simply subtracted from the liquid, for the liquid is not free to become impoverished in any random substance. In general, impoverishment in any substance will cause the liquid to pass within a region of saturation and induce the precipitation of some of the phases with which the liquid is saturated.

Let us take, for example, a magma saturated with biotite, say, a granitic magma. This magma is effectively supersaturated with olivine, pyroxene and amphibole and cannot dissolve them in spite of the marked contrast of composition, which is often supposed to be an aid to the solution of inclusions. But the magma can and will react with these minerals and convert them into biotite, usually by steps. The subtraction of material necessary to produce biotite will cause the precipitation of the minerals with which the magma is saturated until either the liquid or the inclusions are used up or the reaction is brought to an end on account of mechanical obstruction.

---

1 This is only approximately true, for equilibrium is always established at a somewhat lower temperature and the cooling of the mass supplies a little of the heat.

Similarly, granitic magma saturated with an acidic plagioclase cannot dissolve basic plagioclase but can only react with it and convert it into more acid plagioclase.

These remarks are tantamount to the statement that saturated granitic magma cannot dissolve inclusions of more basic rocks. The magma will, however, react with the inclusions and effect changes in them which give them a mineral constitution similar to that of the granite. These changes will often be accompanied by disintegration of the inclusions and the strewing about of the products which may be indistinguishable from the ordinary constituents of the granite. The inclusions may thus become completely incorporated though not in any sense dissolved. It is this action of magmas upon inclusions that makes particularly difficult the problem of distinguishing xenolith from autolith, i.e., accidental inclusion from cognate inclusion. V. M. Goldschmidt had previously deduced from the doctrine of differentiation by crystallization as given by me that salic magmas were saturated with the materials of basic rocks. In applying this fact to the explanation of sharp boundaries he does not appear to have appreciated the full possibilities of reaction. The considerations just outlined show that in all cases sharp boundaries must be due to rapid cooling and consequent inadequate time.[1]

Whatever origin one may assign to a granitic magma—let it be formed by differentiation of more basic magma, by differentiation of syntectic magma or by palingenesis of sediments—there seems no escape from the conclusion that it will normally be saturated. The normal effect of granite upon more basic inclusions should therefore be such as has been outlined above. Thus we find that Fenner, in describing the action of granitic magma on basic bands in an injection gneiss, says, "In other places the dark minerals appear to have been taken up or digested by the magma and to have crystallized out again in large blades. Even in the latter case it is not always certain that perfect solution has been effected at any one time. The process may have been rather in the nature of a chemical reaction with the original minerals or the solution and redisposition of a portion of the material at a time,[2] leaving the general relations undisturbed. This possibility is suggested by the fact that frequently even the coarser micaceous blades or aggregates of dark minerals show evidence of parallelism and this would be difficult to account for under the supposition that solution was so perfect that the original structure was completely wiped out."[3] Here, apparently, we have a good example of the transformation, by reaction rather than solution, of the dark bands into mica-rich material, mica being the dark mineral with

1 *Vidensk. Selsk. Skr. I. Mat.-Naturv. Kl.*, 1916, No. 2, pp. 91-3.
2 Note discussion on p. 192.
3 C. N. Fenner, "The Mode of Formation of Certain Gneisses in the Highlands of New Jersey," *Jour. Geol.*, 22, 1914, pp. 602-3.

which the magma is saturated. The change of serpentine into biotite as observed by Gordon at the borders of granitic pegmatites is precisely the action to be expected.[1]

V. M. Goldschmidt describes the strewing about of the minerals of a basic hornfels in magmas of the Christiania (Oslo) region. He says, "This strewing about hardly has its origin in a solution of the minerals and their later separation. Had solution occurred the grains would not have retained their original forms and they would have differed in composition from the minerals of the hornfels." Near the border of an apophysis of the nordmarkite, grains of diopside from the hornfels are surrounded by a rim of aegirite. In the center of the apophysis the aegirite has no core of diopside.[2]

In the foregoing discussion granitic magma has been taken merely as an example to which the principles developed may be applied. As a further example it may be pointed out that saturated dioritic magma cannot dissolve inclusions of gabbro, peridotite or pyroxenite but, given the opportunity, it will react with those inclusions and convert them into the hornblende and the plagioclase with which it is saturated, at the same time precipitating a further amount of this hornblende and plagioclase from its own substance.

These are but examples of the application of the principle that a saturated magma cannot dissolve inclusions of material farther back in the reaction series (in general more basic) than the crystals with which it is saturated, but the magma can attack these inclusions, reacting with them in such a manner as to convert them into the crystals with which it is saturated.

The dioritic magma we have considered will not remain indifferent to inclusions even of the exact composition of the crystals with which it is in equilibrium, for as the temperature falls it will modify the composition of these inclusions just as it modifies the composition of its own crystals. Indeed this case may be regarded as a special case of that just discussed, for, as the temperature falls, the composition of the liquid changes, and the inclusions then pass into the class of those considered above.

We now come to the case of inclusions of material later in the reaction series than the crystals with which the liquid is saturated. It should be noted that this includes masses of rock of the same composition as the liquid itself, for example its own chilled border phase.

Saturated basaltic magma can react with inclusions of igneous rocks later in the reaction series (in general more acid) in such a way that the inclusions become part of the liquid, crystals of the phases with

1 "Desilicated Granite Pegmatites," *Proc. Acad. Nat. Sci. Phila.*, Part I (1921), p. 169.
2 V. M. Goldschmidt, "Die Kontaktmetamorphose im Kristianiagebiet," *Vidensk. Selsk. Skr. I. Mat. Naturv. Kl.*, 1911, No. 1, pp. 107-8.

which the basalt is saturated being precipitated at the same time. If these crystals are removed by gravity or otherwise, the action on the inclusions may continue, the liquid changing in composition toward the composition of the inclusions and precipitating later and later members of the reaction series until finally it is saturated with precisely the crystalline phases contained in the inclusions. If granitic inclusions, say, were available at the upper contact of a mass of basaltic magma, they would be attacked by the magma in the manner noted and, in a lower layer, accumulation of the precipitated products of the reaction would take place. These would be the early crystals formed in basaltic magma. The upper liquid is thus gradually changed in composition and the crystals precipitated from it are successively later and later members of the reaction series. Attack upon the inclusions continues until finally the upper liquid becomes granitic. All of this depends on a rate of cooling slow enough for free crystal settling to occur. But if the cooling is sufficiently slow for crystal settling all of these results could accrue from the simple differentiation of the basaltic magma. Indeed the principles developed show that the inclusions can become part of the liquid only when they have a composition toward which the composition of the liquid can vary by spontaneous differentiation.

Nevertheless it is apparent that the amount of granitic differentiate might be greatly augmented by this action. It may safely be assumed therefore that in many individual cases considerable quantitative importance in the production of a granitic differentiate of basic magma is to be assigned to the action noted. It is a sort of solution of granitic inclusions though not a simple, direct solution and is in no sense essential to the production of a granitic differentiate.

Daly is of the opinion that many granites are secondary, that is, are formed by solution of granite in basaltic magma and subsequent differentiation.[1] It is seen from the above that theoretical considerations support belief in a process which, in its results at least, is practically that advocated by Daly. The process itself he considers to be rather a simple solution of granite in superheated basaltic magma. We have seen that no superheat is necessary to produce solution by a sort of reactive process. Moreover, we have seen that the incorporated granitic material is to be regarded rather as a contribution to the normal granitic differentiate. There appears, however, to be no reason to doubt that, at times, this contribution might equal or possibly even exceed in amount the granitic differentiate capable of formation from the uncontaminated magma. The limiting factors are principally mechanical rather than thermal or chemical and are very difficult to evaluate. It

1 Indeed, Daly derives in this manner all granites except a supposed original granitic shell of the earth (*Igneous Rocks and Their Origin*, p. 323).

should be noted, in particular, that the combination which is most favorable for significant effects in the way of reactive solution, viz., decidedly acid inclusions and decidedly basic magma, is unfavorable in another respect. The inclusions will be lighter than the magma and will not tend to sink in it, whereas it is the sinking of inclusions through the magma which favors particularly notable reaction effects since it continually brings new magma into contact with the inclusions.

As an example of the effect of basic magma on more acid igneous inclusions basaltic magma and granitic inclusions have been taken. Between such extremes the more marked effects should be obtained, but it cannot be doubted that any basic magma can dissolve, by the same reactive process, inclusions of a rock later in the reaction (crystallization) series. Direct melting of granitic inclusions to masses of liquid by basaltic magma is a definite possibility in small amount but it is a transient condition which must soon give place to the reactive solution process above described.

EFFECTS OF MAGMA ON INCLUSIONS OF SEDIMENTARY ORIGIN

The general problem of the effects of magma upon inclusions of sedimentary origin is much more difficult than the similar problem in connection with igneous inclusions. Sedimentary rocks have their compositions determined by processes wholly independent of igneous action and do not correspond in composition with the products precipitated from magmas at any stage of their career, that is, cannot be placed definitely in the reaction series. However, certain minerals that can be formed in magmas do occur in the sedimentary rocks and often the composition of a sediment is such that by mere heating it can be transformed into an aggregate made up exclusively or almost exclusively of igneous rock minerals. Again sediments exhibit extremes of composition, being very rich in calcium carbonate, aluminum silicate or silica itself, and these present a special problem. Yet it is perhaps not generally realized how much even of these extreme sediments might be incorporated in an igneous rock without changing its mineralogy. The fact is an obvious deduction from the equilibrium diagram of any investigated three component system and it is equally true of a more complex system. Fig. 54 shows the solid phases formed immediately upon complete consolidation of any mixture of CaO, Al_2O_3 and SiO_2. A mixture of composition (A) consists, upon complete consolidation, of anorthite, wollastonite, and gehlenite, one-third of each. One may add to this mixture any amount of CaO up to about 15 per cent of itself, without changing the mineral composition of the consolidated product. Similar amounts of either Al_2O_3 or SiO_2 might be added, the only change in all cases being in the relative amounts of the minerals, not in the kind of minerals. A certain amount of change in the order of separation of the minerals would be effected but the temperature of final consolidation, the composition of the final liquid and the possible differentiates that might be formed by fractional crystallization would in all cases remain as before. Only when amounts are added in excess of those mentioned will a new crystalline phase appear and new fractionation possibilities enter.

There are, however, certain mixtures in the system that will immediately present a new phase upon addition of the slightest amount of CaO, Al$_2$O$_3$, or SiO$_2$. Such is the mixture (B) which consists, on consolidation, of anorthite and wollastonite, one-half of each. The addition of either CaO or Al$_2$O$_3$ will bring in the new phase, gehlenite, and

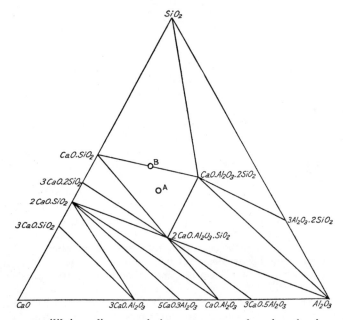

FIG. 54. Equilibrium diagram of the system, CaO-Al₂O₃-SiO₂ showing phases formed upon complete consolidation (after Rankin and Wright with corrections after Bowen and Greig).

of SiO$_2$ the new phase, tridymite, and each will change the course of crystallization. This is because the composition chosen is a limiting case in the ternary system and is in reality of only two components; and the addition of, say, lime carries it out of the two component system. It will be noted that the mixture considered contains the three oxides, though of only two components, and on consolidation only two solid phases are formed. It might seem on first thought that this corresponds with the case of the natural magmas, for these usually form solid phases fewer in number than the oxides present. However, there is another factor, namely, solid solution, that may give rise to this peculiarity and we shall find that it is to solid solution that the limited number of phases formed from magmas is usually to be referred.

Let us now examine a ternary system of oxides that has been completely investigated and in which the factor of solid solution enters. Such is the system CaO : MgO : SiO$_2$ studied by Ferguson and Merwin.[1] A liquid of composition A, Fig. 55, forms on complete consolidation just

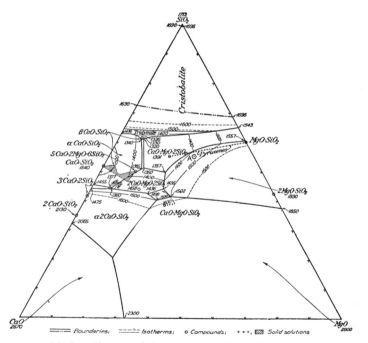

FIG. 55.   Equilibrium diagram of the system CaO-MgO-SiO$_2$ (after Ferguson and Merwin with additions after Greig).

two solid phases, the olivine, forsterite, and clinopyroxene of composition between diopside and MgSiO$_3$. To this liquid any amount of calcite

1 *Amer. Jour. Sci.*, 48, 1919, p. 109.

up to 12 per cent of its weight could be added without changing the mineralogy of the consolidated product. This would still consist of olivine and clinopyroxene but the pyroxene would be richer in CaO, that is, closer to diopside. Only an amount of calcite in excess of 12 per cent would bring in another phase, akermanite, together with the diopside and olivine. It is easy to see from an inspection of the figure that addition of dolomite would likewise have no effect on the kind of phases crystallizing unless more than about 16 per cent were added.

When we have spoken of adding calcite and dolomite to the mixtures mentioned we have imagined that the lime and magnesia have been taken into solution as they might be in a laboratory furnace, where the furnace supplied the requisite quantity of heat. This quantity would be very large, for the conversion of carbonate into silicate is an endothermic reaction and its conversion into silicate in solution is undoubtedly still more strongly endothermic. Now if the liquid were originally a saturated liquid and calcite or dolomite were added without any provision for additional heat other than that already contained in the liquid, the lime and magnesia would not be simply dissolved. Instead, precipitation of olivine (forsterite) would occur, no matter whether calcite or dolomite was added, and as the forsterite separated the liquid would attack the calcite, converting it into lime silicates. Since this is an endothermic reaction the heat for it can be supplied only by crystallization of something from the liquid which will be, of course, the phase with which it is saturated, viz., forsterite. There is another reason why forsterite is precipitated for the formation of lime silicates requires some silica but the liquid cannot be desilicated since it lies on the forsterite saturation surface. Therefore the liquid changes its composition by moving on the forsterite surface toward lower temperature. It does not move directly away from forsterite, however, but curves somewhat in the direction of the composition of the inclusions. If the reaction with the inclusions is strongly endothermic, as it is in this case, the liquid would move down to the forsterite-pyroxene boundary curve where separation of pyroxene would occur and the liquid would be entirely used up without ever getting over to such compositions that akermanite would separate from it. However, some part of the inclusions might have been converted to akermanite. Thus, however large a supply of inclusions were available, even in excess of the 12 per cent mentioned above, the liquid might never get over to such compositions that any new phase is precipitated from it, but would precipitate only forsterite and pyroxene as in normal crystallization, with, however, a certain increase in the lime content of the pyroxene.

These considerations lead us to the conclusion that the liquid we have chosen, even if it has a moderate amount of superheat and therefore is capable of directly dissolving a little lime and magnesia from calcite or dolomite, does not suffer a change in the kind of solid phases capable of forming from it. There results only a modification of the composition of the phase of variable composition (pyroxene). And, as a consequence of the heat effect of the solution of inclusions, saturation shortly ensues and thereafter further action upon the inclusions is accomplished only with concomitant precipitation of the phase or phases with which the liquid is saturated, whereby the liquid is constrained to follow a general course not significantly different from the one it would follow were no inclusions present.

The results obtained in the foregoing enable us to draw certain conclusions as to the effects of natural magmas upon inclusions of various sedimentary rocks. One point that does not seem to be realized is that, when a sedimentary inclusion becomes immersed in a magma, nothing is added that the magma does not already contain. Both belong to the same polycomponent system embracing all the rock-forming oxides. Obviously the effects of all possible sediments cannot be examined, but our purpose will be served if we take the most extreme departure from igneous composition. As representative of this condition, for quartzite we may imagine the addition of pure quartz; for limestone, of pure calcium carbonate; and for shale, of pure kaolin. Any actual sediment would usually contain all of these, together with other constituents that lessen its departure from igneous composition.

Let us take a magma of basaltic composition which, on crystallization with comparatively rapid cooling, would form mainly plagioclase, and clinopyroxene, with some olivine, a little ore mineral and possibly some orthopyroxene. All of these are minerals of variable composition; some of them, in particular the pyroxene, vary with respect to several components and to this is to be attributed the fact that the number of solid phases formed is less than the number of oxides present. This fact permits particularly wide adjustments in the composition of the solid phases without the appearance of new ones. Such basaltic magma, with a little superheat, could directly dissolve a moderate amount of sediments, yet even if these were of extreme composition, the magma would crystallize with the production of the same solid phases as those mentioned above if crystallized under the same conditions.

Normally only saturated magma would be available and the superheated magma mentioned above would rapidly became saturated as a result of solution of inclusions. For the case of such saturated magma it may be stated as a first principle that the sediment would, in so far as its composition permitted, tend to be converted into the phases with which the magma is saturated. And the material necessary for such changes in the sediment would not be merely subtracted from the liquid but adjustments of the composition of the liquid would occur through separation of further amounts of the phases with which the liquid is saturated.

The precise changes in the composition of the solid phases formed cannot be represented graphically on account of the number of components involved, but equations can be written that afford a generalized conception of the possible adjustments for the addition of calcite, silica, and kaolin respectively.

The phases capable of formation from the original unchanged magma are:

PHASE	MINERAL MOLECULES REPRESENTED
Olivine	$Mg_2SiO_4 + Fe_2SiO_4$
Magnetite	$FeO . Fe_2O_3$
Plagioclase	$CaAl_2Si_2O_8 +$ etc.
Pyroxene	$CaMgSi_2O_6 + MgSiO_3 + Al_2O_3* + CaFeSi_2O_6 +$
	$FeSiO_3 + Fe_2O_3*$

* Often written as existing in the Tschermak molecule, for which there is no good reason. See Washington and Merwin, *Amer. Jour. Sci.*, 3, 1922, p. 121.

Upon addition of CaO the following principal adjustments in the proportions of these mineral molecules may occur without the appearance of new phases:

$$CaO + 2FeSiO_3 + Fe_2O_3 = CaFeSi_2O_6 + FeO . Fe_2O_3 ;$$
$$CaO + 3MgSiO_3 = CaMgSi_2O_6 + Mg_2SiO_4 ;$$
$$CaO + Al_2O_3 + 4MgSiO_3 = CaAl_2Si_2O_8 + 2Mg_2SiO_4.$$

Upon addition of SiO_2:

$$SiO_2 + Mg_2SiO_4 = 2MgSiO_3 ;$$
$$SiO_2 + Al_2O_3 + CaMgSi_2O_6 = CaAl_2Si_2O_8 + MgSiO_3.$$

Upon addition of kaolin, which we may regard as $Al_2SiO_5 + SiO_2$, with SiO_2 having the same effect as above:

$$Al_2SiO_5 + CaMgSi_2O_6 = CaAl_2Si_2O_8 + MgSiO_3.$$

The results may be put in words by stating that addition of lime tends to increase the amount of magnetite and olivine, to make the pyroxene more nearly a pure diopside-hedenbergite and to increase the anorthite content of the plagioclase. The addition of silica tends to decrease the amount of olivine and to increase the magnesian content of the pyroxene and the anorthite content of the plagioclase. The addition of kaolin tends to increase the amount of magnesia in the pyroxene and of the anorthite in the plagioclase.

In the case of superheated magma the added material might be directly dissolved and upon solidification the adjustments noted would appear in the crystalline phases. In the case of saturated magma the phases noted would be developed by reaction with the added material and at the same time a further amount of them would be precipitated from the liquid. In neither case would the course of crystallization be fundamentally changed since crystallization produces only the same solid phases slightly modified in composition. If consolidation took place under conditions permitting fractionation by settling of crystals no fundamentally new differentiation potentialities would be introduced by the solution or reaction with foreign material that has been discussed. The magma thus modified could give a diorite-granodiorite-granite sequence, say, only if the original magma could also have done so under the same conditions. It is a question whether, in the case of unsaturated magma, it can be safely assumed that the degree of superheat may be such that an amount of material can be dissolved in excess of that which can be taken care of by the adjustments in composition of existent phases. If this is possible new phases will appear and the course of crystallization and differentiation might be fundamentally modified. In the case of added CaO, for example, the new phase melilite might appear and the differentiates formed might be fundamentally different; might be, say, certain alkaline types as Daly has postulated.

It is especially to be noted that the laboratory demonstration that melting together of a basic rock and limestone will give a melilite rock does not prove that similar results are accomplished in natural magma. It does, of course, show that the possibility of the reaction is to be favorably entertained but one must look to rocks themselves for evidence of its occurrence. Nor does such action appear to be of the sort that is likely to be obscured. Indeed one would expect to find, fairly commonly, a melilite-bearing reaction rim about inclusions of limestone in basalt. The mere metasomatic development of melilite in the limestone[1] is not enough, though, of course, suggestive and justifying diligent search for convincing proof. This must lie in the occurrence of melilite-bearing reaction rims of such a nature as to indicate a rim of hybrid magma capable of precipitating melilite. In this connection it is to be noted, too, that melilite basalts are apparently never basalt plus lime nor yet basalt plus lime and magnesia. If they are the result of solution of limestone or dolomite in superheated basaltic liquid, melilite basalts should show this simple relation in at least some instances.

An alternative method of development of melilite in igneous rocks is discussed on a later page. It consists in the reaction of alkalic liquid with the crystalline phases of a basic rock, especially aguite.

EFFECTS OF BASALTIC MAGMA ON ALUMINOUS SEDIMENTS

Except in the case of excessive superheat, the statement should hold that reaction with foreign material can produce no new differentiation potentialities in the magma. Yet it is certain that some lines of differ-

[1] J. Stansfield, *Geol. Mag.*, 40, 1923, p. 441.

entiation may be emphasized by such agency, that is, that certain types of differentiate should be quantitatively of greater importance. This we have found to be true of reaction of magma with previously solidified igneous material. Thus when basaltic magma reacts with granitic material the tendency is to increase the amount of granitic differentiate capable of forming from the basalt. The effect of reaction with aluminous sediments is of sufficient importance to justify further discussion at this point. The result has been shown to be the emphasizing of more magnesian pyroxene and of anorthite. Now under certain conditions, not well understood, magnesian pyroxene separates from magmas as a distinct phase, an orthopyroxene, and the addition of aluminous sediments should emphasize this tendency. The formation of norite and of pyroxenites characterized by orthopyroxene as differentiates from basaltic magma, may therefore be facilitated by reaction of such magma with aluminous sediments. The relation has been advocated by a number of investigators including A. N. Winchell, A. Lacroix, J. W. Evans, H. H. Read, C. E. Tilley and others. In this case we find that expected effects are amply revealed in the reaction rims of aluminous inclusions.

An example of the reaction of basic magma with aluminous sediments, in which the relations discussed seem particularly clear, is afforded by the so-called Cortlandt series.[1] The early work of Williams and the later work of Rogers on this series treats in some detail the features bearing upon the question here at issue.[2] The Cortlandt series is intrusive into the Manhattan schist (locally into other formations also) and the interaction between schist and magma is in places rather well displayed. The Manhattan schist is a metamorphosed sedimentary rock of the nature of a shale. Its composition, as given by a composite anaylsis of five specimens, is shown in Table XIV under I. No doubt it varies considerably from this average and is sometimes more aluminous, but it is always far from the composition of kaolin, which composition we have used in discussing the effects of basic magma on aluminous sediments. This affords an opportunity of discussing the behavior of an actual example of aluminous sediment. The ordinary Manhattan schist is made up principally of the minerals quartz, biotite, muscovite, orthoclase and plagioclase. These are all minerals of ordinary igneous rocks, particularly of more "acid" types, and correspond to a rather low temperature equilibrium. A glance at the analysis shows, however, that the composition is far from that of an ordinary igneous rock, which means that the minerals are of somewhat different composition and are present in different proportions. Now we have found in our discussion of the re-

1 G. H. Williams, *Amer. Jour. Sci.* (3), 31, 1886, p. 26; (3), 33, 1887, pp. 135-91; (3), 35, 1888, p. 438; (3), 35, 1888, p. 254.
2 G. S. Rogers, *Ann. N.Y. Acad. Sci.*, 31, 1911, pp. 11-86.

TABLE XIV

	I	II
SiO_2	57.94	40.16
Al_2O_3	21.70	29.50
Fe_2O_3	1.57	19.66
FeO	5.90	5.80
MgO	2.49	trace
CaO	.58	.85
Na_2O	1.74	1.46
K_2O	4.68	1.36
$H_2O +$	2.17	....
$H_2O -$	.29	....
TiO_2	1.01	....
MnO	.19	....
S	....	.82
	100.26	99.61

I—Manhattan schist. Composite analysis of five specimens beyond border of the Cortlandt series. Analyst, G. S. Rogers.
II—Manhattan schist on contact of Cortlandt series. Analyst, F. L. Nason in G. H. Williams, *Amer. Jour. Sci.* (3), 36, 1888, p. 259.

action of any saturated magma upon igneous inclusions that if the inclusions belong to a later stage in the reaction series they may become a part of the liquid by causing the precipitation of the phases with which the liquid is saturated. Average Manhattan schist, since it consists of the minerals of an acidic igneous rock, may be regarded as consisting in part of material belonging to a later stage in the reaction series than basaltic magma, but since it does not correspond exactly with any such igneous mass it must be regarded as having a certain amount of surplus material in addition. If we imagine saturated basaltic magma reacting with inclusions or wall rock of schist we may expect the action to be selective. Such substances as may become a part of the liquid would be removed from inclusions or wall rock, with corresponding enrichment in what has been called surplus material. The substances removed would be principally silica, alumina, alkalis and to a minor extent other oxides, all in the proportions in which they enter into some "acid" igneous rock. Our knowledge of the exact proportions may be thus indefinite and yet sufficient for a general solution of the problem. Comparison of the analysis of the Manhattan schist with those of acid igneous rocks gives us a good conception of what the surplus material will be. It will plainly be rich in alumina and iron. A certain stage of the reaction between magma and inclusions or wall rock should exhibit a mass rich in these oxides. This stage is abundantly represented in both wall rock and inclusions by richness in sillimanite, staurolite and other aluminous and ferrous minerals. Chemically it is shown by analysis II in Table XIV which represents wall rock at the margin of the intrusive.

The so-called surplus material does not remain indefinitely as such, but by the time we have obtained inclusions very rich in sillimanite a turning point in the process is reached. Hitherto certain constituents of the schist have become a part of the liquid in virtue of the precipitation of various phases with which the liquid (magma) was saturated. Now reactions between magma and inclusions become of such a nature that the precipitation of phases with which the magma is saturated is the sole process, these phases being appropriately modified by the inclusions. There is now no addition to the liquid. The exact modification of the phases that is produced by sillimanite has already been discussed and equations representing the changes have been written. The net result is an increase in the amount of anorthite and magnesian pyroxene at the expense of lime-bearing pyroxene, and this tends to promote the separation of magnesian pyroxene as a distinct phase, orthopyroxene. Thus the tendency of the magma to give a noritic differentiate is increased, as well as the likelihood of formation of a pyroxenite containing orthopyroxene. These expectations are well matched by the Cortlandt series.

Apart from possible later differentiation this production of norite and related types is the end result of the action of basic (basaltic) magma on sillimanite-rich inclusions. We may with profit examine the details of the action, that is, the processes going on within and immediately around the sillimanite-rich inclusions, immersed in a magma rich in plagioclase and pyroxene. This examination serves to throw some light on the detailed mineralogy of the inclusions, and in particular on the separation of free alumina, as corundum.

If a mass of sillimanite were added to some anorthite just above its melting point, it can be readily seen from examination of Fig. 56 that some of the sillimanite would be converted into corundum. This is because the line joining sillimanite and anorthite passes through the corundum field. We may imagine, for example, that the bulk composition of a layer immediately surrounding the sillimanite is 50 per cent sillimanite[1] and 50 per cent anorthite. The mixture represented by this layer, at 1550°, would consist of mullite, corundum, and liquid. If the original anorthite liquid had an excess of silica amounting to about 10 per cent, no corundum would be formed from sillimanite, because the join of sillimanite with such a composition misses the corundum field. Plainly the freeing of corundum from sillimanite depends, in these liquids, on the liquid being "basic," that is, having not more than a moderate excess of silica over the feldspar composition. Free silica associated with the sillimanite does not have a comparable effect in restricting the forma-

1 Sillimanite is not shown on the diagram because it is not formed under the conditions represented by the diagram. Its composition is approximately 63 per cent Al_2O_3, 37 per cent SiO_2.

tion of corundum; indeed, the silica would require to be nearly equal in amount to the sillimanite in order to neutralize the tendency to form corundum.

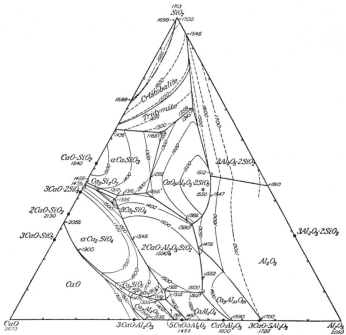

FIG. 56. Equilibrium diagram of the system CaO-Al₂O₃-SiO₂ (after Rankin and Wright with corrections after Bowen and Greig).

The system $MgO : Al_2O_3 : SiO_2$ given in Fig. 15 shows the same general condition. If a mass of sillimanite were immersed in molten $MgSiO_3$ and Mg_2SiO_4 in equal parts, a layer around the inclusion, which might have the bulk composition 75 per cent sillimanite, 25 per cent the above mixture, would consist at higher temperatures of corundum and liquid; at lower temperatures, of corundum, mullite, spinel, and liquid; and at still lower temperatures, of mullite, spinel, and cordierite. Since experiment has not revealed the conditions of stability of sillimanite we cannot state what determines its formation. Presumably it comes in only at lower temperatures or possibly at high pressures.

It should be realized that these conditions we have pictured as occurring about sillimanite inclusions are transient states. We may deduce from an equilibrium diagram the condition of a certain layer about an inclusion, but the system as a whole is not in equilibrium, since there are composition gradients about these inclusions. Nevertheless this changing state might become fixed and be revealed as a result of complete consolidation of the mass.

The actual magma by which the schist inclusions were attacked may be regarded as showing the combined effects of the anorthite liquid and the magnesian liquid of the experimental systems. It is true the magma did not correspond definitely with any such mixture of anorthite and magnesian silicate, but it was closely related, consisting mainly of plagioclase and magnesian silicates, and its effect on sillimanite-rich inclusions might reasonably be similar. The effect was indeed very similar, for the sillimanite-rich inclusions are found to be changed by the magma to masses rich in corundum, spinel, and cordierite, with other related minerals. It is inclusions that have been thus affected that constitute the emery deposits. They represent the reaction between the magma and inclusions arrested midway. No doubt many inclusions completely disappeared, becoming an integral part of the igneous rock in virtue of reactions involving adjustment of composition of existing phases. (See p. 206.) The feldspathic emery and the noritic emery may, from this point of view, be regarded as inclusions approaching their final disappearance.

These transient states in which inclusions may be very rich in certain substances are no doubt of some importance in differentiation. Localized masses in process of reaction with the general mass may move, say in response to gravity, and their accumulation may give rise to bodies rich in minerals formed during the reactions. All up and down the Appalachian Mountain system of Eastern North America there are intrusive masses showing ultrabasic differentiates, with dunite as the extreme and with associated peridotites and pyroxenites frequently rich in rhombic pyroxene, saxonite, websterite and enstatolite itself. These are usually, perhaps always, intrusive into slates and mica schists that were originally aluminous sediments, the Farnham slates of Quebec, the Savoy and Rowe schists of New England, the Manhattan of New York, the Wissahickon of Pennsylvania and Maryland, and the Carolina gneiss of the Southern States. They have perhaps had an important influence in emphasizing differentiates of the types mentioned above. There are, moreover, corundum deposits, either as emery or in purer forms, in frequent association with these ultrabasic rocks and in some cases the origin and accumulation of corundum may be referred to processes outlined above. Gordon and Miss Cobb have demonstrated a different origin for some of them,[1] but the action here described seems unquestionable for the

1 S. G. Gordon, Formation by reaction between pegmatite and serpentine. *Proc. Acad. Nat. Sci. Philadelphia*, Part I (1921), p. 169; and Margaret Cobb, Dissertation, Bryn Mawr, 1924.

Cortlandt emery and the gangue minerals of some of the Carolina deposits strongly suggest a similar origin. These gangue minerals are basic plagioclase, sillimanite and cyanite. That such minerals, together with corundum, could form from basic magnesian rocks by simple differentiation is very doubtful.[1]

Recognition of the probable saturated condition of the magma is of importance because it shows that there could not have been formed, at any time, a liquid whose composition was simply the sum of that of the original magma and the inclusions. Even that portion of the foreign matter which becomes a part of the liquid does so only by precipitating phases with which the magma is saturated and must itself be of a composition toward which the liquid may go spontaneously by fractional crystallization. So in the case of the Cortlandt series various diorites, some syenite and a considerable amount of granite (first recognized by Berkey as a part of the series) were formed by differentiation, as they would have been under the same conditions without reaction with slate material, though presumably the amount of "acid" differentiates was augmented by its addition.

The detailed studies of several norites by Read have revealed the selective nature of the reaction with aluminous inclusions. He notes that the "gabbro magmas become richer in alumina and potash, and poorer in lime and magnesia; iron-oxides and soda appear to play no constant part."[2] On the whole the behavior is about that to be expected from the reactive precipitation discussed in the foregoing.

One of the interesting examples of the effects of basic magma upon inclusions of aluminous sediments has been studied in great detail by H. H. Thomas.[3] The inclusions referred to occur in basic sills of Mull and the action is particularly intensive. It is believed to have occurred in a deep-seated reservoir and to have been more or less perfectly "quenched" upon intrusion of the magma as sills. In these rocks some aluminous inclusions have been converted largely to liquid, with some of the more refractory elements left in excess, and the liquid has cooled to a glass containing crystals of mullite,[4] corundum and rutile. The further reaction of the inclusion with the magma has given corundum-anorthite-spinel rims. Thomas discusses the reaction in terms of experimental results, referring especially to the same diagrams and reaching much the same conclusions as those given in the foregoing.

In connection with the formation of norite through the influence of absorbed slates, it should be realized that the action is probably an emphasis upon normal processes. It may very well be, however, that

1 See Rankin and Merwin, "The Ternary System MgO-Al$_2$O$_3$-SiO$_2$," *Amer. Jour. Sci.*, 45, 1918, p. 325.
2 H. H. Read, *Quart. Jour. Geol. Soc.*, 79, 1923, p. 479.
3 *Quart. Jour. Geol. Soc.*, 78, 1922, p. 229; and Mull Memoir, p. 268.
4 See N. L. Bowen, J. W. Greig and E. G. Zies, *Jour. Wash. Acad. Sci.*, 14, 1924, p. 183.

if there were no argillaceous sediments, norites would be of much rarer occurrence than they are.

THE ACTION OF BASIC MAGMAS ON SILICEOUS SEDIMENTS

A number of examples of argillaceous quartzites are known that have been invaded by basaltic magma and in which it is believed by some investigators that absorption of the siliceous sediments by the magma has occurred. Two carefully studied examples are the Pigeon Point sill of Minnesota[1] and the Moyie sills of British Columbia.[2] Such quartzites may again be regarded as consisting of material belonging to a later stage of the reaction series than basaltic magma, together with a certain surplus. The former, corresponding in composition with some acid igneous rock, should be capable of becoming part of the liquid magma by precipitating its heat equivalent of the phases with which the magma is saturated. There is no theoretical objection, therefore, to the belief that a certain amount of the inclusions could be incorporated in this manner even though the magma is saturated. It should be borne in mind, however, that the material that can thus become a part of the liquid must be of a composition toward which the magma could change spontaneously by fractional crystallization. Once incorporateed, it requires, to produce the acid differentiates that were there formed, the same conditions of crystallization as would have produced an acid differentiate from the uncontaminated magma. In all of these examples the normal course of differentiation is the primary consideration. The extent to which incorporated material contributed to the bulk of the acid differentiate may not have been important in these small bodies even though there is plain evidence of incorporation.

At both localities mentioned evidence of some incorporation is unquestionable. About inclusions of the sedimentary rocks reaction rims of a granitic nature have been formed. We have seen on page 190 that the reaction should emphasize, in the liquid around the inclusion, material belonging to a later stage of the reaction series (i.e., toward which the magma can crystallize), and this should not be intermediate in composition between magma and inclusion. Corresponding with this deduction we find that the rims about xenoliths are not simply melted xenolith but essentially normal igneous material of a late stage of the reaction series. The reaction-rim stage is a temporary one except in so far as it may be preserved about some inclusions by exhaustion of the magma. Others disappear entirely by diffusion of the rim material into the magma and distribution (with possible precipitating effects) of any surplus material. Thus the liquid is pushed onward in the reaction series,

1 W. S. Bayley, *U.S. Geol. Surv., Bull.* 109 ; R. A. Daly, *Amer. Jour. Sci.,* 43, 1917, p. 423.
2 R. A. Daly, *Geol. Surv. Can., Mem.* 38, p. 226. S. J. Schofield, *Geol. Surv. Can., Museum Bull. No.* 2.

not only through addition of the rim material, but also because this necessitates some precipitation of the early-formed minerals from the basic magma. Further fractional crystallization may therefore give differentiates identical with, or closely related to, the reaction-rim material, but normal differentiation might have given it also.

EFFECTS OF GRANITIC MAGMA ON INCLUSIONS OF SEDIMENTARY ORIGIN

In discussing the reaction of magmas with inclusions we have, in the case of basaltic magma, made some reference to the superheated condition. Daly points to basaltic magma as the heat bringer, and has presented evidence that such magma enters into igneous-rock economy on a different basis from all other magmas.[1] If this be true, and his reasons seem to me convincing, basaltic magma is the one magma that may, presumably, be assumed to have superheat on some occasions. All other magmas, whether they may be formed by differentiation of basaltic magma or by differentiation of syntectic magma must usually be saturated, unless it be that locally, at volcanic vents, a special source of heat is available. This possibility has little quantitative petrogenic significance and it is perhaps a realization of the commonly saturated condition of other magmas that has led Daly to adopt basaltic magma as his solvent. Thus all of Daly's syntectics are of *basaltic magma* with various types of foreign matter. We have already seen, however, that the saturated condition is no bar to a *reaction* between magma and inclusions. This fact is as true of intermediate magmas as of any others and the same principles apply to them.

If the foreign material belongs to an earlier stage of the reaction series the tendency is to make it over into those phases with which the magma is saturated and to precipitate a further amount of these phases from the magma itself. If the foreign material belongs to a later stage of the reaction series it tends to become a part of the liquid by precipitating phases with which the magma is saturated.

Sediments do not belong in the reaction series at all and certain types of sediments contain material belonging in both the above classes and both effects may be obtained. Our chief purpose here is to consider principles, and it seems unnecessary, therefore, to discuss individually the action of various intermediate magmas on various foreign inclusions.

It is perhaps desirable, however, to discuss the action of granitic magma on sediments, its action on igneous inclusions having already been described. For the reason mentioned above, only saturated granitic magma will be considered. Quartzites and slaty rocks offer no special problem. They are readily transformed into phases with which the

1 R. A. Daly, *Igneous Rocks and Their Origin*, p. 458.

granite is saturated, an action that any magma will accomplish in so far as the composition of the sediment permits. The conversion of inclusions of such rocks by granitic magma into masses of quartz, feldspars, and micas, in varying proportions, should therefore be the result. A certain amount of mechanical disintegration might cause the strewing about of these products in such a way as to make them an integral part of the mass but there should be no solution. Intermediate steps might see the formation of such minerals as sillimanite, garnet, and others characteristic of contact rocks, but these should be temporary or should survive only because of exhaustion of the liquid.

The kind of effect that sillimanite produces by reaction with basic magmas, namely, the precipitation of orthopyroxene and basic plagioclase is not to be expected in granitic magma. Rather should we expect precipitation of the micas in acid magmas, and formation of orthopyroxene in such magmas is to be referred to other causes than that here adopted for basic rocks.

When we turn to the case of carbonate rocks we find that the reaction with granitic magma is of a different nature. It is often observed that wall rock and inclusions of carbonates are altered to silicate minerals.[1] It has been assumed by some investigators, therefore, that silica has been subtracted from the granitic liquid and that this may occur to such an extent that some of the feldspar molecules are transformed into the less siliceous, feldspathoid molecules with consequent formation of alkaline rocks. This assumed action is said by certain writers to be in agreement with Daly's theory of the origin of alkaline rocks. We have seen above, however, that Daly assumes that superheated basaltic magma is the starting point for all his syntectic magmas and that alkaline rocks are differentiates of some of these syntectics, principally those formed with carbonate rocks. We have pointed out on a preceding page that, if adequately superheated basalt were available, it might form, by solution of carbonates, a melilite basalt and, given the latter, alkaline differentiates seem not impossible and so indeed some alkaline rocks may be formed. Not all alkaline rocks can be so explained, for much nephelite syenite shows intimate genetic relations with granites and on Daly's general theory the original basaltic magma would require to be silicated by solution of acid material to form the granite and desilicated by solution of carbonates to form the nephelite syenite. That the differentiates should show evidence of both seems out of the question. The solution of foreign matter must result in either desilication or silication according to the preponderance of one or the other type of

1 For example in the large-scale production of amphibolites in the Haliburton-Bancroft area. Adams and Barlow, *Can. Geol. Surv., Mem. No. 6,* 1910.

foreign matter and subsequent differentiates should be in conformance with one or the other but not both.

This brings us back to the question whether a nephelite syenite, intimately associated with a granite, could have been formed as a result of desilication of the granite by carbonate inclusions, that is, back to the question of the effect of granitic magma on carbonate inclusions.

We have noted the silication of the inclusions and the consequent supposed desilication of the liquid. In discussing investigated systems we have already found, however, that saturated liquids cannot have ingredients subtracted from them at randon without causing precipitation of other ingredients so that the effect on the liquid is not mere impoverishment in the ingredient subtracted. This may perhaps be made clear by a simple example. If we had a solution of salt at — 20° C, and any hypothetical substance was placed in the solution that withdrew the salt from it, the result would not be simply the leaving behind of liquid water. The reason is simply that liquid water cannot exist at — 20° C and the actual result would be that, as each small amount of salt was removed, a small amount of ice would form and, when all the salt was withdrawn, all of the water would have become ice. For the maintenance of liquidity, the salt and the water are necessary each to the other. And so it must be with a saturated granitic solution. Remove the silica from it and other substances must be precipitated. Now the reaction of granitic magma with inclusions of carbonate rock is not a simple addition of silica to the latter but usually other substances are added as well, these being such as to convert the inclusions ultimately into diopsidic pyroxene or hornblende. The reason for the formation of these phases is that they belong at an earlier stage of the reaction series than the biotite with which granitic magma is normally saturated. The subtraction of the substances necessary to produce these minerals must, for reasons outlined above, cause concomitant precipitation of the other phases normally formed from granitic magma, principally feldspar. Thus the action described must bring about an exhaustion of the liquid by causing precipitation. There seems to be no reason for believing that it could first exhaust the free silica, leaving a feldspar-rich *liquid*, then, upon further action, cause removal of some of the silica from the feldspar *liquid* leaving a *liquid* containing feldspathoid molecules.

A reaction of the kind described, that is, a using up of some silica to form diopside with the consequent precipitation of feldspar and quartz, would seem to be the mode of formation of the diopside-bearing variety of the Beckett gneiss, of which Eskola has written a description and interpretation.[1] The transformation of the solid dolomite into solid diopside, with its effect upon the granitic liquid, was the dominant action in the

1 P. Eskola, *Jour. Geol.*, 30, 1922, pp. 265-94.

production of the types of gneisses there found, and of their banding, rather than actual solution of the dolomite or skarn in the granite and subsequent differentiation of the syntectic liquid. It must be admitted, however, that the solubility of $CaCO_3$ in magmas is probably greater than that of CaO and that under conditions permitting the retention of CO_2 an amount of limestone might be dissolved greater than that suggested by the reaction effects already discussed. The usual free conversion of limestone into silicates indicates, however, that it is not commonly so dissolved.

The formation of basic silicates, without the production of feldspathoids, seems to be the ordinary result of the action of granitic magma on limestone inclusions. Thus, in the granitic portions of the Bushveldt laccolith altered limestone inclusions are surrounded by a halo of dioritic material, but not by alkaline rock.[1] Other examples might be given; in fact, the ordinary effects of granite on limestone seem to be those we have deduced for a saturated granitic magma.

In one locality the alkaline facies of the Bushveldt complex is, it is true, intimately associated with a mass of limestone, and as a result of a study of this locality Shand has concluded that there is some connection between the production of the feldspathoids and the desilicating action of the magma. Apparently Shand does not believe that the entire production of nephelite is due to this action but rather that a nephelite syenite magma becomes ijolite by desilication.[2] This is a quite different matter from the production of the original nephelite syenite magma by such desilication. No theoretical objection can be raised against the belief that interaction with limestone could reduce the amount of feldspar and increase the amount of feldspathoid in a magma already capable of precipitating both of these. Such adjustment of the relative amounts of minerals we have found to be a common effect of inclusions.

It is probable that alkaline rocks are ordinarily produced by crystallization-differentiation from subalkaline magmas. In another chapter we shall discuss a method of production of some leucite and some nephelite rocks which depends upon the incongruent melting of orthoclase for the leucite rocks, and on the pseudo-leucite reaction for the nephelite rocks. The fact that the excess silica must be no more than a small amount should be noted, for this fact renders it possible that limestone may, in spite of the many objections that have been raised above, have some influence in promoting the formation of alkaline rocks. The influence is, however, an emphasizing of a normal tendency rather than a fundamental necessity. This we have found to be a general rule in connection with the effects of inclusions. If the differentiation of the magma

1 Oral communication, Professor Brouwer; but cf. Brouwer, *Jour. Geol.*, 36, 1928, p. 545.
2 S. J. Shand, *Trans. Geol. Soc. South Africa*, 22, 1921, pp. 144-6.

which gave rise to the orthoclase-rich liquid took place in the presence of a supply of limestone inclusions this would tend to reduce to a minimum any excess silica that might otherwise be associated with the orthoclase. Thus the normal tendency of the orthoclase to break down into leucite under the proper conditions would be free to assert itself. We may therefore accept the possibility that reaction with limestone may emphasize the tendency toward the formation of an alkaline differentiate, though it is not essential to it. Other factors, such as the failure of olivine to form at an early stage in the magma's history, or the free resorption of such olivine as does form, may also assure a low excess of silica at a late stage with likelihood of the separation of feldspathoids in cases where the control is the leucite effect above noted.

DEDUCTIONS TO BE COMPARED WITH OBSERVED RESULTS

Throughout the foregoing study of the reactions between inclusions and magma, attention has been directed mainly to its theoretical aspects, that is, to deducing from equilibrium considerations what reactions should occur, together with the effects of these upon the further crystallization of the magma. All of these deductions can be put to the test by observation of what has actually occurred, in particular by a study of the reaction rims formed about inclusions. It should not be expected that each inclusion will tell the whole story, but a general study of inclusions should do so. Not all the differentiates that might later form from the hybrid mass need be shown by the reaction rims, but certainly there should be formed some whose relationship to these possible later differentiates is established by their frequent association in many areas.

In some instances examples have been cited which appear to show that the expected reactions do occur. Such are the formation of granitic reaction rims by the action of basaltic magma on acidic rocks, the making of basic inclusions into biotite-rich masses by granitic magma, and others. The formation of alkaline rocks by the action of ordinary magmas on limestones is, at present, incapable of support on the above grounds: No example is known where inclusions of limestone, contained in an ordinary rock, are surrounded by reaction rims of feldspathoid-bearing rock. It is true that limestones and alkaline rocks are often intimately associated, but there is no assurance that the magma was not already an alkaline magma before it acquired this association. As we have already pointed out, this appears to be the conclusion that Shand reaches concerning the Sekukuniland occurrence, though he favors also the conception that the limestone emphasized its alkaline nature.

In the Fen area of Norway, one of the newer areas to which the limestone-syntectic hypothesis has been applied, there is a very striking association of alkaline rocks and carbonate rocks.[1] However, nothing there displayed demonstrates a change of subalkaline magma to alkaline magma through the influence of the carbonate rock.[2] No support in the way of reaction rims of the appropriate kind has yet been found for the limestone-syntectic hypothesis; with the possible exception of a case described by Eskola in which no actual feldspathoids are developed.[3] More definite support is desirable before the hypothesis can be accepted, even though there is reason to believe, as pointed out above, that the presence of limestone might emphasize the normal tendency of magmas to give an alkaline differentiate.

SUMMARY

The question whether magmas can dissolve large quantities of foreign inclusions is one that has been much debated by petrologists. Some have claimed great powers for magmas in this respect and in addition have assigned a dominant rôle in the production of differentiation to such solution of foreign matter. Others have insisted that magmas have not the necessary heat content to enable them to give significant effects of this kind. A study of some simple equilibrium diagrams, with the object of determining the heat effects connected with solution, gives every reason for believing that the effect is a large absorption of heat, usually of the order of magnitude of the latent heat of melting. For simple solution, then, it is unquestionable that large amounts of heat will be required.

Those who believe in the actuality of the solution of considerable amounts of foreign matter in magmas have usually realized this fact and have sought a source of the heat in magmatic superheat of great amount, that is, in a large excess of temperature of the magma above its crystallization range. A study of the probabilities of the case and of the usual effects of magmas upon inclusions leaves little reason for believing that magmas can ordinarily have any considerable superheat.

Unquestionably, then, the observed effects of magmas upon inclusions are usually to be referred to an action other than the direct solution of inclusions in superheated magma. An application of the conception of the reaction series to the solution of the problem affords an explanation of the effects of magmas, even though saturated. Certain principles governing the effects of liquid upon inclusions belonging to reaction series can be developed by studying the equilibrium diagrams of sys-

1 Cf. W. C. Brögger, "Die Eruptivgesteine des Kristianiagebietes IV," *Vidensk. Selsk. Skr. I. Mat. Naturv. Kl.*, 1920.
2 An alternative explanation of these rocks has been offered which interprets the carbonates as a later introduction having a replacing relation to the silicate minerals. N. L. Bowen, *Amer. Jour. Sci.*, 8, 1924, p. 1; *ibid.*, 12, 1926, p. 499.
3 P. Eskola, "On the Igneous Rocks of Sviatoy Noss in Transbaikalia," *Finska Vetensk.-Soc. Forhändl.*, 43, 1920-2, No. 1, p. 96. See also Brouwer, *Jour. Geol.*, 36, 1928, p. 545.

tems involving both continuous and discontinuous reaction series. In this manner it can be decided definitely that a liquid saturated with a certain member of a reaction series is effectively supersaturated with all preceding members of that series. It cannot dissolve such members but can only react with them to convert them into the members with which it is saturated. The reaction is not a simple subtraction from the liquid of the material necessary for this transformation, but some precipitation from the liquid itself is involved and the liquid ordinarily maintains its position on the same saturation surface. The products of crystallization from the liquid and the possible course of fractional crystallization are thus unaffected.

On the other hand, a liquid saturated with a certain member of a reaction series is unsaturated with all subsequent members of the series. Inclusions consisting of these later members can become a part of the liquid by a sort of reactive solution, the heat of solution of inclusions being supplied by the precipitation of their heat equivalent of the member of the series with which the liquid is saturated. It should be noted that the material that can, by this reactive process, become a part of the liquid must consist of a later member of the reaction series, that is, must be material toward which the liquid could pass spontaneously by fractional crystallization. The net effect upon the liquid is, then, to push it onward upon its normal course.

In Table II the products of crystallization of subalkaline magmas are arranged as reaction series, as definitely as may be in such complex series. The action of magmas upon foreign inclusions of igneous origin may be deduced from this arrangement of the crystalline products as series by application of the principles developed from the above study of simple systems. Thus we find that a granitic magma saturated with biotite cannot dissolve olivine, pyroxene, or amphibole, but can only react with them to convert them into biotite, the phase with which it is saturated. Or, stated more generally, no saturated magma can dissolve inclusions consisting of minerals belonging to an earlier stage of the reaction series (usually more basic).

Saturated basic magma, on the other hand, will react with inclusions belonging to a later stage of the reaction series (more acidic), the reaction being of such a nature that the inclusions become a part of the liquid by precipitating their heat equivalent of the phases with which the magma is saturated (basic minerals). The inclusions, it should be noted, must be of a composition toward which the liquid could pass spontaneously by fractional crystallization. Thus saturated basaltic magma can dissolve granitic inclusions by precipitating basic minerals and the granitic material passing into solution then becomes a contribution to the normal granitic differentiate that may form by fractional crystallization if the conditions are appropriate.

The behavior of inclusions of sedimentary origin is more complicated since sedimentary material does not belong in the reaction series. A consideration of the extent and nature of the variation of composition possible in the crystalline phases formed from a magma shows that the incorporation of considerable amounts of sedimentary material would ordinarily bring about merely an adjustment in the composition and relative proportions of existing phases. As a result of the non-appearance of new phases, the general course of fractional crystallization is unaffected. In general, the adjustment noted takes place through precipitation of the phases with which the magma is saturated. As an example it may be stated that the addition of highly aluminous sediments to basic magma should bring about the formation of anorthite and enstatite molecules at the expense of diopside molecules and should therefore cause the precipitation of crystals rich in anorthite and enstatite. Such action may have been important in the formation of many norites. The foreign material becomes a part of the general mass as a result of reaction and precipitation rather than by simple solution.

The Cortlandt series of New York, with its inclusions, affords an illustration of the behavior of aluminous sediments in basic magma. Such sediments may be regarded as consisting in part of material corresponding in composition with igneous material late in the reaction series, together with a certain excess, which is highly aluminous. The former may become a part of the liquid by the method of reactive solution already described. There results the piling-up of the highly aluminous excess in the inclusions, with formation of such minerals as sillimanite. Moreover, as a consequence of what may be somewhat loosely called the instability of sillimanite in contact with liquid rich in anorthite or magnesian silicates, alumina is set free as corundum. This condition is transient, however, and even these residues from the inclusions may become a part of the general mass as a result of the reactive precipitation noted above. The net result is the formation of noritic material with an increase in amount of the acidic differentiate normally possible.

The addition of limestone to basaltic magma may perhaps give rise to a liquid capable of precipitating melilite in some cases and from such a liquid it is possible that some alkaline rocks may form by further differentiation. It does not seem possible that limestone inclusions can desilicate a granitic magma in such a way as to give rise to a liquid capable of precipitating feldspathoids. However, if limestone inclusions were present during the differentiation of the more basic liquid from which the granitic liquid may have formed, the presence of such inclusions might reduce the amount of free silica associated with the alkaline feldspar in this liquid to such an extent that the normal tendency of orthoclase to break down into leucite would manifest itself. Thus rocks bearing leucite, and possibly other feldspathoids, might form, but influences prevailing during early stages of differentiation, other than the presence of foreign matter such as limestone, may likewise lead to the formation of leucite.

In conclusion, it may be stated, therefore, that magmas may incorporate considerable quantities of foreign inclusions, both by the method of reactive solution and by reactive precipitation, and such action may have been important in connection with the production of certain individual masses. Thus some norites may have been produced as a result of the reactions discussed above, some granites may have had their mass augmented by reactive solution of granitic inclusions in the magma from which they differentiated, some alkaline rocks may have been formed as a result of the presence of limestone inclusions in the liquid from which they differentiated. All of these actions are, however, an emphasizing of normal processes possible in the absence of foreign matter. It is doubtful whether the presence of foreign matter is ever essential to the production of any particular type of differentiate.

PART TWO

CHAPTER XI

THE FORMATION OF MAGMATIC LIQUID VERY RICH IN POTASH FELDSPAR

I N THE study of natural glasses given in Chapter VIII no rocks of that kind were found having compositions corresponding with highly sodic granites and the conclusion was reached that probably highly sodic granites were not formed from liquids of their own composition. On the other hand glassy rocks having the composition of highly potassic granites are found among the pitchstones. Corresponding liquids must therefore be reckoned with in any theory of the derivation of rocks, though it is not necessary to suppose that all potash-rich granites are derived from such liquids.

It is very easy to derive these potash-rich liquids by means of liquid immiscibility, gaseous transfer, or the like, because these processes always do just what one may wish them to do. In the case of fractional crystallization some difficulty is presented. If the feldspars have that relation to each other which is commonly accepted, a liquid very rich in potash feldspar would never appear as a mother liquor of crystallization. We shall therefore offer a modification of accepted views on this relation that may render more credible the derivation of potash-rich liquid by fractional crystallization. In some measure we may be regarded as adopting the same tactics as the proponents of, say, liquid immiscibility: we are attempting to make crystallization do just what we want it to do. But there is this difference. We shall point to certain pertinent characters of rocks as offering at least some support for the fundamental assumptions.

If the relation between the three feldspars $NaAlSi_3O_8$, $CaAl_2Si_2O_8$ and $KAlSi_3O_8$ were that deduced by several investigators of rocks there would be no possibility of the derivation of potash-rich granitic liquid as a mother liquor in the fractional crystallization of basaltic liquid. These investigators have deduced a eutectic relation between anorthite and potash feldspar which gives a ternary diagram of the general form shown in Fig. 57 on the sides of which have been erected the deduced binary diagrams. The incongruent melting of potash feldspar makes it impossible to represent equilibrium in any such manner, nevertheless the relations between the feldspars might be as shown, in a liquid with enough excess silica to neutralize the breaking down of orthoclase. The

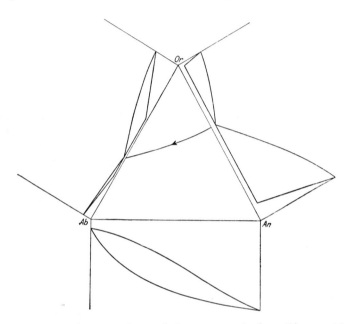

Fig. 57. Binary and ternary figure of the system, orthoclase-albite-anorthite as the relations are ordinarily assumed to be.

diagram might then be regarded as exhibiting equilibrium in such a liquid with the excess silica not shown. There is, however, much reason for believing that the relations in such a liquid are not in accord with these deductions. Upon the potash feldspar-albite diagram there is some difference of opinion as to details but there is little reason to doubt the general correctness of the diagram as given, always, of course, with the proviso that in the pure feldspar liquids the complication due to the incongruent melting of orthoclase must show in all compositions close to orthoclase. But neglecting this factor or considering it in the manner suggested above, potash feldspar and albite are preferably to be regarded as having a eutectic with extensive solid solutions on either side. Many papers have been written on this subject. Vogt has recently attacked the problem again and his paper gives references to all the others.[1] Of the complete solid solution between albite and anorthite there is, of course, no question to be raised.

The relation between potash feldspar and anorthite requires reconsideration. In the first place it is unlikely that the eutectic, if there were one, would lie very far from orthoclase with its melting point (incongruent) at 1170° and that of anorthite at 1550°. In addition there is much evidence in rocks to suggest, if not to prove, that the relation between orthoclase and basic plagioclase, including anorthite, is not of the eutectic kind. Rocks which contain basic plagioclase together with orthoclase are not among the commonest but are nevertheless quite well known and there is in such rocks a very striking tendency for the potash feldspar to occur as shells about the plagioclase. In absarokites it is common, according to Iddings, for the orthoclase crystals to have minute cores of labradorite.[2] In shoshonite and banakite the orthoclase occurs in zones surrounding labradorite, according to the same author (*op. cit.*, pp. 946-8). In various types of rocks in the Roman Comagmatic Region phenocrysts of "labradorite or anorthite are surrounded by mantles of orthoclase."[3] A "syenitic" rock from Korea has feldspars with a core of labradorite-andesine surrounded by a shell of soda-orthoclase.[4] The mantling relation between these two minerals would thus appear to be common in rocks which are cooled in the appropriate manner (rather rapidly) to bring it out.

This formation of orthoclase as a mantle about basic plagioclase casts much doubt on the eutectic relation of these minerals. It is characteristic of pairs of minerals which have a reaction relation such as that exhibited by olivine and pyroxene or by pyroxene and amphibole and is unknown in pairs of minerals such as plagioclase and pyroxene which lack this relation. Incidentally it may be noted that a mantling of the one by the other does not occur with orthoclase and highly sodic plagioclase which is in accord with the deduced eutectic relation at that end of the plagioclase series. A diagram can be constructed for anorthite and orthoclase which shows a reaction relation between them and which has a considerable degree of probability for other reasons. The diagram is as shown in Fig. 58. The fact that there is only very limited solid solution at either end may be regarded as established by the known composition of the natural minerals. A notable feature is that, just as in the albite-anorthite system, the crystals are always enriched in anorthite with respect to the liquid. Unlike the plagioclase system, however, there is no continuous change of composition of the crystals from the one end member to the other. The phase anorthite with only a little

1 *Skr. Norsk Vidensk-Akad.*, I, No. 4, 1926.
2 *Jour. Geol.*, 3, 1895, pp. 940-2.
3 H. S. Washington, *Carnegie Inst. Wash.*, *Pub. No. 57*, 1906.
4 T. Ito, *Jour. Faculty Sci. Tokyo Univ.*, I, 1925, p. 105.

orthoclase in solution separates at all temperatures above the reaction point Z. At this point the crystals react with the liquid and are converted (partly or wholly, depending on the relative amount of liquid) into

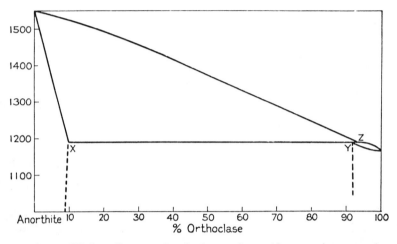

FIG. 58. Equilibrium diagram of orthoclase and anorthite assuming a reaction (not a eutectic) relation and neglecting incongruent melting of orthoclase.

orthoclase crystals with only a little anorthite in solid solution. If this relation holds for anorthite with orthoclase it must hold for basic plagioclase with orthoclase as well. It must disappear before a very sodic plagioclase is reached, for here there is every reason to accept a eutectic relation with orthoclase. The very general tendency for orthoclase to occur as a mantle about anorthite and labradorite in rocks containing both would be accounted for by a relation such as that described, extending from anorthite at least as far as labradorite.

This reaction relation of basic plagioclase and potash feldspar, rendered so highly probable by the frequent mantling relation, carries with it many consequences, the first of which to be discussed will be the possibility of the production of potash-rich liquid of a granitic nature by fractional crystallization. In Fig. 59 is given a ternary diagram of the feldspars which takes account of the reaction relation we have just discussed. Again it is desirable for the sake of simplicity to neglect the incongruent melting of orthoclase or, better stated, to consider the relations in a liquid with excess SiO_2 which neutralizes this effect, and yet not to plot this excess SiO_2.

The boundary curve between the fields of orthoclase and plagioclase must occupy some such position as ENR extending from the eutectic, orthoclase-albite at E to the reaction point orthoclase-anorthite at R. The curve OXP indicates the limit of solid solution of orthoclase in

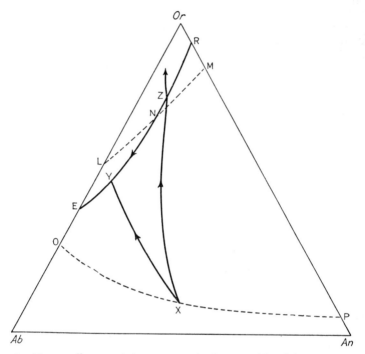

FIG. 59. Ternary diagram of the system, orthoclase-anorthite-albite with assumed reaction relation of orthoclase to basic plagioclase and incongruent melting of orthoclase neglected.

plagioclase and the curve LNM indicates the limit of solid solution of plagioclase in orthoclase. Various points along these curves of saturated solid solutions thus indicate the compositions of the two kinds of feldspar crystals jointly in equilibrium with liquids represented by points along the boundary curve ENR. It will be noted that the solid solution curve LNM cuts the boundary curve ENR at N. The liquids along the boundary curve are thus divided into two classes. In the first class are those extending from E to N (approximately)[1] which are in equilibrium

1 The word "approximately" is inserted here to indicate the fact that, while the liquids are divided into the two classes noted, there are liquids close to N that present even more complex relations. These need not be further discussed in the present connection.

with two kinds of crystals having compositions lying on opposite sides of the boundary curve. These crystals (orthoclase solid solution and sodic plagioclase) are therefore simultaneously subtracted from the liquid. In the second class are those extending from R to N (approximately) which are in equilibrium with two kinds of crystals having compositions lying on the same side of the boundary curve. The one kind of these crystals (orthoclase solid solution) is subtracted from the liquid but the other kind (plagioclase solid solution) reacts with the liquid to produce orthoclase solid solutions and thus gives the mantling of calcic plagioclase with orthoclase which is their characteristic relation.

In the discussion of fractional crystallization of plagioclase liquids in Chapter VII it was unnecessary for the purpose then in view to make any definite assumption as to the position of the orthoclase-plagioclase boundary curve. Now that a conclusion has been reached as to the general position of this boundary curve (ENR of Fig. 59) we may enlarge upon that discussion to some extent, especially by way of pointing out the possibility of the development of potash-rich granitic liquids by fractional crystallization.

In the crystallization of the liquid X (Fig. 59) there are two principal possibilities. The first involves relatively high fractionation (low reaction) giving a curve such as XY which meets the boundary at a point (Y) lying between E and N. When the point Y is reached rather sodic plagioclase is joined by orthoclase and if there is further fractionation the liquid may move somewhat along the boundary curve towards E. There are thus produced liquids which in their feldspar proportions are analogous to the largest class of granitic glasses, namely, the obsidians. They have rather balanced proportions of potash and soda feldspar. If, therefore, this fractionation took place in basaltic liquid and was accompanied by that impoverishment of femic constituents and those effects that lead to the formation of free silica, we appear to have the definite possibility of the development of residual liquids of the nature of obsidian. The details of this process have already been discussed.

The second possibility involves relatively low fractionation (high reaction), at least at early stages, whereby the liquid follows a course such as XZ, encountering the boundary curve at the point Z when the plagioclase crystals in equilibrium with the liquid are still (in virtue of low fractionation) rather highly calcic.[1] The point Z lies between N and R and the orthoclase solid solution which now begins to separate has a reaction relation to the plagioclase and the liquid. If the reaction

---

[1] A course of the liquid such as XZ is definitely assured in any part of a mass where there has been enrichment in calcic plagioclase crystals relative to liquid. A mass in which plagioclase crystals have accumulated by settling or a mass from which some liquid has been squeezed out would thus have interstitial liquid that would follow a course such as XZ, whereas the squeezed-out liquid would follow a course such as XY.

can take place freely the liquid will again move somewhat along the boundary curve towards E, but if reaction is prevented for any reason, such as mantling of the plagioclase or squeezing out of residual liquid, the liquid Z will crystallize in total disregard of past plagioclase crystals. Thus this liquid, which is already very rich in orthoclase molecules will leave the boundary curve and pass into the orthoclase field with separation of orthoclase solid solution and still greater enrichment of the liquid in orthoclase molecules in a manner analogous to that more simply shown in the binary system, anorthite-orthoclase as depicted in Fig. 58. There would thus appear to be the definite possibility of producing from basaltic liquid, by appropriate fractionation, a potash-rich granitic liquid corresponding with the smaller class of natural glasses represented by pitchstones.

CHAPTER XII

THE ALKALINE ROCKS

GENERAL NOTE

THE alkaline rocks constitute a group that is difficult to mark off sharply from their more abundant sub-alkaline relatives. It is equally difficult to single out any definite character that may be said to distinguish the alkaline rocks. Perhaps the outstanding feature of the more notably alkaline types is that they contain one or more members of the group of minerals known as feldspathoids.

The most prominent feldspathoidal rock is nephelite syenite and in a former publication a mode of origin of nephelite syenites was suggested in which the corresponding magma was regarded as the residual liquid of granitic magma, crystallized in the appropriate manner.[1] Reliance was placed upon the indications, furnished by the presence of biotite in granite, that, in the presence of water, there is a tendency towards the breakdown of the polysilicate molecules of the feldspars into the ortho-silicate molecules of the micas, with setting free of SiO_2. The mica and some of the quartz of this stage are thus accounted for and equations have been given on p. 83 to indicate the general type of reaction referred to. This method of origin of nephelite syenite is, therefore, very similar to that which had previously been suggested by Smyth in that it assigns to the volatile components an important rôle in the origin of alkaline rocks.[2] In the above-mentioned suggestion as to the origin of nephelite syenites the molecule $NaAlSiO_4$ was regarded as one of the molecules originating from this action, and it was considered that, as the result of precipitation of the other molecules in the form of mica and quartz, there would be a concentration of the $NaAlSiO_4$ molecule in the residual liquor at a certain stage. The separation of the residual liquor, probably by a squeezing-out process, might then occur and thus an independent phonolitic liquid might originate. On the other hand if the liquor remained in contact with the granitic minerals, crystalline

1 N.L. Bowen, *Jour. Geol.*, 23, Suppl. 1915, p. 55.
2 C. H. Smyth, Jr., *Am. Jour. Sci.*, 36, 1913, p. 46.

nephelite (unlike biotite) being incompatible with quartz, the liquid would not precipitate nephelite but would react with quartz to form albite from the nephelite molecules. Against this mode of origin of nephelite syenite it may be urged that in some cases, where the final cooling had been too rapid to permit this final reaction, one would expect to find phonolitic interstitial material in granite just as one may find, from similar causes, quartzose interstitial matter in olivine-bearing rocks. But the two cases are not strictly parallel, for the amount of interstitial liquor of the hypothetical phonolitic character would be very small compared with the crystals with which it may react and it exposes a large surface to these crystals. Moreover, the condition then existing could have originated only as the result of very slow cooling and this slow cooling must continue, for the mass is so near complete consolidation as to render it impossible to inject it in bulk into a small fissure where alone it might experience rapid final cooling. Only the interstitial liquor is capable of such injection.

Another objection that may be raised to the stated method of origin of nephelite syenite as a residual from granite is that there is a tendency for granites to give quartzose pegmatite and finally quartz veins as residuary material. It is possible, however, that this running on into quartz veins is prominent only when the feldspar fractionation as outlined in Chapter XI has been of the kind which produces the potash-rich granitic material, granitic pegmatite, as originally constituted, being of such a nature. On the other hand it appears that when the fractionation is such that a more sodic granite is produced, the liquid may push on into alkaline and definitely feldspathoidal syenites. This matter is more fully discussed and field examples cited in the original paper.

Other modes of development of nephelite syenite are suggested in the following pages. One of these depends on the demonstrated fact of the incongruent melting of orthoclase. In this method of development granite and nephelite syenite, while intimately related genetically, would never have a reaction relation to each other; neither would ever occur as interstitial material in the other. On the other hand in developing this hypothesis it will be shown that interstitial substance with free silica might occur in closely related leucite rocks. These conditions seem to be definitely in accord with the actual findings in rocks.

Our present problem, then, is to consider how rocks containing feldspathoids and rocks transitional towards these may be developed from magmas which ordinarily give rise only to quartzose late differentiates.

Much discussion has already been given to the factors which control the development of free silica in magmatic liquids and which lead to its precipitation at late stages. Among these factors was the early separation of olivine in excess of its stoichiometric proportion and when this factor is operative it would appear that the possibility of the subsequent development of feldspathoidal and related rocks is not to be expected. There are, however, probable conditions of cooling under which this factor may not be operative and which therefore open up the possibility that differentiation may proceed towards a syenitic rather than a granitic final liquid.

TRACHYTIC ROCKS

In describing Fig. 14 there has already been discussed the olivine control over the development of free silica in late liquids. In this system we are concerned with the reaction relation between forsterite and clino-enstatite. Development of free silica as a result of similar factors was discussed in connection with Fig. 20 also. There the reaction relation obtained between forsterite and a series of pyroxenes which may be called enstatite-diopsides. But of this series of pyroxenes only some show

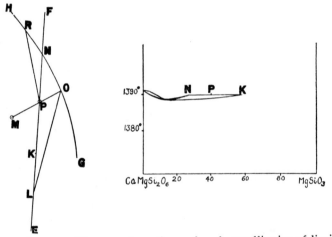

Fig. 60.   Diagram to illustrate alternative modes of crystallization of liquids of the composition of some of the pyroxenes of the enstatite-diopside series.

the reaction relation. Diopside itself and those extending as far as 80 per cent diopside-20 per cent $MgSiO_3$ do not have this relation to forsterite. Pyroxenes lying just beyond this range present interesting possibilities of alternative behavior (depending on the conditions of cooling) which we shall now discuss. The relations will be shown in an enlarged diagram, Fig. 60, dealing particularly with the area close to the point of intersection of the pyroxene conjugation line (of which FE is a part) and the boundary curve (of which GNH is a part). A liquid of the composition of the pyroxene P begins to crystallize with separation of forsterite and the liquid changes in composition to O. When the temperature and composition O are reached pyroxene of the composition L begins to crystallize. The composition of the liquid now changes along the boundary curve, forsterite reacting with liquid and pyroxene separating until the liquid has the composition N. At this temperature all the forsterite has disappeared and the pyroxene has the composition K. Thenceforth the mass behaves as a simple binary mixture and its behavior is expressed by the binary diagram of Fig. 60, in which the points N, P and K correspond with the points so lettered in the ternary figure. This simple, binary crystallization need not be discussed. Suffice it to say that the final product is simply the pyroxene P.

Such is the behavior of the liquid P when perfect equilibrium and no fractionation obtain, but it is plain that the liquid, when it had the composition O, might fail to react with forsterite crystals whether as a result of bodily removal or armoring of these crystals. In such a case the liquid O would not follow the boundary curve but would cross the pyroxene field and reach the pyroxene-silica boundary curve where free SiO_2 crystallizes as tridymite. Failure to pass over to liquids from which SiO_2 crystallizes may be accomplished, as has been noted, by complete reaction of the liquid with the early-crystallized forsterite, but it may be accomplished in another manner as well. If the liquid cooled so rapidly at first that no crystallization took place till the temperature of the point N was attained then no crystallization of forsterite would occur but crystallization would begin with the separation of pyroxene. Fractionation might occur during this crystallization but the possibility of passing over to liquids containing free silica would not enter. Thus if pyroxene of that nature occurred in a melt with plagioclase the fractional crystallization might give rise to liquids ever richer in alkaline feldspar without developing any free quartz. A syenitic (trachytic) differentiate is thus a possible alternative to the granitic differentiate if the pyroxene lies anywhere between N and K. As shown in the figure this is a very wide range of pyroxene compositions extending from about 80 per cent diopside-20 per cent enstatite to 40 per cent diopside-60 per cent enstatite. Pyroxenes outside this range do not show the same alternative possibilities. Those from N to diopside never precipitate excess olivine and therefore can not be the cause of development of free silica in a late liquid. Those from K to enstatite always precipitate excess olivine but the liquid in equilibrium with both pyroxene and olivine always contains excess SiO_2 at every stage of crystallization. At no time could fractionation, resulting, say, from armoring of olivine or otherwise, give rise to any type of liquid except that with excess SiO_2.

Those pyroxenes extending from N to K are thus the only pyroxenes whose presence introduces the possibility of the alternative development of either a syenite-like late differentiate or a granite-like late differentiate. The question now arises whether the natural pyroxenes of basaltic magma or even of some basaltic magmas have the same characters as these synthetic enstatite-diopsides (N — K). It has been well established as the result of the work of Wahl, Holmes, Washington and others that the pyroxenes of plateau basalts are the so-called enstatite-augites or hypersthene-augites. These pyroxenes are comparatively free from alumina, indeed are essentially hypersthene-diopsides. The only significant difference between these natural minerals and the synthetic pyroxenes whose relations have been completely studied lies in the replacement of some of the MgO by FeO. What modification of the physicochemical properties of the pyroxenes will be introduced by this substitution can not be predicted. A small amount of FeO can, of course, only modify the relation somewhat. It is possible that the relation may disappear when a considerable amount of FeO is present but it is equally possible that it is emphasized by the presence of FeO, that is, that an even wider range of pyroxenes can induce this alternative behavior.

The enstatite-augites present in many plateau basalts are the most likely pyroxenes to show these properties as being the nearest relatives of the investigated enstatite-diopsides. If it can be assumed that the pyroxenes of the plateau basalts of Mull and the Hebridean area in general are among those having these properties then a straightforward explanation of the two lines of descent there exhibited is forthcoming. The gabbro-inninmorite-granophyre sequence, on the one hand (Mull Memoir, p. 14), and the gabbro-mugearite-syenite sequence (Mull Memoir, p. 26), on the other, may be the result of alternative lines of descent, rendered possible by a composition of the parental magma such that, when very quickly cooled, it gives enstatite-augites having physicochemical characters like those of the enstatite-diopsides.

Some consideration may now be given to the probable control which determines whether the trachytic or the granophyric material shall be the late differentiate. It has already been suggested that cooling sufficiently slow to permit the complete resolution of olivine (or of excess olivine) would determine a trachytic differentiate. Again very rapid cooling, such that the liquid was cooled through the temperature range in which the excess olivine separates, might permit a like result. The alternative involving slow cooling requires the additional assumption that some sort of stirring effect prevented segregation of the olivine, a possibility which can not be denied and may have been operative in some cases. But this combination of conditions does not seem as likely on general grounds as the rapid cooling of a rather small mass in its early stages, followed by a sufficient slowing-up of the cooling at later stages to permit some fractionation. The relative volumes of the rock types concerned are in accord with this general expectation. The trachytic differentiates of the Hebridean rocks are of insignificant volume and are therefore preferably to be referred to the differentiation of rather small masses of basaltic magma. The granophyric differentiates, on the other hand, occur in some volume and their content of free silica is to be referred to the separation of excess olivine from more considerable bodies of basaltic magma at early stages, and failure of the resorption of olivine because of armoring or segregation to form bodies of peridotite, allivalite, etc. It is to be noted that, even in a very large mass whose liquid course was dominantly towards granitic composition, some part of the mass might have just the right amount of olivine to give barely complete resorption. The course of the liquid might thus locally be directed towards a trachytic differentiate.

In the typical basalt-trachyte association the differentiation may proceed still further in the alkaline direction and give phonolitic trachyte with a deficiency of SiO_2 below that necessary to convert all the alkalis and alumina into feldspars. The development of phonolite probably does not in this case depend upon the preliminary formation of leucite as it does in other cases to be discussed later. If this is true, there is nothing in the results of systems yet investigated to explain the passage onward to phonolite in this particular association. Experimental results have yet thrown light only on the relations involving diopside and the enstatite-diopsides. They indicate that, of these, only diopside should occur together with feldspathoid and that if one begins with any plagioclase and enstatite-diopside the extreme possible limit would be albite with a little diopside, i.e., a trachyte-like product, not phonolite-like. But systems yet investigated throw no light on the factors that control the development of the alkalic molecules in pyroxene or on equilibrium between the iron oxides and the iron silicates. Probably if these were known it would be apparent that crystallization-differentiation permits the possibility of passing on to phonolitic compositions from the trachytic.

A possible solution of another of the many problems raised by Hebridean rocks may lie in this same flexibility that is introduced by variation of the behavior of enstatite-augite with different rates of cooling. It has been noted that the plateau basalt magma of Mull, when rapidly cooled, did not develop a siliceous final residuum but rather tends to develop analcite.[1] This is quite in accord with the possible effects of liquids giving enstatite-augite when rapidly cooled. In such a case there may be no *excess* separation of olivine and an originally undersilicated

1 Mull Memoir, p. 30. See also p. 75 of the present volume.

liquid will remain undersilicated until the latest stages of crystallization. At such a stage and in such a liquid the formation of analcite is a likely development if adequate water be present.

THE BASALT-TRACHYTE ASSOCIATION OF OCEANIC ISLANDS

It is a well-known fact that the association of basalt and trachyte is very common in oceanic islands.[1] The trachyte is of insignificant bulk. Ordinarily rhyolites are lacking. If our general thesis is correct the trachyte may be regarded as a derivative of the basaltic magma, differentiation having occurred through fractional crystallization of quite small bodies. The general tendency towards a trachytic differentiate might then be referred to a similar tendency towards small dimensions of the individual bodies of basaltic magma. The mechanics of the suboceanic crust are presumably at all times analogous to those of continents (especially their margins), during times of continental fragmentation. Intrusive activity finds its expression almost exclusively as dikes with occasional, relatively insignificant, plug-like expansions. There is never that tendency towards the formation of great cake-like masses of magma which is characteristic of continental mechanics. And there is consequently little tendency to form those differentiates of basaltic magma which result from slow cooling of the basaltic magma even in the very early stages of cooling. The dominance of trachytic material among the salic differentiates and the general lack of rhyolitic material would thus appear to be connected with the mechanics of the suboceanic crust.

FELDSPATHOIDAL ROCKS

In the foregoing discussion of the origin of potash-rich granitic liquids and of trachytic rocks evidence has been offered of the existence of a reaction relation between basic plagioclase and orthoclase. In drawing the diagrams necessary to illustrate this condition the relations in the neighborhood of orthoclase have been simplified because the true relations would complicate the matter unnecessarily for the purpose then in hand. In this simplification the incongruent melting of orthoclase has been neglected and the diagrams therefore give the relations developed in a liquid with enough excess silica to neutralize this effect. For our present purpose it will be necessary to consider the probable relations freed of any such simplifying assumptions. The fundamental binary diagram showing incongruent melting of orthoclase is given in Fig. 61. Since the demonstration of the incongruent melting of orthoclase a number of rather peculiar diagrams have been published which were de-

1 Daly, "Geology of Ascension Island," *Proc. Amer. Acad. Arts Sci.* 60, 1925, p. 75; Lacroix, "La Constitution Lithologique des Iles volcaniques de la Polynesie Australe," *Mem. Acad. Sci. Paris,* 59, 1927.

signed to show the relation of orthoclase to one or more other compounds and at the same time to show the incongruent melting of orthoclase. The full relation between orthoclase and either albite or anorthite can be

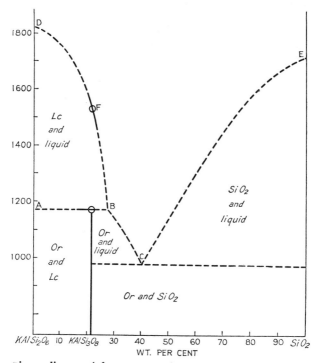

FIG. 61. Binary diagram (after Morey and Bowen) illustrating the incongruent melting of orthoclase.

properly shown only on a three-component diagram and that between orthoclase and plagioclase only on a four-component diagram. Diagrams of the kind given in Fig. 62 are necessary in order to show the orthoclase-anorthite equilibrium. In Fig. 62 (a) the equilibrium is of the kind which, in the simplified diagram, neglecting the incongruent melting of orthoclase, shows as a eutectic relation between anorthite and orthoclase (Fig. 57). In Fig. 62 (b) the equilibrium is of a reaction nature which shows in a simplified two-component diagram in the manner of Fig. 58. We have already given reasons for preferring the reaction relation between anorthite and orthoclase so that Fig. 62 (b) may be taken as the preferred form. A study of the figure will make plain the advantage of presenting the relations first in a simplified binary diagram even although this is not strictly accurate. In the binary diagram (Fig.

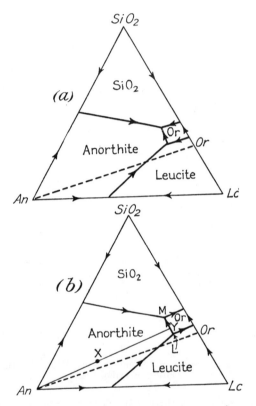

FIG. 62.   Ternary diagram, leucite-anorthite-silica. (a) Assuming eutectic relation between anorthite and orthoclase. (b) Assuming reaction relation between anorthite and orthoclase.

58) the fact is clearly brought out that the liquid is always enriched in orthoclase with respect to the crystals, because the composition of the crystals in equilibrium with each liquid is shown. In the ternary figure the same fact is shown, to be sure, but in a much less obvious form. The fall of temperature from the ternary reaction point, anorthite-orthoclase (solid solution)-silica, to the binary point, orthoclase-silica is the only

expression of this fact in the ternary system. It will be noted that the fall of temperature is in the opposite direction in Fig. 62 (a).

Fig. 62 (b) need not be described in detail. It may be mentioned, however, that in any mixture of anorthite and orthoclase only (i.e., without excess SiO_2) the separation of anorthite is always followed by the separation of leucite, not orthoclase, though this leucite will, under appropriate conditions, be converted to orthoclase at a subsequent stage. The addition of albite can not instantly destroy this relation. It may therefore be stated that in any mixture of orthoclase with a basic plagioclase the separation of basic plagioclase is followed by the separation of leucite, not orthoclase, with subsequent conversion of the leucite to orthoclase if perfect equilibrium is attained. There is no necessity that the addition of albite should ever destroy this relation, but there are grounds for believing that it does and that the fields of leucite and albite never come together. At some intermediate plagioclase, then, this relation disappears, but it probably extends from anorthite through labradorite and perhaps to some andesine. The separation of the feldspathoid, leucite, is therefore to be expected during the fractional crystallization of a magma containing basic plagioclase with some orthoclase (say ordinary basaltic magma) if the pyroxene-olivine reaction has been such that the liquid is not endowed with enough excess silica to neutralize this effect. (See the discussion on pp. 236-7.) If there is this amount of excess silica, which may be a very slight amount, the line indicating the course of crystallization of the mixture, when the plagioclase is pure anorthite, will not coincide with the join anorthite-orthoclase of Fig. 62 (b) but will appear in that figure as a line such as XY, encountering the boundary of the orthoclase field at Y. Leucite is thus missed altogether. A similar condition may readily be pictured for the case where the plagioclase is not pure anorthite but of intermediate composition, remembering that the relation disappears before sodic plagioclase is reached.

One other point may be noted. The boundary curve LM represents the compositions of liquids in equilibrium with both anorthite and orthoclase (solid solution) the relation being a reaction one. Again this relation can not be instantly destroyed by the addition of albite. It must still obtain for basic plagioclase close to anorthite, the indication of natural rocks being, as we have seen, that it extends at least as far as labradorite. The full facts could be accurately expressed only with the aid of a tetrahedron erected on the triangle of Fig. 62 (b), but by plotting the equilibrium as it exists in a liquid with a little excess silica the relations can be shown as they have been in Fig. 59. In that figure the solid solution line and the boundary curve cross each other, the relation being of the reaction type on the anorthite side and eutectic-like on the albite side. The advantage of showing the equilibrium with the aid of this simplified diagram is very apparent.

Recourse must be had, however, to a tetrahedral figure in order to show adequately the factors sometimes involved in the production of feldspathoidal rocks, not only leucite-bearing varieties but nephelite-bearing varities as well. The four faces of the tetrahedron are shown in Fig. 63. The base or central triangle has as its corners $NaAlSiO_4$,

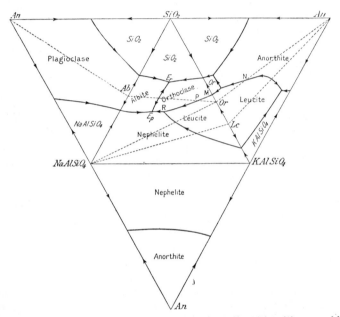

FIG. 63.   Skeleton quaternary diagram, nephelite-kaliophilite-silica-anorthite.

$KAlSiO_4$ and SiO_2. On the potash side there are the two compounds leucite and orthoclase. On the soda side there is but one compound albite. Jadeite, the chemical analogue of leucite, is excluded as being formed under conditions different from those concerned in the ordinary crystallization of the more familiar igneous rocks. When the question arises as to how this central triangle is to be divided into subsidiary triangles certain alternatives are presented. These triangles must, of course, have at their apices those compounds capable of existing together at equilibrium. About the joining of albite with orthoclase and of leucite with $NaAlSiO_4$ there can be no hesitation. The only question, then, is as to how we shall divide the quadrilateral, albite-orthoclase-leucite-NaAlSiO$_4$. The alternatives are the joining of orthoclase with NaAlSiO$_4$ or of leucite with albite. There can be no question as to the choice to be made. Orthoclase and nephelite are well known to occur together under conditions proving their equilibrium with each other, but that albite and leucite can so occur is very doubtful. Leucite occurs, under conditions which suggest equilibrium, only with intermediate to basic plagioclase. Orthoclase and NaAlSiO$_4$ have therefore been joined. This decision necessitates the existence of the two quintuple points, the one having the solid phases, leucite, orthoclase and nephelite, the other orthoclase, nephelite and albite. These are joined by the boundary curve, orthoclase-nephelite but the fields of leucite and albite never come together and there is no boundary curve, leucite-albite. There is no assurance as to the exact position of these points and curves, but there is no great latitude in the choice of their general position. The relation between orthoclase and leucite in the binary system is known from experimental work to be of a reaction nature and the boundary curve orthoclase-leucite must be of that nature for part of its length and probably for all of it. In the figure it is taken to be of this nature throughout. The quintuple point, leucite-orthoclase-nephelite (R) has been taken as lying in the triangle, albite-nephelite-orthoclase and is a reaction point. The quintuple point albite-nephelite-orthoclase (E$_p$) is placed in the same triangle and is therefore a eutectic. The system, albite-orthoclase, is thus a simple binary system for all mixtures extending from albite to the point P. Only the mixtures between P and orthoclase require a ternary figure for their treatment. There is, moreover, an ordinary binary eutectic between albite and orthoclase (solid solutions, of course) which is a point of maximum temperature on the orthoclase-albite boundary curve. The relations shown in the triangles other than the central triangle are comparatively simple and need not be discussed, though certain features will be pointed out in the application of the diagram now to be given.

Such an arrangement of the boundary curves and invariant points is a perfectly reasonable one on the basis of those facts which have been ascertained experimentally and will at the same time offer an explanation for many of the known relations of feldspathoid-bearing rocks. The nature of the reaction occurring at the point R is especially to be noted. At this point the liquid reacts with leucite with formation of orthoclase and nephelite. Since those pseudomorphs after leucite which are known as pseudo-leucite are typically constituted of orthoclase and nephelite we may call this point the pseudo-leucite reaction point. From this point there rises into the tetrahedron a curve which represents the same equilibrium occurring in the quaternary mixtures. It is the pseudo-leucite reaction curve. This curve is of great importance in the explanation of the relations of the feldspathoidal rocks.

The orthoclase-leucite boundary curve is also of great importance and requires some description. It is in all parts a reaction curve and represents reaction of liquid with leucite to form orthoclase. The point P, where this boundary curve crosses the orthoclase-albite conjugation line, is the limit of those mixtures of orthoclase and albite that can be treated as binary and has a special character that can best be brought out by description of the crystallization of certain mixtures. The point P may be called the limit of congruency in the system, orthoclase-albite. Any mixture lying between P and orthoclase begins to crystallize with the separation of leucite. The composition of the liquid therefore changes along a line pointing directly away from leucite until the leucite orthoclase boundary is reached, when reaction occurs between liquid and leucite to form orthoclase. The composition of the liquid moves along the boundary curve towards P and when that point is attained the leucite is completely transformed. The liquid (P) then continues to deposit orthoclase and its behavior is thenceforth binary. It leaves the orthoclase-leucite boundary curve and passes along the orthoclase-albite join to the binary eutectic orthoclase-albite where albite separates and crystallization is complete. If during the separation of leucite there was some relative motion of crystals and liquid then those parts of the liquid impoverished in leucite would have a total composition lying on the silica side of the orthoclase-albite join. In those parts the resorption of leucite would be complete before the point P was attained and then the composition of the liquid would cross the orthoclase field to the orthoclase-albite boundary which would be followed to the ternary eutectic orthoclase-albite-silica. In those parts of the mass enriched in leucite the total composition would lie in the albite-orthoclase-nephelite triangle, or, with still greater enrichment in leucite, in the orthoclase-leucite-nephelite triangle. In both cases there will be some leucite left when the composition of the liquid reaches the point P and reaction of liquid with leucite will continue with change of composition of the liquid towards R. At R liquid and leucite react to produce not only orthoclase but nephelite as well (the pseudo-leucite reaction) and if the total composition is such that it lies in the orthoclase-leucite-nephelite triangle the liquid will be used up by this reaction while some leucite still remains. On the other hand, if the total composition is such that it is represented by a point in the orthoclase-albite-nephelite triangle then leucite is entirely used up while some liquid of composition R remains. This liquid then proceeds to crystallize along RE_p with ordinary separation of both orthoclase and nephelite until the ternary eutectic orthoclase-albite-nephelite is reached, where albite separates in addition and complete solidification occurs.

It is thus plain that the period of leucite crystallization, which is a transient one, the leucite being completely transformed to orthoclase with perfect equilibrium, may nevertheless have an important effect if there is relative movement of liquid and leucite crystals. This action may bring about the survival of leucite in some parts of the mass concerned and in addition the formation of nephelite in some parts of the mass.

If the liquid is separated from crystals at any time after its composition has passed P, that is, when the composition of the liquid lies between P and R, the pseudo-leucite reaction point R will never be attained. The liquid will, in fact, leave the boundary curve and precipitate orthoclase, albite and nephelite without any pseudo-leucite reaction effect. It is important to note the properties of any mass whose total composition is represented by a point lying in the triangle constructed by joining P, R and Or. Such a mass would consist of any liquid lying along the boundary curve MPR with some suspended leucite crystals and it is noteworthy that some of these liquids, those from M to P, would actually contain excess SiO_2. The mass would crystallize in such a way that leucite would disappear entirely when the liquid had a composition between P and R and the final product would consist only of orthoclase, albite and nephelite without pseudo-leucites and without any evidence that any leucite had ever formed. Yet the precipitation of nephelite would have depended upon the former presence of these leucites and upon the existence of a curve with the properties of the curve PR, ending in the pseudo-leucite reaction point R. So in natural magmas the fact that pseudo-leucites are known, necessitates the existence of a curve with the properties of PR and permits the same method of derivation of a nephelite rock with, in many cases, lack of evidence of its derivation being dependent upon the former presence of leucite.

It is to be carefully borne in mind that there rises into the tetrahedron, from the orthoclase-leucite boundary curve in the ternary system, an orthoclase-leucite boundary surface which represents the same reactions as they occur in quaternary liquids. There is thus a curve rising from the point P which represents the intersection of this boundary surface with the plane, orthoclase-albite-anorthite, and all points on this curve have the same general properties as the point P. This curve is the limit of congruency in the ternary system anorthite-albite-orthoclase just as the point P is the limit of congruency in the binary system albite-orthoclase. Any mixture of anorthite, albite and orthoclase from which leucite has crystallized will thus, with perfect equilibrium, accomplish complete resolution of leucite when this curve is attained and will thereafter behave as a simple ternary system, leaving the boundary surface, orthoclase-leucite, and depositing only plagioclase and orthoclase. But at any place where there is local enrichment in leucite crystals the liquid will continue to follow the leucite-orthoclase boundary surface towards the pseudo-leucite reaction curve which rises from R. If there is an adequate amount of leucite this curve is attained and transformation of leucite to orthoclase and nephelite begins. An excess of leucite gives complete solidification with survival of some leucite, and the residual liquid then passes along the orthoclase-nephelite boundary surface to the plagioclase-nephelite-orthoclase boundary curve which rises from E_p. If the amount of leucite had been insufficient to determine the reaching of the pseudo-leucite reaction curve, the liquid would finally precipitate plagioclase, orthoclase and nephelite without intervention of the pseudo-leucite reaction, as in the parallel case discussed fully for ternary liquids. And for these liquids as for the similar ternary liquids there would be no evidence in the final product that leucite had ever formed or that upon its formation depended the development of nephelite.

There is still to be pointed out the fact that this leucite-orthoclase boundary surface is limited upward in the tetrahedron by a plagioclase-leucite-orthoclase boundary curve, part of which lies on the one side and part on the other side of the anorthite-albite-orthoclase plane. On this account it has the same general properties as the orthoclase-leucite boundary curve in the ternary system or the corresponding boundary surface in the quaternary system. Thus in these quaternary liquids from which plagioclase is separating in addition to leucite, accumulation of crystals, among them leucite crystals, will have the same effect as that just described in detail. The liquid will proceed towards the pseudo-leucite reaction curve where crystallization will be complete if there is an excess of leucite but if there is an excess of liquid this will continue to crystallize with separation of plagioclase, orthoclase and nephelite until crystallization is complete.

It is, indeed, in this separation of leucite within the tetrahedron which is accompanied by and has even been preceded by separation of plagioclase that we are particularly interested when we are considering the question of the development of feldspathoidal rocks from ordinary basaltic magma. It is plain that the primary controlling factor, without which no feldspathoidal rocks will develop, is that which determines whether or not the leucite field will be attained, and to this question we will now turn.

If we consider the subsidiary tetrahedron, anorthite-albite-orthoclase-silica, we find that the leucite field encroaches very little upon it. The crystallization of nearly all mixtures in this tetrahedron will never attain the leucite field but will proceed toward their common goal, the ternary eutectic, orthoclase-albite-silica. For certain mixtures lying in or close to the anorthite-albite-orthoclase face (i.e., with little or no free silica) appropriate fractionation may induce the attainment of the leucite field.

The manner in which fractionation may control the attainment or non-attainment of the leucite field may be more easily visualized if we take the triangle albite, anorthite, orthoclase out of the tetrahedron and consider it separately. Fig. 64 shows the triangle which is, of course,

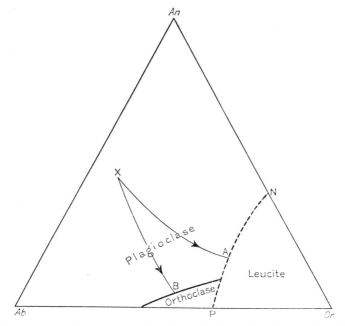

FIG. 64. Ternary diagram, anorthite-albite-orthoclase with the hypothetical extension into it of the leucite field.

the familiar one already considered in other connections. (Figs. 28 and 59.) Here the orientation is different, being made to correspond with its position as it stands in the general tetrahedron constructed with the aid of the triangles of Fig. 63 and the fact of the incongruent melting of orthoclase is not now neglected as it was in Figs. 28 and 59. In the orthoclase corner the field of leucite has been marked off by the broken curve joining the points P and N, which correspond with the points similarly lettered in the tetrahedral diagram (Fig. 63). The behavior of mixtures in the corner marked off by PN can not be adequately represented by the triangular figure but the factors which determine whether leucite shall or shall not form are brought out quite clearly. The crystallization of a liquid such as X may take place with little or with much reaction between the plagioclase crystals and their mother liquor. With little reaction a course such as XB is followed and the plagioclase is joined by orthoclase. The plagioclase-orthoclase boundary is then followed, no leucite separating at any time. The whole course of crystallization is in this case represented by the ternary figure. With greater reaction a more strongly curved course such as XA is followed. When the point A is attained the plagioclase is, in virtue of high reaction, a much more basic plagioclase than when the course XB is followed, and it is then joined by leucite. The well-known fact of the much more frequent association of leucite with calcic plagioclase than with more sodic plagioclase is thus a necessary consequence of the relations here presented. On an earlier page it has been shown that another consequence of a strongly curved course (approaching XA) is the development of orthoclase-mantled plagioclase. We should expect to find a tendency towards association of rocks bearing orthoclase-mantled plagioclase with rocks bearing leucite and the expectation is justified in nature.[1] Italian rocks already mentioned and Javanese rocks described by Iddings and Morley are particularly instructive in this connection.[2] How leucite can ever occur with more sodic plagioclase can be explained only by considering further crystallization, after the appearance of leucite which can not, of course, be represented in the ternary figure. The tetrahedral figure must be returned to and this will be done shortly.

It is especially to be noted that none of the facts just deduced is in any way contradictory to those deduced on earlier pages with the aid of a triangular diagram having the same three compounds as its apices (Figs. 28 and 59), because it was there expressly stated that the incongruent melting of orthoclase was neglected or that the relations were considered as they occurred in a liquid with enough excess SiO_2 to neutralize this effect. A still better form of statement would be that the relations are as they occur in a liquid with enough excess silica to determine that the course of crystallization shall miss the leucite field. The small skeleton tetrahedral figure (Fig. 65) brings out clearly how a little free silica may lead to the avoidance of the leucite field if that result is not attained otherwise, such as in the manner discussed above. The small tetrahedron is, of course, the same as the subsidiary tetrahedron anorthite-albite-orthoclase-silica of the larger tetrahedron (Fig. 63). It is here turned so that its base is the anorthite-albite-orthoclase triangle and is converted into a regular tetrahedron. The figure is

---

1 It has been pointed out in connection with Fig. 59 how a strongly curved course such as XA is locally insured by any process leading to the local enrichment in plagioclase crystals relative to liquid. See foot-note on p. 232.
2 *Jour. Geol.*, 23, 1915, pp. 231-45.

presented merely as an aid to the visualization of the short distance the leucite field extends into it, i.e., the small amount of free silica which will insure avoidance of the leucite field. In the presence of this

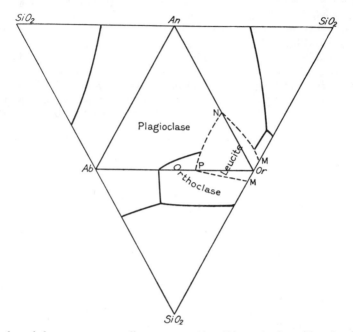

FIG. 65. Skeleton quaternary diagram, anorthite-albite-orthoclase-silica showing limited extension into it of the leucite field.

amount, or a greater amount, of free silica the orthoclase field is always encountered and the leucite field missed whether the plagioclase fractionation is strong or weak.

We must now return to Fig. 63 in order to discuss the behavior consequent upon the separation of leucite in the fractional crystallization of the liquid X of Fig. 64 which has already followed the course XA and has encountered the leucite field. Leucite then begins to separate together with the basic plagioclase with which the liquid A is in equilibrium. Until this point the liquid has always been a pure feldspar liquid like the original liquid X with neither excess nor deficiency of silica. The boundary curve, leucite-anorthite (Fig. 63), is an ordinary subtraction boundary curve representing the separation of leucite and anorthite side by side. The boundary surface, plagioclase-leucite, which passes out into the tetrahedron from this curve must be of a similar nature in part at least. If the plagioclase in equilibrium with the liquid is calcic, and we have seen that only when it is calcic is the leucite field encountered, then leucite and basic plagioclase will separate side by side and the liquid will move along the leucite-plagioclase boundary surface. This surface, like the leucite-anorthite boundary curve, passes into the region of excess silica and there is a definite period during which the leucite is in equilibrium with liquid whose composition lies in the tetrahedron, orthoclase-albite-anorthite-silica. Motion on this surface leads to the encountering of the plagioclase-orthoclase-leucite boundary curve which passes into the tetrahedron from the quintuple point anorthite-orthoclase-leucite. When this boundary curve is encountered leucite begins to react with liquid to form orthoclase. We have already noted that the plagio-clase-orthoclase-leucite boundary curve lies partly on the one side and partly on the other side of the albite-orthoclase-anorthite plane and the consequences of this fact have been pointed out. We may merely recall here that with perfect equilibrium and no segregation of crystals the leucite will be completely resorbed when the limit of congruency is reached and thereafter crystallization takes place entirely in the ternary system anorthite-albite-orthoclase which has already been discussed. The period of existence of leucite crystals is thus transient even when the appropriate fractionation has occurred (in both the olivine-pyroxene pair and the plagioclase reaction series) to ensure that the leucite field is encountered at all. But, given the appropriate conditions, there is a definite period during which leucite crystals occur in a liquid which has an excess of silica. Rapid cooling might give a rock having leucite crystals in a groundmass containing excess silica and it is now well established that such rocks exist.

Cross has described a rock from Wyoming which contains leucite. Analysis of the rock shows that the silica present is entirely adequate to have formed orthoclase with the potash and alumina present. The glassy groundmass[1] must therefore contain excess silica. This condition is entirely in accord with experimental results.

A shoshonite from Yellowstone Park, described by Iddings, has more than enough silica to convert all the bases into their most siliceous compounds (3 per cent normative quartz), yet in this there occurs an "impure leucite."[2] Likewise a banakite from the same locality has, according to one analysis, enough silica to form feldspar molecules, according to another analysis, only a slight deficiency, yet it contains a considerable quantity of leucite.[3]

1 W. Cross, *Amer. Jour. Sci.*, 4, 1897, pp. 122, 132.
2 Iddings, *Jour. Geol.*, 3, 1895, p. 945.
3 Iddings, *loc. cit.*

There is, in addition, the leucite granite porphyry from Brazil which contains phenocrysts of leucite (now pseudo-leucite) in a groundmass of orthoclase and quartz. To this we shall revert when discussing the formation of pseudo-leucite in rocks.

So much for the evidence of rocks revealing the existence of a period of separation of leucite, crystals of which may be preserved in a groundmass with excess silica. This is the period when the liquid is following that part of the plagioclase-leucite-orthoclase boundary curve which lies on the silica side of the orthoclase-albite-anorthite plane. Indeed the liquid, as it follows the boundary curve, will never cross this plane unless there has been local enrichment of certain parts of the mass in leucite crystals. But if there is such local enrichment, whether from gravitative or deformative influences, in certain parts there will be leucite crystals left when the limit of congruency is reached. The liquid in those parts will then proceed along this curve to the pseudo-leucite reaction curve or on to the plagioclase-orthoclase-nephelite boundary, depending on the proportions of the leucite crystals in a manner already discussed.

All of these are possible courses in the fractional crystallization of the original feldspar liquid X of Figs. 64, 28 or 59. The curves of feldspar crystallization which finally encounter the leucite field are merely the extreme members of the family of curves of alternative liquid lines of descent that diverge from the point X. All of these are what would be termed sub-alkaline lines of descent but in the case of the extreme members which encounter the leucite field, accumulation of leucite crystals may cause a switching-over of the liquid into an alkaline line of descent in the parts of the mass where this accumulation has occurred. We may consider this feldspar associated with pyroxene to make a basaltic liquid and that these same types of fractionation of the feldspar can go hand in hand with the tendency towards a general impoverishment of the liquid in pyroxene as the more alkalic mixtures are approached. We then have a picture of how, in the differentiation of such a mass, certain parts of it may proceed ever towards the phonolitic eutectic E_p and certain other parts towards the rhyolitic eutectic E_r.

THE PSEUDO-LEUCITE REACTION AND THE DEVELOPMENT OF SOME NEPHELITIC ROCKS

Pseudomorphs after leucite that are typically constituted of orthoclase and nephelite (or alteration products thereof) are a well-recognized constituent of igneous rocks. In fact, in coarsely crystalline rocks true leucite is not often found, its place being ordinarily taken by the above-mentioned pseudomorphs which have been called pseudo-leucite. A few coarse-grained rocks with leucite are known, however, so there is nothing impossible about its occurrence in that manner but it is in aphanitic or quenched rocks that true leucite is ordinarily found. In these rocks it may occur as quite large phenocrysts or again as a constituent of the groundmass.

It has been suggested by Knight[1] that pseudo-leucites are the result of the breakdown of a mineral analogous to leucite but containing a considerable amount of soda, a soda-leucite. The slow cooling of coarse-granular rocks would favor the breakdown whereas quick cooling would prevent it and the common contrasts between aphanites and phanerites would therein be accounted for. However, it would appear that, if the above explanation were correct, some of the leucites, indeed most of the leucites of aphanites should be this soda leucite. Analyses of leucites have failed to reveal any such composition. They are potash leucites with quite moderate amounts of soda. On this account it is preferable to believe that pseudo-leucites are the result of what we have called the pseudo-leucite reaction, a reaction of liquid with leucite to give ortho-clase and nephelite. The usual contrast between aphanites and phanerites would then be accounted for by the failure of this reaction with rapid cooling. Inadequate time need not, however, be the only cause of failure of reaction. Absence of liquid or of liquid of the right kind would have a like effect so that coarse-grained, true leucite rocks remain a possibility.

We may consider the pseudo-leucite reaction more closely by comparing it with, say, the pyroxene-hornblende reaction. They differ, of course, in the fact that the product of reaction is a single mineral in the one case and an aggregate of two minerals in the other. In other respects there is much similarity. Sometimes pyroxene is transformed to hornblende which retains the form of the original mineral, but this change has involved interchange of substance with the liquid. Again the hornblende may form as rims about pyroxene and there may be continued outgrowth of the hornblende material. And with little doubt, under conditions of very slow cooling, pyroxene passes into solution and hornblende is precipitated about entirely separate nuclei so that evidence of the reaction relation may be suppressed. In the case of the pseudo-leucite reaction we probably have similar gradation of development. Under certain conditions the leucite is transformed into an aggregate of orthoclase and nephelite which retains the form of the original leucite. Under other conditions there is further outgrowth of the ortho-clase and nephelite beyond the limits of the original leucite, giving indefinite rounded patches of pseudo-leucite material. In yet other conditions there is probably solution of leucite and precipitation of nephelite and orthoclase about quite independent nuclei. The evidence that some of the nephelite and orthoclase were formed by reaction with leucite

1 C. W. Knight, *Amer. Jour. Sci.*, 21, 1906, p. 293.

would in the last case be entirely suppressed. It is quite possible then that in a great many cases the evidence might be lacking that the formation of nephelite and of nephelite-bearing liquids depended upon the preliminary formation of leucite which was followed by the pseudo-leucite reaction. The finding of evidence that the formation of nephelite-bearing liquids depended on the separation of leucite would be even less likely in those cases where a liquid which had attained a subsilicic composition in virtue of reaction with leucite crystals is then separated from these crystals, say, by squeezing-out of the liquid. Such a liquid would form only nephelite, orthoclase and albite without giving any suggestion of ever having been associated with leucite.

However, in some cases individual rocks or again rock associations have retained the evidence that the formation of nephelite and of nephelite-bearing liquids depends upon the leucite relation we have just attempted to discuss in all its complexity. There may be evidence, in a single rock series, of the control of leucite-formation over the development of free quartz in those rocks of the series from which the leucite has been removed and concomitant development of nephelite in those rocks of the series to which the leucite has been added.

Near Loch Borolan in Scotland there is a differentiated laccolith which is described by Shand as being made up of the following in stratiform arrangement, as stated below, and shown in Fig. 66 :[1]

  I. Quartz syenites (nordmarkite with 12 per cent quartz and other more quartzose types).
 II. Transition zone of quartz-free syenites.
III. Feldspathoid-bearing syenites.
IV. Probable ultrabasic zone (noted in one locality).

Some of the syenites of III contain rounded masses interpreted as pseudo-leucite. The interpretation has been questioned but the descriptions seem to warrant it and to render any other rather doubtful. If it can be imagined that the laccolithic chamber was filled with a magma very rich in alkaline feldspars, and with not more than a moderate excess of free silica, this magma might, as our results show, begin to crystallize with separation of leucite. The actual proportions of the various rock types in the composite mass (see Fig. 66) show that the general liquid would be very rich in feldspar, principally orthoclase, and excessively poor in the molecules that go to make up the heavy minerals. The density of such a liquid as a glass at ordinary temperatures would probably be not far from that of rhyolitic obsidian (2.37)[2] The density of leucite at ordinary temperatures is 2.46. At higher

1 S. J. Shand, *Trans. Edin. Geol. Soc.*, 9, pt. III, 1909, p. 202, and pt. V, 1910, p. 376. See also Horne and Teall, *Trans. Roy. Soc. Edin.*, 37, pt. 1, 1892, p. 163.
2 H. S. Washington, "Rhyolites of Lipari," *Amer. Jour. Sci.*, 50, 1920, p. 449.

temperatures, then, there would probably be a definite though small margin of density in favor of leucite crystals, especially since the liquid would, no doubt, contain a fair amount of volatile substances, whereas

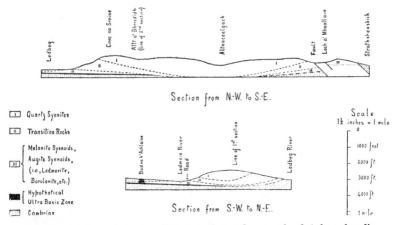

Section from N.-W. to S.-E.

Quartz Syenites

Transition Rocks

Melanite Syenoids,
Augite Syenoids,
(i.e., Ledmorite,
Borolanite, etc.)

Hypothetical
Ultra Basic Zone

Cambrian

Section from S.-W. to N.-E.

Scale
1½ inches = 1 mile

1000 feet
2000 ft
3000 ft
4000 ft

1 mile

FIG. 66.   Sections of the laccolith at Loch Borolan, Scotland (after Shand).

the obsidian noted above is nearly free from these. Presumably the leucite crystals could settle under the influence of gravity; indeed, the general arrangement of zones can scarcely leave room for doubt that they did and gave a lower zone (III) much enriched in that mineral. When the time of reaction of the leucite with liquid arrived there would be an amount of leucite above that requisite for the reaction and some would be left in excess. The excess leucite was, during the further crystallization of the mass, transformed into pseudo-leucite (orthoclase + nephelite) as a result of the pseudo-leucite reaction. On the other hand, in the zone from which the leucites were entirely removed (I) there were no crystals to react with the liquid, and it crystallized appropriately with an excess of free silica.

The Ilimausak batholith in Greenland is stratified in a manner closely related to that shown by the Loch Borolan mass. At the top is a quartzose phase, arfvedsonite granite, which passes downward into quartz-free syenite and finally into sodalite and nephelite syenites of great variety.[1] In one of these, the so-called lujavrites, which are the lowest exposed rocks of the mass, there are large crystals of analcite which Ussing presents reasons for believing were formerly leucite.[2] The nature of the

[1] N. V. Ussing, "Geology of the Country around Julianehaab, Greenland," *Med. om. Grönland,* vol. 28, p. 322, Fig. 29, 1911.
[2] *op. cit.,* pp. 164, 165.

stratification and the existence of these pseudo-leucites (?) is so strikingly similar to the conditions at Loch Borolan that one must consider the possibility that the early separation of leucite has been a factor controlling the differentiation of the mass and the development of nephelite in it.

The igneous massif of Bezavona in Madagascar, as described by Lacroix, consists of quartz syenites, nephelite syenites, monzonites, and other types, including various dike and flow rocks. In the coarse granular rocks there is apparently nothing suggesting the formation of leucite, but in quickly chilled facies, which may be regarded as quenched, leucite appears. Thus there are micro-syenites with leucite and also leucite phonolites.[1] These facts again suggest the possibility that the formation of leucite may have been an intermediate step in the genesis of the nephelite rocks.

Still less definite as evidence, and yet worthy of consideration, are certain structures observed in nephelite rocks that suggest the former presence of leucite. In some members of the Ice River complex of British Columbia there are spots of a "finger-print-like" intergrowth[2] of orthoclase and nephelite that is practically identical with the "dactylotype" intergrowth of these minerals when they form pseudo-leucites. Structures that are perhaps of similar origin are described by Lacroix from the nephelite syenites of the Los Archipelago[3] and by Brouwer in rocks from the Transvaal.[4]

An intimate relation between leucite and nephelite rocks is observed in many fields. A striking example is shown by the Magnet Cove complex. In particular it is noteworthy that a leucite porphyry and a foyaite, associated there, have nearly identical chemical composition, at least in some specimens. Washington has called attention, also, to a similar relationship of a leucite-rich rock (leucite phonolite) of the Sabatinian district, Italy, with a nephelite syenite of Beemerville, New Jersey.[5]

In volcanic fields the frequent intimate association of trachyte, leucite-trachyte and phonolite is suggestive when it is recalled that upon the completion of the reaction whereby leucite is transformed into nephelite and orthoclase any liquid that may be left proceeds towards the phonolite eutectic E_p of Fig. 63.

---

1 A. Lacroix, "Les roches alcalines d'Ampasindava," *Nouv. arch. du muséum*, Paris, Série 4, 5, 197 and 207, 1903.

2 J. A. Allan, "Geology of the Field Map-Area, B.C. and Alberta," *Geol. Surv. Can.*, Mem. 55, pp. 133, 285, 1914, Plate XVIIB.

3 A. Lacroix, "Les syénites néphéliniques de l'archipel de Los," *Nouv. arch. du muséum*, Série 5, 3, 1911, p. 53.

4 H. A. Brouwer, *Transvaal Nephelien-Syenieten*, p. 40, Pl. I, Fig. 1.

5 H. S. Washington, "Igneous Complex of Magnet Cove, Arkansas," *Bull. Geol. Soc. Amer.*, 11, 1900, p. 399; and "Roman Comagmatic Region," *Carnegie Inst. Wash.*, Pub. No. 57, 1906, p. 47.

CHAPTER XIII

LAMPROPHYRES AND RELATED ROCKS

GENERAL CHARACTERS

A SPECIAL study of lamprophyres has been made by Niggli and Beger.[1] They conclude that these rocks are to be accounted for by the local accumulation of early crystals which have then remelted or redissolved and given a liquid of lamprophyric composition. Perhaps lamprophyres are too broad and ill-defined a group to enable one to make a general statement concerning them to which no exception might be taken. But after all, lamprophyres are characteristically porphyritic rocks, the porphyritic elements highly femic (olivine, hornblende, mica) and the groundmass notably alkalic. In default of a porphyritic structure they have "a rather exceptional structure due to a strong tendency to idiomorphism of all the constituent minerals."[2] The mere existence of a porphyritic structure may mean nothing in itself, but if a group of rocks is characteristically porphyritic or has the special panidiomorphic structure[3] one may reasonably entertain some doubts as to the existence of liquids corresponding with the rocks in bulk composition. It is impossible to be sure from existing descriptions of lamprophyric dikes whether they ever have tachylitic, spherulitic or uniformly aphanitic selvages of the same composition as the main mass. If they have not, their characters would appear to be unfavorable to the hypothesis of remelting, at least in its simple form. We shall not further pursue the problem of lamprophyres in general but attention will be directed to the olivine-bearing varieties.

OLIVINE-BEARING LAMPROPHYRIC TYPES

Olivine is often an important constituent of lamprophyres and in certain important representatives, mica peridotites, it is usually the most abundant constituent. These peridotites occur in dikes, often quite small

1 *Gesteins und Mineralprovinzen*, I, 1923, pp. 217-574, especially 571-4.
2 Harker, *Petrology for Students*, 1919, p. 134.
3 On p. 153 a suggestion is made as to the origin of a structure in which the bulk of the rock is made up of nearly idiomorphic crystals.

dikes, but are perhaps represented in greater volume in the form of volcanic necks or pipes. Very intimately related to the mica peridotites are the alnoites. They contain in addition to olivine, augite, and mica, characteristically melilite, often with alkalic minerals of different kinds, which have been the occasion of the introduction of a great number of rock names. These rocks are so intimately interrelated that there can be little doubt that light on the origin of any particular member has an important bearing on the origin of the group. The general character of the group is a richness in ferromagnesian constituents combined with a richness in alkalis, which combination of characters is, indeed, rather distinctive of lamprophyres in general. The question with which we are now concerned is whether all of these constituents have been liquid together or whether the femic constituents were in large part present as crystals before intrusion, the liquid which permitted intrusion thus being a notably alkaline and presumably highly mobile menstruum.

On this question some recent studies of alnoite and related rocks may throw some light. Rocks of this character from Isle Cadieux, Quebec, have constituent minerals plainly divisible into two genetic classes.[1] In the first class are augite and olivine which originally made up nearly the whole of the rock and were in large crystals from $\frac{1}{2}$ to 1 cm. in diameter. In the second class are principally biotite, monticellite, melilite and perovskite which now make up more than one-half the rock, the liquid which deposited them attacking and replacing large amounts of olivine and augite. An alnoite from Montana shows very similar relations, as given by Ross.[2] In this rock the original minerals are olivine and nephelite, with possible augite, and these have been attacked by liquid which partly replaced them and deposited principally melilite, hauynite and phlogopite. The action in the case of the Isle Cadieux rock has been attributed to an alkalic liquid which attacked augite and olivine and precipitated the other minerals as a result of this reaction, both the solid minerals and the liquid contributing to the composition of the new minerals. The observations of Ross have led him to a quite similar general conclusion. There is thus no evidence from either rock that there was ever a liquid corresponding approximately in bulk composition with the final product; indeed, the evidence is such as to cast much doubt on the existence of such a liquid. The original olivine and augite of the Isle Cadieux rock may have been a crystal accumulation from a liquid of more complex composition. Even the liquid which permeated and replaced this largely crystalline mass has not left a collection of minerals of the same composition as itself but has left only minerals produced by interaction of liquid and crystals.

1 N. L. Bowen, *Amer. Jour. Sci.*, 3, 1922, pp. 1-33.
2 C. S. Ross, *Amer. Jour. Sci.*, 11, 1926, pp. 218-27.

That an alkalic liquid, especially a nephelite-rich liquid, could produce the effects observed in these rocks is rendered likely by certain experimental observations now to be noted.

The system $CaO-MgO-SiO_2$ has been investigated by Ferguson and Merwin and their results are expressed in the equilibrium diagram, Fig. 55. A liquid made up of 50 per cent forsterite and 50 per cent akermanite begins to crystallize with deposition of the olivine, forsterite, an end member of the chrysolites; when the temperature falls to 1450° monticellite begins to crystallize; and at 1436° akermanite, an end member of the melilites, begins to crystallize and monticellite redissolves until finally monticellite is entirely replaced by akermanite and the whole mass has completely solidified. In this case, then, we have the succession, forsterite (chrysolite)-monticellite-akermanite (melilite), the akermanite having a replacing relation to monticellite. This liquid and the adjacent liquids which show the same effects are, of course, themselves very rich in lime and magnesia, but when other components are added the same relations must persist at first, though they will be modified and may finally disappear as more and more of other components are added. The effect of the addition of nephelite to some of these mixtures has been investigated. The complete quantitative solution of the problem can not be expressed in terms of less than five components, and this is, of course, not possible without a vast amount of preliminary work on the simpler systems. However, an approximate statement is possible; in other words, a statement of the kinds of mineral phases present at various temperatures may be made without giving the precise composition of the phases of variable composition or the exact relative proportions of the phases.

The results of the investigation of equilibrium in mixtures of diopside, $CaMgSi_2O_6$, and nephelite, $NaAlSiO_4$ are given in Fig. 67, which, it must be realized, is not a two-component diagram, though having the same general form. It gives the names of the phases present at various temperatures in mixtures whose total composition can be expressed in varying proportions of diopside and nephelite, but the composition of some of the individual phases can not be so expressed.

Since both components melt congruently there is, at either end, a portion which behaves at higher temperatures as an ordinary two-component mix but towards the middle the relations are much more complicated. In particular it is to be noted that the early crystals are olivine and melilite, that is, nephelite reacts with diopside in such a way as to produce these very basic molecules which are then precipitated. The separation of these necessarily renders the liquid with which they are in equilibrium correspondingly more siliceous. Another important general feature is the very low temperature at which there is still some liquid in the intermediate mixtures. Thus at temperatures in the neigh-

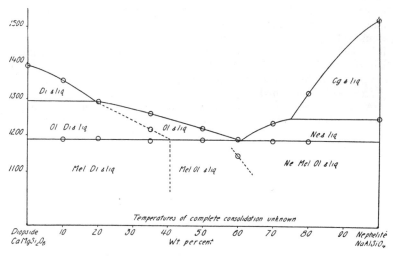

FIG. 67. Pseudo-binary diagram of equilibrium in mixtures of nephelite and diopside.

borhood of 1000° some liquid remains and in many mixtures it is impossible to tell at what temperature complete solidification occurs, on account of the slowness with which crystallization proceeds at such temperatures.

The behavior of any particular mixture is given by the diagram, but will be put in words for certain of these that are of special significance. A mixture of 30 per cent nephelite, 70 per cent diopside begins to crystallize at 1270° with separation of forsterite. This continues to separate until a temperature of 1240° is reached, when diopside begins to separate and forsterite to redissolve. The re-solution of forsterite continues until the temperature has fallen to about 1180° and at about the same temperature melilite begins to separate. At temperatures below this and as far as 1100° the mass consists of diopside, melilite and liquid. Below 1100° the crystals formed are so small and the attainment of equilibrium so slow that further changes can not be definitely described, the temperature of final consolidation being also unknown. No nephelite is formed at any temperature where the identification of phases is certain; indeed the reaction which forms melilite presumably destroys the nephelite and part of the diopside.

A mixture of 50 per cent nephelite and 50 per cent diopside begins to crystallize at 1220° with separation of forsterite which continues down to about 1180°. At this temperature melilite begins to crystallize, and below this temperature the mass consists of melilite, forsterite and liquid down to temperatures where the crystals formed are no longer identifiable. Here again the temperature of final consolidation is unknown. Neither nephelite nor diopside appears, at least not at any temperature where the identification of phases is possible. They mutually destroy each other in the reaction which produces melilite.

A mixture of 70 per cent nephelite and 30 per cent diopside begins to crystallize at 1235° with separation of nephelite. At about 1180° the nephelite is joined by olivine and melilite, and down to temperatures where the crystals formed are no longer identifiable the mass consists of nephelite, melilite, forsterite and liquid. No diopside appears, it being used up by reaction with nephelite to form melilite.

NATURE OF THE MELILITES

The melilites formed from the mixtures studied do not correspond with the end member, akermanite, whose equilibrium relations are given in Fig. 55, but are strictly analogous to natural igneous melilites. Some of the crystals, particularly those formed at higher temperatures, are positive like akermanite but of much weaker birefringence. With increasing richness in nephelite in the total mixture the melilites separated become isotropic and finally pass over into negative melilites. The decrease of birefringence is accompanied by a slight decrease of refringence which is about 1.630 in the sensibly isotropic member. The series is quite different, therefore, from the akermanite-gehlenite series of Ferguson and Buddington in which the refringence increases as the birefringence decreases on passing away from akermanite. The present artificial melilites are, however, readily explained and correlated with natural melilites in the light of the recent work of Buddington.[1] The relation to Buddington's mixtures is brought out clearly by adding nephelite directly to akermanite and crystallizing the fused mixture at a low temperature. When 10 per cent nephelite is added to akermanite the product is nearly homogeneous melilite of somewhat lower refringence and distinctly lower birefringence than akermanite. There is, however, a very little forsterite in excess and also minute stringers of glass. As more nephelite is added the forsterite and glass increase somewhat and the birefringence of the melilite further decreases. When

1 These melilites fall under the subdivision humboldtilite as used by Buddington, *Amer. Jour. Sci.*, 3, 1922, p. 75.

25 per cent nephelite is added the melilite is sensibly isotropic and its index about 1.630 and further addition of nephelite produces negative melilite. The series of melilites is evidently strictly analogous to the akermanite: sarcolite: soda-sarcolite series of Buddington (p. 56). An equation can be written showing the formation of these molecules by reaction of akermanite and nephelite (the akermanite being in excess) with the formation of some forsterite and a more siliceous alkalic liquid that remains as glass. The equation is as follows:

$$\underset{\text{nephelite}}{15(NaAlSiO_4)} + \underset{\text{akermanite}}{6(2CaO . MgO . 2SiO_2)} = \underset{\text{sarcolite}}{4(3CaO . Al_2O_3 . 3SiO_2)} +$$
$$\underset{\text{soda-sarcolite}}{2(3Na_2O . Al_2O_3 . 3SiO_2)} + \underset{\text{olivine}}{3Mg_2SiO_4} + 3NaAlSi_2O_6$$

The melilites formed in the nephelite-diopside mixtures are exactly like those formed in the nephelite-akermanite mixtures except that they do not show as great range of composition. As noted, they may be either positive or negative but are never far removed from the isotropic member; in other words they never approach the pure akermanite end member but appear to center around the low melting mixture which Buddington has decided is the case in natural melilite (p. 87). The reactions taking place when nephelite is added to diopside may be pictured by imagining that the first step is the production of forsterite and akermanite from diopside by desilication as follows:

$$4CaMgSi_2O_6 + 3NaAlSiO_4 = 2(2CaO . MgO . 2SiO_2) + Mg_2SiO_4 + 3NaAlSi_2O_6$$

The akermanite then further reacts with nephelite after the manner indicated in the first equation to produce the other molecules of the typical melilites. The forsterite and the melilites resulting are the first crystals precipitated from the intermediate mixtures.

The correspondence with the melilites of the rocks described is quite marked for, in the portion of the rock where reaction with the alkalic liquid has not proceeded far, the melilite is positive (akermanite-rich) whereas in the biotite-melilite seams, where reaction has gone farther, the melilites have isotropic cores and negative outer portions. A zoning of this particular nature has been noted in melilite from San Venanzo.[1] In the natural rock ordinary olivine was not one of the products of the reaction for the reason that the reacting liquid differed from the experimental liquid, notably in the potash content, with the result that mica was formed instead.

SPACE RELATIONS OF THE EQUILIBRIUM FIELDS

The separation of melilite and olivine from liquids whose total composition can be expressed as a mixture of nephelite and diopside has already been discussed in some detail and pictured in Fig. 67. As an

1 Rosenbusch, *Mikroskopische Physiographie*, I, 2, 1905, p. 71.

aid to the understanding of equilibrium relations in the rock we have described, it is perhaps desirable to present the results in a somewhat different manner, that is, in terms of a composition tetrahedron having as its base the CaO-MgO-SiO$_2$ triangle and nephelite at the other corner. The compounds and their equilibria with liquids for all compositions lying on the base have been studied by Ferguson and Merwin, and our present purpose is to describe how their fields are affected as we pass out into the tetrahedron, that is, as nephelite is added. It is apparent from the equations that have been given above that even four components are not sufficient to express accurately the composition of all the phases present but we could, as in the pseudo-binary diagram (Fig. 67), mark the tetrahedron out into fields showing the nature of the solid phase in equilibrium with liquid, the *total composition of the mixture* being given by a point in the tetrahedron.

No attempt has been made to map out the fields throughout the whole tetrahedron, which would be a labor of many years. Attention has been concentrated upon the position of the melilite, olivine and pyroxene fields as having particular interest in the present connection.

The fields that have been delimited for the MgO-CaO-SiO$_2$ mixtures constituting the base will pass upward into the tetrahedron, the boundary curves becoming boundary surfaces. The position of these surfaces as they are encountered by the lines indicating the changing composition of any crystallizing liquid mixture may be shown approximately by projecting upon the base the surfaces where thus encountered. The position of the point on the nephelite-diopside join (A) serves as a reference or indicating point.

In the CaO-MgO-SiO$_2$ system itself as shown in Fig. 55 all of the liquids in equilibrium with monticellite are more siliceous than monticellite itself. On the other hand diopside lies in the pyroxene field, that is, the liquids with which it is in equilibrium surround it and are both more siliceous and less siliceous. When nephelite is added the pyroxene field is pulled over toward the silica corner and with about 19 per cent nephelite the boundary surface pyroxene-forsterite passes through the indicating point (A) which thereafter lies in the forsterite field. At 20 per cent nephelite the section through the tetrahedron shows the projected fields in the relative positions of Fig. 68, the arrows indicating falling temperature. This shows that forsterite crystallizes first, pyroxene second, and then melilite from a mixture given by the indicating point.

With 40 per cent nephelite the relative positions of the projected fields are as shown in Fig. 69. Thus forsterite crystallizes first and is joined simultaneously by both pyroxene and melilite from a mixture of diopside and nephelite as given by the indicating point. With 55 per cent nephelite the fields as projected are shown in Fig. 70.

The manner in which the fields of forsterite and melilite are drawn

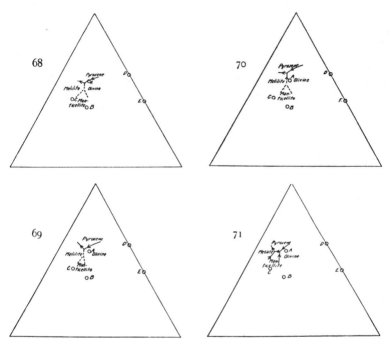

FIGS. 68, 69, 70 and 71. Projections of fields of stability in nephelite-diopside mixtures upon the CaO-MgO-SiO₂ plane. Same orientation as Fig. 55. Projected point for mixtures of nephelite with diopside. A; with monticellite, B; with akermanite, C; with enstatite, D; with forsterite, E. Fig. 68, 20 per cent nephelite; Fig. 69 (lower left), 40 per cent; Fig. 70 (upper right), 55 per cent; Fig. 71, deduced relations in magma.

farther over toward the silica corner as more and more nephelite is added is well shown by this series of figures (55, 68, 69 and 70).[1] A simple, if only qualitative explanation of this is that equilibrium in the liquid is such that the nephelite in the liquid takes silica to itself and leaves relatively little for the lime and magnesia, thereby increasing the concentration in the liquid of the low-silica, lime and magnesia compounds and causing their precipitation at an early stage. Indeed the

[1] In Fig. 55 the point corresponding to (A) is the diopside point itself.

series of figures is but a graphical and qualitative expression of the equations given on page 263. This affords an example of the manner in which both affinity and solubility combine to determine the solid phases that shall separate from a liquid, the affinity of nephelite for silica being unquestionably an important factor in the present instance.

On account of the impossibility of distinguishing between forsterite and monticellite in the small crystals obtained at the relatively low temperatures, it is impossible to speak with certainty regarding the behavior of the monticellite field. It is apparently pulled over in the same manner as the other fields, indeed even in the ternary system itself (Fig. 55) it is displaced in that direction, so the addition of nephelite presumably emphasizes this natural tendency. Monticellite may therefore separate from liquids not particularly rich in lime and magnesia, providing they are alkalic. The manner in which the monticellite field separates a part of the forsterite field from part of the akermanite field, with a consequent reaction relation, has already been pointed out. This separation is apparently more marked in the alkalic liquids and in the case of the natural rock described it appears that, for the system it represents, the fields must be considered to have the relative positions shown in Fig. 71, which gives their projection upon the $CaO-MgO-SiO_2$ plane in a manner analogous to that of Figs. 68, 69 and 70. Such a figure will, at any rate, afford an explanation of the paragenesis of minerals in the rock and is but a further modification of Figs. 68, 69 and 70 in the direction in which modification has been demonstrated in the presence of alkali-rich liquid. The figure (71) shows that the crystallization of a liquid given by the indicating point may begin with the formation of olivine, which is followed by pyroxene, then monticellite, then melilite, the monticellite and melilite having a reaction relation to pyroxene and olivine; that is, the latter are resorbed and replaced by monticellite and this in turn by melilite. This is the relation that the actual minerals of the rock show, with biotite as an additional replacing mineral, the investigation of which would require the addition of several other components. The occurrence of melilite as reaction rims about augite or olivine has been noted in other rocks of this area[1] and indeed a great many of the occurrences of melilite suggest its origin by reaction of pyroxene with alkalic liquid.

FORMATION OF LIME-RICH MINERALS IN ALKALIC ROCKS

The series of figures (55, 68, 69 and 70) serves to emphasize the fact that in the presence of considerable nephelite the very basic silicates forsterite, melilite and possibly monticellite separate from liquids well over on the silica side of the nephelite-diopside join. The presence in a

1 Robert Harvie, *Trans. Roy. Soc. Can.*, 3rd Ser., Vol. 3, sec. IV, 1910, p. 261.

rock of the lime-rich mineral melilite need not, therefore, be considered as indicating the existence of a magma particularly rich in lime but rather of an alkali-rich magma from which melilite-bearing facies might be formed by the reaction effects noted. Moreover, the intimate association of melilite and nephelite rocks in nature is to be regarded as further evidence of the importance of crystallization-differentiation, melilite being a normal cafemic constituent of nephelite-bearing rocks.

Even such minerals as garnet and vesuvianite, closely related as they are to melilite,[1] may, perhaps, be capable of formation as normal constituents from alkalic magmas and are not necessarily an indication of contamination of the magma by lime-rich rocks such as limestone. These very basic silicates may conveniently be regarded as the normal basic minerals of rocks (mainly metasilicates) desilicated in the presence of alkalic molecules such as nephelite. It will be noted that this action, which has been demonstrated for the artificial mixtures, is exactly the opposite of that which Daly supposes to take place in magmas, viz., the desilication of feldspathic molecules by lime, with formation of feldspathoids. This fact should not, however, in itself, be urged as necessarily disproving Daly's assumption, for it is conceivable that we are dealing with a condition of equilibrium whose position may be varied under different conditions.

The information obtained from the investigated liquids shows that the mineral association of the Isle Cadieux rocks can originate through the action of alkalic liquid on augite and olivine. There is no necessity for the existence of a lime-rich liquid and in particular no evidence of the existence of a liquid of the total composition of the rock. Concerning the related Montana rock Ross says that it "is not associated with limestone beds; it contains no calcic inclusions, and is low in calcium since it contains less than 9 per cent CaO. It is evident that an alnoitic rock can develop without the assimilation of limestone and in the absence of an excessively large lime content."

On the other hand Scheumann,[2] studying very similar rocks from Bohemia, and Stansfield,[3] studying rocks near Isle Cadieux, have concluded that assimilation of limestone has occurred and has produced lime-rich liquids. Scheumann believes also in the re-solution of early sunken crystals to form liquid strongly enriched in these constituents. From this action he excludes olivine, believing that the re-solution of constituents occurs only "bis auf dem Olivinbestandteil" in his rocks. Even Scheumann, then, does not advocate the existence of liquids of the total composition of his lamprophyric rocks though he goes much farther in that direction than the facts seem to warrant in the other cases that

1 The sarcolite molecule of melilite is also a garnet molecule.
2 K. H. Scheumann, *Centralbl. Min.*, 1922, pp. 495-545.
3 J. Stansfield, *Geol. Mag.*, 40, 1923, p. 433.

have been cited. It is not impossible that each conclusion is correct for the individual example. As Fig. 67 shows, complex liquids rich in lime and magnesia, which are at the same time rich in alkaline compounds of the right kind, can exist at comparatively low temperatures so that we have not the same barrier to the acceptance of such liquids as we have in the case of liquids rich in lime and magnesia alone. The comparatively simple liquid made up of 60 per cent $NaAlSiO_4$ (nephelite) and 40 per cent $CaMgSi_2O_6$ (diopside) begins to crystallize only at 1190° and is partly liquid at temperatures well below this. This liquid has the oxide composition SiO_2 47.6, Al_2O_3 21.5, MgO 7.4, CaO 10.4 and Na_2O 13.1. Related liquids of more complex character (i.e., containing additional components such as iron and potash compounds) could undoubtedly exist at still lower temperatures especially when some volatile constituents are present. At the same time it should be realized that increasing the content of MgO or CaO of this liquid must raise the temperature at which the mass will be entirely liquid, for this liquid (60 Ne 40 Di) is saturated with olivine and melilite. Melilite is thus not to be expected as inherently present, in the stoichiometric sense, in the low melting liquid though it may form from such a liquid just as olivine may form from many liquids (including these liquids) whose composition is such that, stoichiometrically, they contain no olivine. We have thus additional reasons for believing that melilite, occurring in the rocks discussed, is for the most part formed either by reaction between alkaline liquid and already crystallized pyroxene or is precipitated from a liquid rich in feldspathoid molecules and pyroxene but at the same time not rich enough in lime to contain melilite in the stoichiometric sense. Its presence does not, therefore, indicate the existence of a liquid very rich in lime though a liquid comparable with the synthetic liquid (60 Ne 40 Di) in this respect is presumably formed in some cases. This matter is treated further in connection with the development of certain liquids by the fractional resorption of hornblende.

CHAPTER XIV

THE FRACTIONAL RESORPTION OF COMPLEX MINERALS AND THE FORMATION OF STRONGLY FEMIC ALKALINE ROCKS

ACCORDING to the conclusions reached in the foregoing discussion, the olivine-bearing lamprophyres have not been formed from liquids of their own bulk composition. Their constituents may, in general, be separated into a group of notably femic minerals which were acted upon by an alkalic liquid with resultant deposition of another group of minerals, and the only liquid immediately concerned in their formation was this alkalic liquid. This association of accumulated femic crystals with an alkalic liquid is not, however, to be regarded as fortuitous. It is believed to be due to the fractional resorption of complex minerals, notably hornblende, which will now be discussed.

In our study of rocks formed by the accumulation of crystals of olivine and of basic plagioclase we found reason to believe that there was no significant re-solution of these minerals. In stating this conclusion we had no desire to set up such behavior as a standard to be applied to any mineral which accumulates in a similar manner. We merely examined rocks enriched in one or both of these two minerals and found in their characters the evidence of lack of re-solution. Olivine and basic plagioclase are minerals of comparatively simple constitution and it may be safely said that, since they are early members of the reaction series, each crystal of either that is added to any layer of magma raises the crystallization temperature of that layer, not only the temperature of beginning of crystallization but also the (equilibrium) temperature of completion of crystallization. It is this fact that acts as a bar to the re-solution of crystals of these minerals. The same statements are, in general, true of pyroxenes, certainly of orthopyroxene and of much clinopyroxene. Some of the more complex clinopyroxenes may exhibit relations similar to those of hornblende. In the case of hornblende we have a mineral of very complex constitution. The usual igneous hornblende is a very basic mineral, a fact which was pointed out in discussing the methods whereby the residual liquids from fractional crystallization of basic magma are enriched in free silica. But hornblende is basic in a very complex way and, whereas its addition to a certain layer will raise the temperature of beginning of crystallization of that layer, it may at the same time lower the temperature of completion of crystallization and affect the course of crystallization in important respects.

The problem is best attacked by considering the behavior as a manifestation of fractional resorption. We should expect hornblende, sinking into a region of somewhat hotter liquid, not to be redissolved as such, except in quite small amount, but rather to enter into reactions which precipitate, as crystals, earlier minerals of the reaction series. These will be olivine and pyroxene and perhaps some calcic plagioclase; for the most part minerals of simple constitution. If a large amount of hornblende suffers such action there should result an enrichment of the liquid in all the constituents of hornblende other than those precipitated during the reaction. We may ascertain just what these constituents will be by studying the norm of hornblende-rich rocks such as hornblendite, for in the norm the rock analysis is converted into simple minerals, some of them the same as those we have noted as the earlier members of the reaction series which are precipitated by this reaction. In Table XV we give, therefore, the normative composition of ten hornblendites taken from Washington's Tables of 1917. Of the normative minerals listed, albite and orthoclase are late members of the ordinary reaction series and their tendency would be to become part of the "hot" liquid. It will be noted, too, that nephelite is a rather important constituent of nearly all the norms. As shown by Fig. 67, nephelite is capable of forming with diopside a liquid of quite low melting temperature, so it is to be expected that nephelite together with a nearly equivalent weight of diopside could enter into the liquid. Probably the small amounts of leucite shown in some of the norms and some of the calcium orthosilicate of two of them would tend to pass into the liquid also. On the other hand, olivine, excess of pyroxene and perhaps basic plagioclase would tend to be precipitated by the "hot" liquid. It should be noted that by "hot" liquid we do not mean superheated liquid but merely liquid not yet saturated with hornblende. It may be saturated with olivine and/or pyroxene and/or basic plagioclase; indeed, superheated liquid would rapidly become saturated with these as a result of the reaction. Plainly only a thoroughly basic, presumably basaltic liquid is to be considered effective. If the mass whose total composition has been thus affected by the addition of hornblende cools slowly the liquid would eventually become saturated with hornblende, the reactions noted would be reversed and a mass rich in hornblende would result. Such a rock as a cortlandite might then form. But let the mass be injected as a small dike into cold rocks before the formation of hornblende sets in and the rapid cooling might give such rocks as the olivine-bearing lamprophyres we have been discussing. If the injection is accompanied by a filtration effect which tends to separate crystals and liquid the liquid portion would crystallize

TABLE XV

NORMATIVE COMPOSITION OF HORNBLENDITES

	A						B		
	Ab	Or	Ne	Di	Lc	Cs	An	Hy	Ol
1	19.39	2.78		26.88			21.13	9.30	9.47
2	19.18	7.23	10.79	26.57	3.92				8.05
3	3.14	10.01	7.10	14.36			26.41		18.12
4			8.52	25.52	4.36	6.19	23.91		16.75
5			6.53		3.49	14.62	30.58		27.29
6	7.07	6.67	8.09	51.94			7.51		11.56
7	11.00	1.11	4.26	31.04			15.01		28.76
8	2.62	7.23		47.92			18.35		9.25
9	2.62	8.34	3.98	26.60			18.36		14.70
10	5.24	2.22		14.39			20.29	22.65	25.29

Norms of ten hornblendites from Washington's Tables, 1917. (Magnetite, ilmenite, apatite, etc., omitted.)

The normative constituents under A tend to become part of the liquid when hornblende reacts with "hot" liquid. We must except that part of the diopside in excess of the figure for Ne (approx.) and perhaps also some calcium orthosilicate.

The normative constituents under B tend to be precipitated by the reaction. We must except some normative anorthite that may go into pyroxene molecules.

to a mass rich in nephelite and pyroxene especially if crystallization took place under surface or near-surface conditions. Thus may originate the nephelinites or, when some olivine crystals have been carried along, the nephelite basalts.

The sinking of biotite, either alone or with hornblende, and its reactive solution in "hot" liquid should be regarded in the same general way. Biotite may be regarded as made up of highly femic molecules, mainly olivine, that would be precipitated and alkalic molecules that would in large part enter into the liquid. The separation of liquid from crystals, say by a squeezing-out process, would appear to be the most promising method of developing liquids that are strongly alkaline and at the same time rich in ferromagnesian constituents, the latter largely as metasilicates (pyroxenes) but not as orthosilicates (olivine). In this direction lies what at present appears to be the most promising method of development of nephelinite, leucitite, nephelite basalt, leucite basalt, and nephelite-melilite basalt. The leucite-bearing types are to be regarded as developed from liquids enriched principally in certain of the constituents of biotite, and the nephelite and nephelite-melilite types as developed from liquids enriched in some of the constituents of hornblende, with, of course, every gradation between the extremes. It is especially noteworthy in this connection that a porphyritic nature is so definitely characteristic of nephelite basalt, leucite basalt, and melilite basalt that the designation "porphyritic" is part of the definition of each of these types as given by Rosenbusch. Yet this is not a mere matter of definition for there are no non-porphyritic rocks of corresponding composition. In other words it is an inherent characteristic of material of the composition of these rocks that it should exhibit phenocrysts. The phenocrysts are olivine and augite and one can conclude only that there were no liquids of the bulk composition of these rocks. The only liquids concerned were notably poorer in olivine and augite substance. The closely related nephelinite, leucitite, etc., ordinarily lack these phenocrysts but only because their total composition is such that they are impoverished in the substance of the phenocrysts. Nephelinite and leucitite thus correspond in general with the groundmass of nephelite basalt and leucite basalt respectively. They represent the liquid portions of the corresponding basalts with the crystals removed in some manner and are consequently richer in alkaline minerals or, better, in the low-melting combination nephelite-pyroxene on the one hand and leucite-pyroxene on the other.

When the crystallization of these femic-alkalic mixtures takes place under certain conditions there is a tendency for two very basic minerals, viz., olivine and melilite to separate in excess, if we may reason by analogy with the simple system of Fig. 67. As a result residual liquids are enriched in silica and impoverished in mafic constituents. Crystallization would thus proceed towards a yet lower-melting liquid of the general composition of phonolite, that is, rich in alkaline feldspar (feldspathoid + silica) and feldspathoid. In this tendency there probably lies the explanation of the natural association of nephelite-basalt, melilite-basalt and phonolite. The fractional resorption of accumulated hornblende thus permits a switching over of the liquid into an alkaline line of descent. It is probable that, among deeper-seated rocks, such types as theralite, basanite, etc. belong in this line of descent.

If we examine a table of igneous rocks which divides them into plutonic rocks and their effusive equivalents we find an effusive equivalent opposite any plutonic rock that lies in or approaches any of the liquid lines of descent. Thus such rocks as diorite, granodiorite, nephelite syenite all have effusive equivalents. If we turn to rocks which we have reason to believe are formed by accumulation of crystals especially where this must be very marked we find no effusive equivalents. Thus peridotite and anorthosite have no effusives to correspond. Lack of an effusive equivalent of hornblendite will also be found in such a table, yet the condition is not as true of hornblendite as of the other types mentioned. To be sure, there is no mineralogical equivalent of hornblendites among effusives, nor yet is there an exact chemical equivalent. This is because hornblende, like olivine and basic plagioclase, fails to suffer simple re-solution or remelting to give a liquid of its own composition. But unlike the very "refractory" olivine and basic plagioclase, the less "refractory" hornblende does suffer some action of this kind, but it is of the selective nature that we have pointed out. The result is that hornblendites, while not having equivalents among the effusives, have what might be termed representatives among these and they are nephelite basalts and related rocks. The dike representatives are the olivine-bearing lamprophyres, alnoites and related types. These dike and effusive representatives show some departure from hornblendites in composition on account of the selective nature of the re-solution. The tendencies brought to light in such a survey of rocks leave no doubt as to the restrictions that must be imposed, as well as the liberties that may be allowed in any hypothesis of re-solution of accumulated minerals.

CHAPTER XV

FURTHER EFFECTS OF FRACTIONAL RESORPTION

REVERSAL OF NORMAL ORDER OF ZONING

THE possibility of fractional resorption is not, of course, confined to any particular mineral or minerals, though some are more susceptible. It is not to be doubted that, during the slow crystallization of any large mass of magma, any crystal formed in a cooler upper portion may sink into a lower stratum at least somewhat hotter and that there some resorption would occur. Two principal cases may be distinguished in both of which we assume that the lower liquid has the same composition as the upper liquid had before it began to crystallize. In the one case the lower stratum may be regarded as still somewhat above the temperature of saturation for the sinking mineral, and in the other it may be regarded as saturated with that mineral phase. If the latter condition exists and the crystalline phase is a definite compound, no solution can occur, but since most minerals are solid solutions the warmer lower stratum might be saturated with a crystalline phase but not saturated with exactly the same composition of that crystalline phase. For example, the upper part might have become saturated first with Ab_1An_4 and have cooled further until it is now saturated with Ab_1An_2. If the first crystallization had been rather rapid the crystals would be zoned, with composition ranging from Ab_1An_4 to Ab_1An_2. Now if we may suppose that these crystals sink into a hotter layer which, because it is hotter, is saturated with a more calcic plagioclase, say Ab_1An_3, then the liquid will react with crystals in the effort to transform them into Ab_1An_3. Since the action must take place at the interface between liquid and crystals, in any moderate period only an outer shell of the crystal will have been transformed, so that there will be an outer zone exhibiting a reversal of the normal order of zoning. In the meantime even the deeper layer in which the crystals are now suspended may have experienced sufficient cooling that it is now passing into a condition of equilibrium with less calcic plagioclase and the deposition of less calcic zones may be resumed, with the normal order of zoning again in evidence. The temporary reversal of zoning may thus be explained as due to the sinking of crystals into a hotter zone. A carrying forward of the whole mass to a higher level, with institution of a new period of cooling and sinking, might bring about a repetition of the same effect with a general tendency towards more sodic shells but with another reversal of the normal zoning. Oscillatory zoning of plagioclase may thus find a probable explanation in several periods of sinking with intermittent surging forward of the mass. It may also be that the forcing of a hotter liquid magma through a crystal mesh, repeated several times, is a common cause of oscillatory zoning. Fenner has expressed the opinion that oscillatory zoning, or indeed a wide range of zoning even of the normal kind, testifies to a long period of suspension and disproves crystallization-differentiation.[1] His conclusion regarding a long period of suspension is undoubtedly correct if it means that not all the crystals sank rapidly to the "bottom" and stayed there, but if it carries the implication that they did not move downward at all then it is decidedly questionable. A long period of suspension with continued sinking is to be expected of crystals that differ as little in density from the liquid as the plagioclases ordinarily do. It would be possible at any time during the process to drain off a portion of the liquid showing suspended crystals of a certain kind with a wide range of zoning or perhaps with oscillatory zoning. But it might be equally possible to draw off another portion of the liquid from which all of the crystals, at least of that kind, had been removed, and yet another portion from which they have been partly removed. The existence of the one condition, locally, is no bar to the existence of the other conditions, locally. Indeed it should be fairly obvious that a long period of suspension of crystals may be synonymous with a long period of sinking of crystals. The occurrence of a long period of suspension (sinking) certainly cannot be taken as proof that crystallization-differentiation does not occur, for, though still suspended in one part of the liquid, the crystals may have come from another part which suffered the appropriate change of composition consequent upon their subtraction. Or again, if the oscillatory zoning is caused by the second method suggested (repeated filtering) each repetition of this action is a crystallization-differentiation. A wide range of zoning, of the normal kind, is itself a form of crystal fractionation even in the absence of any sinking of the crystals, for in virtue of such zoning the liquid is able to attain compositions it would not otherwise attain.

LIMITS OF RESORPTION

If, in a mass in which crystal settling is taking place, the lower liquid is unsaturated with the sinking crystal phase some direct solution will

1 C. N. Fenner, *Jour. Geol.*, 34, 1926, p. 703.

occur, but various factors combine to limit the scope of this action. As the concentration of the dissolving substance increases in the liquid the temperature at which saturation will occur becomes higher. Moreover, in the act of solution, heat is absorbed and the temperature of the liquid is lowered. Soon the temperature is reached at which saturation occurs for the new concentration, and simple solution ceases. Thereafter the only action will be the reactive effect already described.

It is impossible to say what limit is to be placed upon the simple solution of sunken crystals in any individual instance, but the general survey of the limits of composition of aphanites, given in another chapter, clearly demonstrates the general limits. Apparently solution of olivine never occurs beyond an amount sufficient to endow the liquid with 10 to 15 per cent normative olivine. Likewise solution of calcic plagioclase never occurs beyond an amount sufficient to endow the liquid with normative plagioclase more calcic than Ab_1An_2. It is not impossible that some liquids containing the higher amounts of olivine or of anorthite may have been formed by solution of crystals of the one or the other in liquids less rich in these high-melting compounds. For example, the plateau basalts appear to have, in general, plagioclase approaching Ab_1An_1 in composition. It is possible that some basaltic liquids having the normative plagioclase Ab_1An_2 are derived from plateau magma by a process of solution of sunken basic plagioclase. On the other hand it is possible that original basaltic magma varies notably at its source, wherever that may be. The process of solution of sunken crystals can not therefore be appealed to as the necessary explanation of the composition of any individual liquid, though it must occur in some instances. But, as we have seen, a general upper limit can be placed beyond which we may be sure the action has never gone.

LOCALIZED RESORPTIVE EFFECTS

Although the total composition of a liquid is ordinarily changed but little by solution of crystals[1] it is possible that somewhat stronger effects can occur locally about sinking crystals and crystal groups when these constitute but a small proportion of the mass. In such circumstances the relatively free flow of heat from the large body of liquid may keep the reacting system in the immediate vicinity of the crystals at sensibly the same temperature as the whole mass even though endothermic changes may be going forward there. Such a condition is rendered possible by the fact that the diffusivity of temperature is much greater than the diffusivity of concentration, so that the reaction goes on only locally and affects the composition of the liquid only locally, whereas the heat necessary for the reaction is derived from a large mass

1 The exceptions are those to whose discussion Chapter XIV has been devoted.

of surrounding liquid whose temperature is changed inappreciably. The most important result of such action from the viewpoint of crystal fractionation will be the possible production of a crystalline phase which is an earlier member of the same reaction series to which the sinking crystal belongs. To some extent such action has already been described in discussing the reversal of the order of zoning in plagioclase, though there no limitation was assumed as to the proportion of sinking crystals. More notable effects of the kind described may be obtained on a few crystals than on a larger proportion.

FORMATION OF SPINEL IN ULTRABASIC ROCKS

Results of greater consequence are to be obtained when the crystals concerned are not simply members of a continuous reaction series (solid solution series) like the plagioclases but are members of a discontinuous reaction series. We shall discuss this feature of resorption with special reference to the probable mode of development of spinel in ultrabasic rocks, and, to illustrate the principles involved, the diagram of the system, anorthite-forsterite-silica (Fig. 53, p. 194) will be used.

A mixture of the composition X is made up of anorthite 59 per cent and enstatite 41 per cent. The liquid becomes saturated at approximately 1305°, anorthite and forsterite separating together. We may suppose that a large mass of liquid of this composition and at a temperature of 1325° is placed in such surroundings that it cools principally from its upper surface. The upper portion would cool quickly to 1305° and separation of anorthite and olivine would begin. Crystals of these would grow and, by hypothesis, would sink. We shall suppose that rare crystals and perhaps glomeroporphyritic groups might eventually reach a layer of magma still uncooled; i.e., at a temperature of 1325°. Crystals would begin to dissolve and in the immediate vicinity of a crystal group there would be a localized system, with liquid varying in composition from X, the unchanged liquid, to a composition, in contact with the crystal group, which has dissolved all the anorthite and forsterite possible at 1325°. But, since the anorthite-forsterite boundary curve is cut off at 1320° by the spinel field, the liquid which has dissolved as much crystals as possible will lie on the 1325° isotherm in the spinel field, that is, it will be saturated with spinel, not with anorthite or forsterite. In short, the liquid will transform some of the crystals into spinel. It is an endothermic reaction and no great quantitative proportion of spinel could be formed without cooling any but a very hot liquid to a temperature at which it would be saturated with anorthite and olivine, whereupon action would cease. Nevertheless a small amount of spinel might be formed by this method from a liquid only slightly above its temperature of saturation with anorthite and forsterite.

In order to introduce the above discussion in the simplest manner, we have taken a liquid which becomes saturated with both anorthite and forsterite simultaneously, and have regarded it as somewhat super-heated. For a liquid of such composition a certain amount of superheat is essential to the reaction but it is to be noted that there are any number of liquids of the system that require no superheat in order to produce spinel in the manner described. We may take, for example, any liquid which lies in the forsterite field and becomes saturated with forsterite at a temperature above $1320°$, the lowest temperature of the spinel field *in this system*. Such a liquid, even although it is already saturated with forsterite, so long as it is not yet saturated with anorthite in addition, will produce spinel from any crystal group of forsterite and anorthite that might be placed in it; say, might sink into it from a cooler part of the liquid. Likewise liquids which first become saturated with anor-thite, but are not yet saturated with forsterite, can produce similar effects upon such crystal groups.

There is thus furnished a possible method of formation of spinel from a liquid which, in the ordinary course of cooling, is incapable of forming spinel. Spinel so formed should be a transient phase and should dis-appear in virtue of a reversal of the same reaction on cooling. Forma-tion of a corona about it might prevent its disappearance or the same result might be accomplished by local accumulation of crystals (spinel and its associates) in such quantity that the local supply of liquid was inadequate.

Although we have taken a specific system of very simple type to illustrate this action, it should not be regarded as possible only in this system. Indeed it is possible in any liquid which can precipitate olivine and basic plagioclase for such a liquid must belong to a system which has the forsterite-anorthite system as one of its binary limits. There must be a spinel region close to this binary limit in any such system, however complex, and any liquid that becomes saturated early with an olivine close to forsterite or a plagioclase close to anorthite is capable of transforming these into spinel if the liquid is at a temperature within the spinel region for that system. In general the presence of some olivine in a liquid which is otherwise a mixture of plagioclase with metasilicate (approximately) will tend to move that liquid closer to the spinel field.

A very fine example of spinel reaction rims has been described re-cently by Barth.[1] The spinel occurs between olivine and plagioclase in rocks which Barth interprets as formed by crystal accumulation (sink-ing).[2] The spinel may be interpreted as formed by reaction effects between plagioclase and olivine sinking into a liquid as yet saturated with only one of them, a rising temperature effect. The subsequent

1 Tom. Barth, *Skr. Norske Vidensk. Akad. Oslo I. Mat.-Naturv. Kl.*, 1927, No. 8, p. 15.
2 Barth, *op. cit.*, p. 29.

partial replacement of spinel by hypersthene (or augite) may be regarded
as a reversal taking place with falling temperature. Barth cites numerous examples of spinel-bearing reaction rims in rocks. In each case the
essential feature appears to be juxtaposition of olivine and plagioclase.
There is nothing incompatible between these minerals. They occur side
by side in many rocks without interposition of any other material. We
therefore believe that the action of a liquid not yet saturated with both
olivine and plagioclase is to be invoked in those cases where spinel has
been formed. This liquid forms with the crystals a localized system
which will develop spinel at the temperature concerned though the liquid
alone is incapable of developing spinel at any temperature.

Some of the spinel formed during this reaction may participate in the
process of crystal accumulation either independently or as a constituent
of crystal groups. The ultrabasic rocks of Skye, like many other occurrences of such rocks, sometimes carry spinel[1] (pleonaste) and the association is such as to suggest that spinel may have been formed by the
action described. These rocks have been formed, if our conclusions are
correct, by the accumulation of early crystals, principally olivine and
bytownite-anorthite, from a liquid of the composition of olivine basalt,
the plateau magma of that region. Hot, but not necessarily superheated magma in the deeper layers of a large mass may have acted on
these early crystals to produce spinel. It is, of course, an action not unrelated to the melting-up of crystals in deep-seated layers, an action
which some have proposed as an explanation of the production even of
extreme liquids such as peridotite or even dunite. Reasons have already
been given for believing that action of such intensity is both unlikely on
general grounds and apparently unwarranted by any evidence in rocks
themselves.

ORIGIN OF PICOTITE AND CHROMITE

The minerals of the spinel group, picotite and chromite, are more
commonly developed in ultrabasic rocks than is the ordinary spinel,
pleonaste. In the investigated system cited in connection with the origin
of spinel only the pure magnesian spinel could develop. In a related
natural system, containing other oxides, no doubt other spinel-forming
oxides would enter into the spinel. Among these would be iron and a
little chromium, possibly nearly all the chromium of the liquid; but it is
very doubtful if a spinellid as rich in chromium as picotite could be
formed directly by that process. The belief in the remelting of sunken
crystals to give ultrabasic liquids has the advantage that a rather simple
method of developing picotite and chromite can be based thereon. Given
the necessary temperatures it is also a not improbable method. It is only

1 Harker, *Tertiary Igneous Rocks of Skye*, p. 71.

necessary to assume that there is further crystal fractionation and re-melting of sunken early crystals, among which would be a spinellid. By repetition of this process the spinellid might eventually become strongly enriched in its high-melting components including, presumably, chromium oxide. Vogt has proposed such a process for the production of chromite.[1] The difficulties involved are that it would require very high temperatures not only to produce such liquids but to keep them liquid, so that we should have evidence wherever they have been intruded that they were at such high temperatures. There is no such evidence; furthermore the absence of aphanites corresponding to these chromite-bearing rocks renders it unlikely that any such liquids are to be reckoned with. Indeed it seems necessary to find some method of developing picotite and chromite by crystal accumulation and such reactions as may be consequent thereon without appealing to the extreme temperatures necessary for remelting. A possible method will now be suggested. The first step in the process is the development of a spinel by the method of reaction of olivine and basic plagioclase with liquid of non-extreme (basaltic) composition. This spinel may be assumed to carry some chromium. Spinel so developed is, as we have pointed out, a transient phase. When the mass cools the reaction is reversed, the liquid reacts with the spinel, producing basic plagioclase and olivine. But basic plagioclase and olivine are incapable of containing any significant quantity of Cr_2O_3. Not until the time of precipitation of monoclinic pyroxene does a phase appear into which Cr_2O_3 can enter in important amounts. By hypothesis that stage has not yet arrived, so that, as the reaction described goes on, the spinel may be pictured as continually losing $MgO \cdot Al_2O_3$ with consequent enrichment in Cr_2O_3. According to the extent of the reaction picotite or chromite is formed. They too would disappear later, by entering into clinopyroxene, but in the meantime further crystal fractionation may occur and the picotite or chromite may be concentrated locally, together with olivine and other minerals of the same period. There may, therefore, be an inadequate supply of liquid to effect incorporation of the chromium into other minerals, and picotite or chromite may persist. These chrome minerals are thus to be regarded as released minerals of this reaction just as magnetite is sometimes a released mineral of the pyroxene-hornblende reaction.

It is especially noteworthy in this connection that local richness of spinel of adequate degree may bring about local complete (or nearly complete) disappearance of liquid during the reaction of spinel with the liquid. Basaltic liquid may be regarded as pyroxene plus plagioclase plus olivine. But pyroxene plus spinel may give only anorthite and

[1] J. H. L. Vogt, *Vidensk. Selsk. Skr. I. Mat.-Naturv. Kl.*, 1924, pp. 85, 86.

olivine as may be illustrated roughly by an equation where iron oxides are omitted for simplicity:

$$CaMgSi_2O_6 + 2MgSiO_3 + MgO . Al_2O_3 =$$
$$CaAl_2Si_2O_8 + 2Mg_2SiO_4$$

The result is that basaltic liquid, reacting with spinel in appropriate quantity, may give only or practically only plagioclase and olivine. In such a case the Cr_2O_3 would of necessity be thrown out in some form and its occurrence as picotite grains embedded in olivine and bytownite would seem likely. A mineral that is thus thrown out in small amount as a sort of excess material in a reaction tends to occur as small well-formed crystals. Magnetite is so found in many rocks where it has such origin and it tends to occur in clusters which are, as it were, a sort of pseudomorph of the resorbed mineral. Perovskite, too, may so occur when formed in a similar manner in alnoite. The tendency of chromite to occur in clusters may be due to similar causes and not to a "swimming together" as Vogt has supposed.

The varying aspects of the spinel reaction may thus be summarized in order. Spinel may be formed by the reaction of liquid with anorthite and olivine (rising temperature). When so formed it is ordinary spinel and may persist as such as a result of failure of the reverse reaction. The reverse reaction may take place with falling temperature giving, locally, basic plagioclase with olivine and a chrome-rich spinellid. With further opportunity for reaction the chrome spinellid may disappear by incorporation in a notably chromiferous clinopyroxene. Finally, with perfect redistribution of products and unlimited opportunity for reaction—an unlikely case—the result would be the same as if spinel had never formed and the Cr_2O_3 would occur generally distributed as a minute proportion of the whole mass, presumably in the pyroxene.

CHAPTER XVI

THE IMPORTANCE OF VOLATILE CONSTITUENTS

INTRODUCTION

IT IS well known that all magmas contain volatile constituents. The most abundant of these is undoubtedly water, with carbon dioxide, chlorine, fluorine and others probably always present in small amounts. The influence of any such substance upon the properties of a silicate melt is unquestionably greater than that of a corresponding quantity of any of the ordinary rock-forming oxides. This fact does not, however, justify the attitude that a great many petrologists have towards the volatile components. The properties of many "dry" melts have been determined in the laboratory. The majority of petrologists accept these determinations and are willing to believe that the addition of small quantities of any ordinary oxide to these melts would modify these properties a moderate amount, but let there be a spectroscopic trace of a volatile component and the liquid is assumed to acquire properties wholly unrelated to those of a "dry" melt. To many petrologists a volatile component is exactly like a Maxwell demon; it does just what one may wish it to do. The facts are that volatile components in small amounts can only modify the characters appropriate to the dry melt and the modification must be progressive, a very small amount producing a proportionately small effect. This question will be considered in greater detail with the aid of diagrams. I have profited much from the writings of Morey and of Niggli on the subject of volatile components and am especially indebted to my colleague Morey for much helpful discussion of the factors involved.

SYSTEMS WITH WATER

Many binary systems consisting of a certain compound with water have been completely studied throughout the range of compositions from pure water to the pure compound. As a particularly simple example we may note the system NH_4NO_3-H_2O, of which the equilibrium diagram is given in Fig. 72. With the exception of the minor complications introduced by the fact that NH_4NO_3 occurs in several crystalline forms, it is an ordinary eutectic diagram, the melting-point curves of water and of NH_4NO_3 meeting at the eutectic at about — 20° C.[1] This system

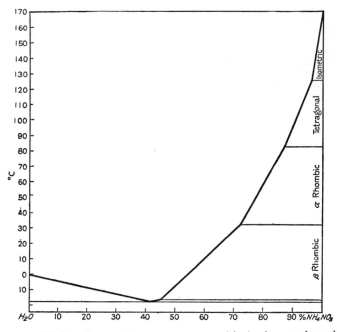

FIG. 72.   Equilibrium diagram (temperature, composition) of ammonium nitrate-water (after Millican, Joseph, and Lowry).

is particularly simple because the pressure (of water vapor) never rises above a value which permits investigation in an open vessel. Even though the temperatures of fusion of those mixtures rich in NH_4NO_3 is well above the boiling point of water, the pressure of water vapor in equilibrium with these solutions (melts) is always well below one atmosphere, a fact which is emphasized by the pressure-temperature curve of saturated solutions[2] (Fig. 73). The curve has a characteristic shape. The pressure rises with increase of temperature but this effect is finally overbalanced by the decreasing concentration of water so that the curve passes through a pressure maximum and falls to zero at the

1 Millican, Joseph and Lowry, *Jour. Chem. Soc. Lond.*, 121, 1922, p. 959.
2 Prideaux and Caven, *Jour. Soc. Chem. Ind.*, 38, 353 T, 1919.

melting point of the anhydrous substance. The maximum is in this case at less than ½ atmosphere.

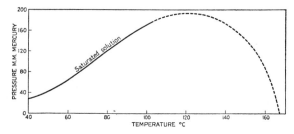

FIG. 73. Pressure-temperature diagram of the system, ammonium nitrate-water (after Prideaux and Caven).

Solutions exhibiting such low pressures are to be obtained only with very soluble compounds. For most substances the pressure of solutions saturated at the higher temperatures rises to one atmosphere and the solution boils. Investigation of solutions of higher concentration must be carried on in a closed container capable of withstanding the pressures developed, but with such a provision it may be possible to determine the

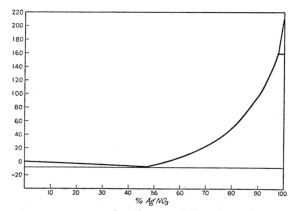

FIG. 74. Temperature-concentration diagram of the system, silver nitrate-water.

solubility (or melting) curve throughout all compositions. A simple example is afforded by $AgNO_3$ which shows a eutectic with water at — 7° (Fig. 74).[1] Since the pressure rises above one atmosphere in certain concentrations, part of the investigation must be carried on in a closed

1 Landolt-Börnstein Tabellen.

vessel. In this case, as a result of the high solubility and low melting point of $AgNO_3$, the maximum pressure is only about 1 1/3 atmospheres.[1] (Fig. 75.) There is obviously no difference in principle between the case

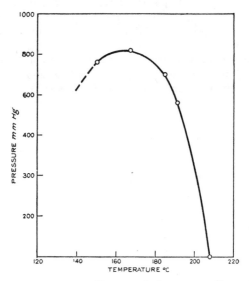

FIG. 75.   Pressure-temperature diagram of the system, silver nitrate-water.

of $AgNO_3$ and that of NH_4NO_3. It merely happens that the atmospheric pressure is such that a certain difference of experimental procedure is necessary.

The phenomenon of the "second boiling-point" would be shown by saturated $AgNO_3$ solutions at 1 atmosphere pressure. If a line is drawn on the p.t. diagram (Fig. 75) at a pressure of 1 atmosphere it cuts the vapor pressure curve at two points, the one at 131°, which is the first boiling point, and the other at 191°, which is the second boiling point. The second boiling point is observed on cooling in the open air, from, say, 200°, a molten charge of $AgNO_3$ containing a little water. As it cools $AgNO_3$ separates as crystals and the pressure rises until at 191° the pressure attains one atmosphere, whereupon the liquid boils. The temperature remains constant, boiling away of the water and crystallization of $AgNO_3$ taking place, until both actions are complete. It is plain that the second boiling point, like the first, is defined only when the pressure is stated.

1 Roozeboom, *Die heterogenen Gleichgewichte*, II, 1, 1904, p. 352.

When we pass on to substances of lower solubility and higher melting point, among which substances are the silicates, the same type of equilibrium diagram may be found in some cases. Morey and Fenner have investigated potassium silicates with water and their results will now be considered.[1] Since the silicate compounds studied have high melting points, the pressures obtained in some saturated solutions rise to quite high values so that closed containers of great strength (steel bombs) were necessary for their investigation. The equilibrium diagrams for the binary systems of H_2O with K_2SiO_3 and with $K_2Si_2O_5$ are shown in Fig. 76. They are the same general type as the diagrams for

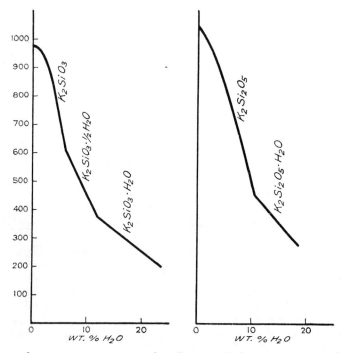

FIG. 76. Temperature-concentration diagrams of the systems, H_2O-K_2SiO_3 and H_2O-$K_2Si_2O_5$.

the nitrates already given, except that each of these silicates forms one or more compounds (of incongruent melting points) with water. Apart

[1] G. W. Morey and C. N. Fenner, "The Ternary System H_2O-K_2SiO_3-SiO_2," *Jour. Amer. Chem. Soc.*, 39, 1917, p. 1173.

from these complications water lowers the melting point of these silicates in a progressive manner just as it does the melting points of the nitrates. The pressure-temperature diagrams for these silicates (Fig. 77) are only slightly more complicated than the corresponding dia-

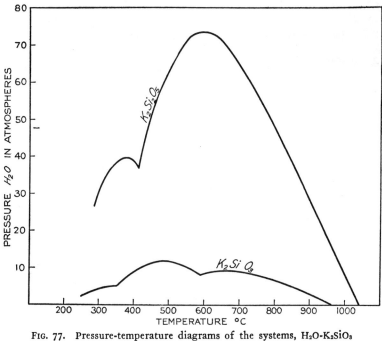

FIG. 77.   Pressure-temperature diagrams of the systems, H_2O-K_2SiO_3 and H_2O-$K_2Si_2O_5$.

grams for the nitrates, the complications again being due to the formation of these compounds, so that there is more than one pressure maximum. It will be noted that the maximum pressure of solutions saturated with $K_2Si_2O_5$ is about 75 atmospheres. This obtains at a temperature of about 600° and over solutions containing about 8 weight per cent H_2O.

·The phenomenon of the second boiling point would be shown by saturated $K_2Si_2O_5$ solutions under any external pressure less than 75 atmospheres. Thus if one had a solution (melt) of $K_2Si_2O_5$ containing, say, one per cent H_2O and under an external pressure of 50 atmospheres and if this melt were cooled, crystallization of $K_2Si_2O_5$ would begin at about 1020°. With further cooling and crystallization the pressure would rise until at about 800° it would reach 50 atmospheres and the melt, which now contains about 5 per cent H_2O, would boil. Upon further abstraction of heat the temperature would remain constant, with crystallization of $K_2Si_2O_5$ and evolution of water vapor until complete crystallization and dehydration had occurred. On the other hand, if the external pressure had been more than 75 atmospheres progressive cooling would occur with continued crystallization of $K_2Si_2O_5$ and enrichment of the residual liquid in water. We may state this in terms of geological environment. If the melt pictured were cooled under a cover of about 600 feet of rock it would crystallize in the ordinary way until a temperature of about 800°, whereupon vesiculation would occur and crystallization would then proceed to completion as a result of loss of water as gas without further cooling. After final crystallization the mass would then cool to the temperature of its surroundings. On the other hand, if the melt were cooled under a rock cover of 1000 feet or more it would crystallize gradually as cooling proceeded, the pressure rising at first and then falling off even though the concentration of water in the liquid increases continually. When it reached the temperature of its surroundings at, say, 100°, the residual liquid would be simply a hot aqueous solution exerting a moderate pressure and capable of escaping into surrounding rocks as such, i.e., as a liquid.[1]

A fact of interest may be noted in connection with the first case described, viz., that in which cooling took place under 600 feet of rock. If in any way the mass was able to expand its chamber during the stage of vesiculation and the water vapor, instead of escaping, remained disseminated through the porous crystalline mass, then when this mass reached a temperature of 450°, the external pressure being still the weight of 600 feet of rock, some of the crystalline $K_2Si_2O_5$ would remelt (redissolve) and form with the water vapor a liquid containing approximately 10 per cent H_2O. This action would continue at constant temperature until all the water vapor dissolved, an appropriate amount of $K_2Si_2O_5$ melting simultaneously. Further cooling and crystallization would then occur in exactly the same manner as in the case where the water remained in solution throughout the whole process.

Of the magnitude of the lowering of the melting point of these silicates it is perhaps well to make a note. In each case 10 per cent water induces a lowering of somewhat more than 500°, an average of about 50° for each unit per cent of water. It is a notable effect but it is not miraculous. The melting point of albite is not significantly different

1 In making this statement we are assuming that the liquid is incapable of reacting freely with the crystalline $K_2Si_2O_5$ as a result, say, of the formation of narrow reaction rims of $K_2Si_2O_5 . H_2O$. If complete equilibrium occurred the liquid would be entirely used up at 420° and the whole mass would consist of the crystalline hydrate $K_2Si_2O_5 . H_2O$.

from that of $K_2Si_2O_5$ and if one were called on to make an estimate, on the basis of present knowledge, of the amount of lowering of the melting temperature of albite (or orthoclase) by water, a figure of the same order would be a natural choice. Thus it seems likely that a lowering of $100°$ in the melting temperature of alkaline feldspars would require perhaps 2 per cent H_2O in the melt. The pressure required to hold such a quantity of water in the melt would probably be much higher in the case of the feldspar melts than in the case of the melts of the more soluble alkaline silicates. Indeed, all the common rock-forming silicates are so insoluble in water that critical phenomena in saturated solutions would probably be shown in the binary systems of any one of these with water and it will be necessary now to consider such modification of the relations as may be introduced in such circumstances.

In all of the systems hitherto considered the solubility of the substances in water at the critical temperature of pure water has been sufficiently great that the critical temperature of the solution is raised to a value well above that at which any crystallization occurs. Critical phenomena are, therefore, shown only in unsaturated solutions at temperatures well above those at which any crystallization (saturation) occurs. A general diagram which shows the composition of the gas phase as well as the other phases has, for such pairs of substances, the form of Fig. 78. In this figure there is a critical curve K_A-K_B which joins the critical temperatures of the two substances and all parts of it lie above the saturation or melting curves. The rest of the diagram is an ordinary eutectic diagram such as we have been considering, with the composition of the vapor phase shown in addition. Thus the curve B_SE_2 is the melting point curve of solid B, and A_SE_2 is the melting point curve of solid A, which curves meet at the eutectic E_2. The curve A_SE_1 gives the composition of the vapor phase in equilibrium with the solid A and the liquids A_SE_2. The curve B_SE_1 gives the composition of the vapor phase in equilibrium with solid B and the liquids B_SE_2. It will be noted that the vapor is always much richer in the more volatile component A than is the liquid. The systems already considered are of this type, the vapor consisting, indeed, of practically pure water.

The corresponding pressure-temperature diagram is like the actual pressure-temperature diagrams already given for definite systems with a critical curve added. It takes the form of Fig. 79 where K_A-K_B is the critical curve lying entirely above the curve of vapor pressures of saturated solutions $S_B + L + G$. This latter curve is of course the same type of curve with a maximum that we have already discussed. It is plain that no critical phenomena will occur during the crystallization of mixtures of such a system.

We pass now to a consideration of systems of the other type, viz., that in which, by virtue of low solubility, critical phenomena may occur in saturated solutions. The diagram for this case takes the form

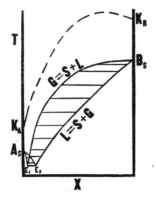

FIG. 78. Ideal T-X projection (after Morey and Niggli); the critical curve is not intersected by the lines representing two-phase equilibrium in coexistence with solid.

K_A represents the critical point of A (the more volatile component, e.g., H_2O), K_B that of B (the less volatile component, e.g., silicate); the broken line $K_A K_B$ the critical curve.

A_S and B_S represent the melting points of A and B, respectively.

$A_S E_1 B_S$ gives the composition of the vapor in coexistence with liquid and solid; $A_S E_2 B_S$ that of the liquid in coexistence with the vapor represented by $A_S E_1 B_S$ and with solid.

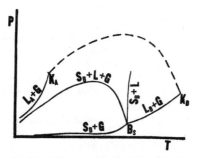

FIG. 79. Ideal P-T projection (after Morey and Niggli); the critical curve is not intersected by the lines representing two-phase equilibrium in coexistence with solid.

K_A represents the critical point of A, K_B that of B; the broken line $K_A K_B$ the critical curve; B_S the triple point of B (melting point under its own vapor pressure). The eutectic relations at low temperatures are not indicated.

of Fig. 80. The saturation curve of the less volatile substance B, instead of being a continuous curve passing from the eutectic to the melting point of B, is now discontinuous, being made up of the two sections

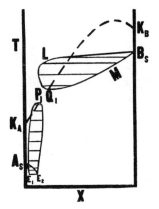

Fig. 80.   Ideal T-X projection (after Morey and Niggli); the critical curve is intersected at P_1 and at Q_1 by both lines of two-phase equilibrium in coexistence with solid.

K_A represents the critical point of A (the more volatile component, e.g., H_2O), K_B that of B (the less volatile component, e.g., silicate); the broken line K_A P_1-Q_1K_B the critical curve; P_1 and Q_1 the critical end points.

A_S represents the melting point of A, B_S that of B; the curves $A_SE_1P_1$ and Q_1LB_S give the composition of the vapor in coexistence with liquid and solid; $A_SE_2P_1$ and Q_1MB_S that of liquid in coexistence with the above vapor and with solid.

Between P_1 and Q_1 the only phase which can coexist with solid is a fluid (gaseous) solution.

E_2P_1 and Q_1B_S. P_1 and Q_1 are the points where the critical curve intersects the saturation curve. The curve P_1E_2 gives the composition of liquids in equilibrium with solid B and vapor of compositions lying along the curve P_1E_1. Similarly B_SMQ_1 gives the composition of liquids in equilibrium with solid B and vapor of compositions lying along B_SLQ_1. In the simpler case of Fig. 78 the vapor is always much richer in the more volatile component than is the liquid, and this is true in the present case except at the two critical end points P_1 and Q_1 where the liquid and vapor in equilibrium with solid B become identical in composition and, indeed, in all properties. Between P_1 and Q_1 only one phase, a fluid (gas) phase, can exist in equilibrium with solid B. The corresponding pressure-temperature diagram has the form of Fig. 81. The critical region between P and Q has, of course, no curve of invariant equilibrium crossing it.

It would be profitless for our present purpose to discuss all the possible complications of a system of this type. It is sufficient to point out that, in the cooling of any liquid mixture initially rich in the less vola-

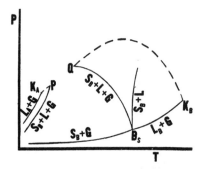

FIG. 81.   Ideal P-T projection (after Morey and Niggli); the critical curve is intersected at P and Q by the two-phase lines in coexistence with solid.

K_A represents the critical point of A, K_B that of B; the broken line K_A P-QK_B the critical curve; P and Q the two critical end points; B_S the triple point of B.

Between P and Q only fluid (gaseous) solutions are stable in coexistence with solid; the eutectic relations at low temperatures are not indicated.

tile component, there will always be a period during which the system consists of solid and gas without liquid. The gas may show an indefinite degree of approach to the liquids in composition and properties and if conditions are such that the critical end-points are attained the gas (fluid) phase becomes identical with the liquid. It is necessary, however, to bear in mind the position which these critical end-points would occupy in a binary system of components showing such enormous contrast of properties as water and silicate. Since the silicate will ordinarily be but very sparingly soluble in water at its critical temperature the first critical end-point P_1 must lie at a temperature and pressure but little above the critical temperature and pressure of pure water and at a composition but little removed from pure water. The second critical end-point Q_1 must also lie at compositions very rich in water. For all practical purposes connected with the crystallization of a deep-seated mass (high external pressure) the melting point curve of B (B_SMQ_1, Fig. 80) is not different from the simpler case of Fig. 78 where the melting point curve merely falls continuously to the eutectic. Only when the liquid has attained a composition such that the concentration of water is many times that of silicate may critical phenomena be expected even in those cases most favorable for their development. The main course of crystallization of a molten silicate containing a little water will therefore be the same whether the system be of the type of Fig. 78 or of Fig. 80. The only difference will come at a very late stage when the residual liquid is a dilute aqueous solution. At this stage critical phenomena may occur in the one type. But throughout the whole course of crystallization the gas phase, if any is formed, must be very rich in the more volatile substance and very poor in the less volatile substance.

Now, while it is probably true of most rock-forming silicates that they would form, with water, a binary system of the second type, in so far as the system remained binary, it is not probable that a molten *mixture* of silicates containing a small amount of a *mixture* of volatile components would behave in a similar manner. In such a complex melt, and natural magmas are such, the progressive crystallization of silicates would lead to continual concentration of the volatile components in the residual liquor until a stage came when there would be a noteworthy concentration of comparatively soluble compounds formed by what may be termed a decomposition of the silicates. This decomposition would be the result of the action of H_2O, CO_2, Cl, F, etc., on the silicate compounds, and the soluble hydrates, carbonates, chlorides, etc. formed would probably be adequate to eliminate critical phenomena altogether. It is doubtful, therefore, whether the complex systems constituted by magmas are such as to permit the existence of critical phenomena. In any case it is certain that if critical phenomena occur at all they occur at a stage where the liquid is to be described as an aqueous solution rather than a magma and that in the general problem of the differentiation of igneous rocks critical phenomena need not be considered, even though it is not impossible that they may have a place in the problem of vein-forming solutions. This does not mean, of course, that gas can not form at some stage of the crystallization process. It may, indeed, form copiously at any stage if the external pressure be adequately reduced, say, by injection of the mass into a position where it has but a moderate cover. All the information we have on the subject would lead us to believe that the proportion of what we may term solids dissolved in the gas phase would be very small. In the slow crystallization of a deep-seated mass, there is every reason to believe that the dominant rôle of the volatiles is the passive rôle of becoming increasingly concentrated in the residual liquid. With this increasing concentration goes hand in hand, of course, an increasing effect on the properties of the residual liquid.

GASEOUS TRANSFER

An action frequently appealed to in present-day writings as a cause of differentiation of igneous rocks is gaseous transfer. Most petrologists leave one very much in the dark as to what they mean by the terms.

Fenner has, however, recently advocated a process of gaseous transfer as the principal factor in igneous differentiation, listing it before all others, and giving a more definite picture of the process. He therefore offers something concrete for discussion in his process of a streaming through of gas bubbles which brings about differential transfer of materials.[1] That gases as such, that is as definite bubbles, do stream through magmas when the magma body is open to the air is well known and that a differential transfer of material is effected by the gas is indubitable. From his own experience and from the literature Fenner offers many examples of sublimates formed by this differential transfer.[2] It is, however, scarcely to be questioned that surface lavas and volcanic conduits are at best very unimportant seats of igneous differentiation. If this bubbling through of gases is to be the principal factor in igneous differentiation it must occur in deep-seated bodies. Vesiculation should be a common condition, perhaps the normal condition, of deep-seated masses. It might, of course, be transient, but deep bodies have sent dikes into surrounding rocks at all stages of their careers and these dikes are often, at least at their margins, aphanitic to glassy. They have thus crystallized so rapidly that if the liquid were vesicular the aphanite would be vesicular also. Yet vesicular dikes or dike selvages are exceedingly rare and are apparently always found in association with flows which indicate for them a shallow depth. Besides this definite evidence from rocks themselves of normal lack of vesiculation of magmas in depth, certain theoretical considerations have been brought to bear on the problem. Shepherd and Merwin have calculated, with the aid of certain assumptions as to saturation that can not be far wrong, that it would require very great quantities of volatiles in a magma to produce vesiculation at any considerable depth.[3] It is very much to be doubted, therefore, that any process which involves vesiculation can be the principal or even an important factor in igneous differentiation. But even on the supposition that vesiculation is a common condition in deep-seated magma and that for some reason the evidence of this condition is concealed from us, the quantitative adequacy of the process is seriously in question. We have already seen that the concentration of "solids" in the gas phase must be small. The actual proportion of volatile substances in magmas is itself small. If any part of an igneous mass is to be endowed, by differential gas-bubble transfer, with an oxide composition notably different (say as different as granophyre is from diabase) the proportion of the mass so endowed could not be more than a fraction of one per cent of the total. Such quantities are certainly not adequate to explain the ascertained facts.

1 C. N. Fenner, *Jour. Geol.*, 34, 1926, pp. 743-4.
2 *op. cit.*, pp. 718-45.
3 Shepherd and Merwin, *Jour. Geol.*, 35, 1927, pp. 114-15.

There is another form of so-called gaseous transfer that appears to be in the minds of many petrologists. Strictly speaking it is not gaseous transfer at all but is a supposed differential transfer taking place through the intervention of the volatile components without any actual separation of these as gas. There is a continual tendency towards an escape of volatile components into the more or less porous surrounding rocks. They carry at least some "solids" with them and these may be deposited in surrounding rocks. But the innumerable, well-established cases in which igneous masses have introduced great quantities of material into adjacent rocks can scarcely be accepted as proof that the introduction was effected principally in this manner, and especially at this stage. The loss of the volatiles and their load from border portions of the magma would set up a flow of these materials from central portions. But, all being in the liquid phase, this flow must be merely a diffusion, and diffusion through significant distances is an excessively slow process. Nor can appeal be made to unlimited time for, slow though the cooling of a large mass of magma must be, the rate of diffusion of heat is many times the rate of diffusion of substance.

But even if we admit for volatiles a miraculous rate of diffusion through the liquid and suppose that at the stage of complete liquidity the principal introduction of material into surrounding rocks occurs, the next step in the reasoning is decidedly difficult to accept. It is the supposition that notable differentiation of the liquid igneous mass would accompany this action. There would be an undoubted flow of the volatiles towards border portions, but this flow would be a consequence of impoverishment of the borders by escape into surrounding rocks. It would continue only so long as this condition persisted and if the action were interrupted at *any* stage we should find the borders poor in the moving materials as compared with central portions. Yet it has sometimes been contended that this diffusion towards the border of an entirely liquid mass accounts for certain border facies which, in their type of crystallization or their mineral content, give evidence of special richness in mineralizers. Plainly the hypothesis encounters insuperable difficulties at every turn.

The alternative picture is one in which there is only a very limited escape of volatiles into surrounding rocks at the stage when the magma is completely or almost completely liquid. As crystallization proceeds there is a building-up of the concentration of volatiles in the residual liquid. The growth of crystals involves diffusion through moderate distances, but relative movement of crystals and residual liquid, whether by gravity or by deformation of the mass, may bring about a transportation of the volatiles dissolved in the liquid through great distances. In this manner a mass might come to give evidence of special richness of volatiles in its outer and upper portions. It is a case of silicate transfer of the volatiles. And when the mass approaches complete crystallization there is a long period of very slow cooling during which escape of the relatively concentrated aqueous residues into the pores and channels of surrounding rocks may occur. This would appear, on general grounds, to be the promising period for notable introduction of material into surrounding rocks. It is the period, too, to which those who have studied them most have assigned the principal examples of contact metasomatism.

Whatever may be the appropriate designation of the transfer of substance in adjacent rocks, the principal action in which volatiles are involved in the igneous mass itself is silicate transfer of the volatiles, a direct and indirect consequence of the progress of crystallization.

PROPORTIONS OF THE VOLATILE CONSTITUENTS AND THE PROBABLE EFFECTS OF SUCH PROPORTIONS

In many analyses of rocks only the one volatile constituent has been determined. This is, of course, water. Lack of detailed information of the quantities of all the volatiles is, however, not a serious matter for our present purpose. It is generally accepted that water is the principal volatile. Shepherd has studied the gases of rocks by heating them to high temperatures in vacuo and then separating the various gases. Of their proportions he says: "Regardless of origin, the volatiles which can be obtained from lavas seem to agree upon one thing, and that is that the water content shall be about 80 per cent of the total. There are exceptions, of course, but for active lava the figure seems quite general."[1] The determination of water may, therefore, be taken as a sufficient indication of the quantity of volatiles. It is commonly assumed that crystalline rocks have lost the greater proportion of these substances in the act of crystallization and in many examples this is probably true. We may turn to glassy rocks with some hope of finding an approach to the maximum quantity of water. In many cases these, too, may be impoverished but in others there are indications that nearly all the volatiles must have been retained. There are, for example, many occurrences in which the top of a glassy flow is coarsely vesicular. This passes downward into material with fine vesicles, which in turn gives place to non-vesicular lava. The obsidian of Obsidian Cliff in the Yellowstone Park furnishes an example. There appears to be adequate reason for assuming that the compact obsidian still retains all the volatiles it had when it arrived at the surface. The actual amount is less than 1 per cent. It is hardly possible to escape the conclusion that many deeper-seated rocks of like composition crystallized in the presence of no

1 E. S. Shepherd, *Bull.* 61, *National Research Council*, p. 260.

greater original quantity of volatiles. Yet the fact that this mass is found at the surface might be regarded as indicating that it had occupied a near-surface position for some time and there lost much of its volatiles. Nevertheless, the non-vesicular condition of the lower portions of the flow shows that the mass was non-vesicular when it reached the surface and could not have been losing gas by vesiculation in any postulated near-surface position. Diffusion through the mass and into the surroundings thus offers the only means of escape of volatiles while it occupied this supposed position and this action is too slow, relative to rate of cooling, to account for notable losses. It therefore seems safe to accept the conclusion that the compact obsidian referred to contains practically all the volatiles that were ever contained in the magma from which it formed. The same conclusion can probably be drawn from many other compact obsidians and the general conclusion is strengthened that many granitic rocks have crystallized in the presence of no greater original quantity of volatiles than is shown in the analysis of many obsidians.

We are led now to examine the analyses of obsidians in order to see what proportion of volatiles they commonly display, taking the determined water as an adequate indication. Of 44 obsidians (or equivalents such as rhyolite glass) in Washington's Tables of 1917, the first seven in point of content of water show that constituent in the following quantities in weight per cent: 8.7, 7.2, 6.5, 5.1, 3.8, 2.4, 2.2. Six contain between 1 and 2 per cent and the rest contain less than 1 per cent. The few with high values and the rare rocks known as pitchstones, some of which contain even greater amounts of water, demonstrate the existence of liquids of a granitic character with such quantities of water. But the great preponderance of glasses with much lower quantities, viewed in the light of the limitations that are imposed on the loss of volatiles from an entirely liquid mass, indicates that the normal water content of granitic magmas is much less.

The salic magmas have been dealt with in the above discussion because it is generally believed that, with the possible exception of some alkaline magmas, they contain the highest amounts of volatiles. The evidence of this fact is derived principally from the more notable effects of such magmas in the introduction of materials into surrounding rocks. It is probable, therefore, that most magmas normally contain much less water and other volatiles than those moderate amounts we have found reason to believe are normally present in granitic magma.

The hypothesis of fractional crystallization as here conceived leads to belief in the derivation of granitic magma as a late crystallization from more basic, probably basaltic, magma when that crystallization has been adequately fractional. Granitic magma so derived would be the natural home of a relatively high concentration of volatiles, their proportion in the liquid being ordinarily increased continually as crystallization proceeds. Their abundance in granite is thus the result, not of some mysterious process of gaseous transfer of silicates, but of the simple process of silicate transfer of the volatiles, a melt of general granitic composition, i.e., rich in alkaline feldspars and free silica, being the natural residue of appropriate fractional crystallization even of a dry melt of the general composition of basaltic magma. On account of the well-demonstrated properties of the plagioclase solid solution series petrologists have not, in general, found it difficult to believe that the course of crystallization would be such as to lead to enrichment of alkaline feldspars in the residuum. Some have, however, expressed the opinion that free silica is not naturally to be expected in the low melting residuum, pointing to its high melting-temperature as a reason for their opinion and suggesting that volatile constituents account in some manner for its presence in this residuum.[1] The facts are that, in spite of the high melting-point of silica, every dry system investigated in which SiO_2 is one of the components, whether the system is binary, ternary or quaternary, has shown free silica as one of the solid phases separating at the lowermost eutectic. It is true of the binary systems of SiO_2 with any one of the following, Al_2O_3, CaO, MgO, K_2O and Na_2O. It is true of all three ternary systems of SiO_2 with any two of CaO, MgO, Al_2O_3. It is true of the ternary systems of CaO and SiO_2 with either Na_2O or K_2O. It is true of every binary and ternary section through quaternary or quinary systems that has yet been made. Among these may be mentioned the two ternary systems of forsterite and silica with either diopside or anorthite. In many of these the eutectic mixture which has silica as one of its solid phase is the only eutectic, the relation between other compounds being of the reaction type, so that with fractional crystallization this composition acts as the eutectic for the whole system, including even mixtures which originally have no free silica, stoichiometrically. This reaction relation between compounds is, of course, due to the incongruent melting of one of them and in every case yet noted the compound breaks up in such a way as to cause a separation of crystals of a more "basic" compound and consequent throwing of free silica into the liquid. The opposite condition has never been encountered, that is, a case in which a compound breaks up with separation

[1] Petrologists have had some justification for doubting that a granite liquid would be the residuum from even a dry basaltic liquid. This lay in the supposed fact indicated by old experiments, that dry granite melts at a higher temperature than dry basalt. Since this volume went to press, it has been demonstrated experimentally in the dry way by Shepherd and by Greig that granite is converted into glass at a temperature well below that at which basalt begins to show any change. This result was no surprise to them, indeed was a foregone conclusion to anyone familiar with the thermal properties of the mineral constituents of the two rocks. The demonstration should, however, change the opinion of those who have placed reliance in the old experiments.

of crystals of a more siliceous compound or of free silica with consequent throwing of excess of "base" into the liquid.[1]

In view of these well-established facts for dry melts, it is idle to state that the presence of volatile components is necessary to the appearance of free silica in the low melting residuum from fractional crystallization of a basic magma. There is undoubtedly some formation of free silica as a result of the feldspar, mica, quartz, water equilibrium; but that the presence of water is essential to the carrying on of free silica into the late differentiate can not be reasonably maintained in the face of the facts just pointed out for dry melts. In the very late stages of the crystallization of the granite itself, when the concentration of water may become very high, it is not improbable that continued action of water upon the alkaline silicates gives alkaline solutions from which the quartz of veins and perhaps some of the very latest quartz of many granites is precipitated. But at this stage the course of what may be appropriately termed igneous differentiation is complete, and, whatever actions may then go on, in which water and other volatiles play the dominant rôle, we need not recede from our position that the whole course of differentiation and the attainment of a granitic differentiate had long since been determined by the properties of an essentially "dry" melt only slightly modified by the presence of small amounts of volatile components. If it be true that, in the genesis of some nephelite syenites, volatile constituents have played a fundamental part, this fact need not vitiate the general conclusions stated, for nephelite syenites as a whole are but a small factor in igneous geology and it is more than probable that only some of them are so formed.

It is a very useful exercise to compare the crystallization of an anhydrous molten mixture of $K_2Si_2O_5$ and K_2SiO_3 and the crystallization of the same mixture with 1 per cent H_2O added. This can be done with the aid of Fig. 82. A melt containing 50 per cent of each silicate is represented by the point M. It begins to crystallize at 900° with separation of $K_2Si_2O_5$ which continues until 780° when the liquid has the composition E. Eutectic separation of $K_2Si_2O_5$ and K_2SiO_3 then occurs at this temperature until crystallization is complete.

If 1 per cent H_2O is contained in the original mixture (M′) crystallization begins at a considerable lower temperature, 850°, with the separation of the same crystals ($K_2Si_2O_5$) as in the case of the anhydrous melt. The liquid changes in composition to E′, which is attained at a temperature of 730° and here K_2SiO_3 begins to separate. The liquid E′ has almost exactly the same proportion of K_2SiO_3 and $K_2Si_2O_5$ as the anhydrous eutectic E, the only significant difference being the pres-

---

[1] A possible exception may be noted in the case of manganese and silica. It can not yet be regarded as definitely established and in any event can have no significance for igneous rocks.

ence of water. As $K_2Si_2O_5$ and K_2SiO_3 separate, the temperature falls and the liquid changes in composition towards R. The proportions of the two compounds separating are almost identical with those of the

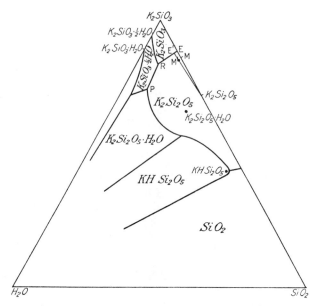

FIG. 82. Equilibrium diagram of the ternary system, K_2SiO_3-SiO_2-H_2O (after Morey and Fenner) redrawn in weight per cent.

anhydrous eutectic and the liquid R contains the two compounds in almost exactly the same proportion. At the point R, which is attained when the temperature reaches 550°, nearly 90 per cent of the original mass has crystallized, and up to this rather advanced stage of crystallization there is no difference in the nature of the compounds separating and a scarcely perceptible difference in the proportion of the compounds as compared with an anhydrous melt. At R action of the liquid upon K_2SiO_3 begins with its partial conversion to $K_2SiO_3 \cdot \frac{1}{2}H_2O$ and this action continues, with perfect equilibrium, until the liquid is exhausted. Or if only partial reaction is possible the liquid will move along RP and even during this course the proportions of K_2O and SiO_2 in the liquid are not changed notably. The relatively insignificant effect of the water upon phase equilibrium until an advanced stage of crystallization is sufficiently clear. The effect upon temperatures of phase changes and their facilitation by lowering of viscosity may, however, be very important.

Now although we have taken a specific system for which actual temperatures and concentrations have been determined it should be plain that if one takes any determined anhydrous binary system and pictures the probable effect of addition of 1 per cent of water the conclusion can hardly be very different in a qualitative way from that reached in the investigated system. From each quadruple point of the binary system a boundary curve must pass out into the ternary system which curve represents univariant equilibrium between liquid and the same solid phases as are present at the quadruple point in the anhydrous system. At a short distance from the binary side-line this boundary curve can not be displaced notably from the position of the quadruple point in the binary melt. There is no escaping the conclusion that in any system containing a small amount of a volatile component, the early stages of crystallization must be concerned with the same solid phases in approximately the same proportions as the early stages of the crystallization of the anhydrous system. To be sure, one can not claim, by analogy with the investigated example, that not until some 90 per cent of the mass has crystallized do notable effects of the volatiles appear in the way of changing the nature and proportion of solid phases. Notable effects may appear sooner in the complex examples presented by natural magmas. On the other hand they may not appear till later. In any case it can be stated that a considerable proportion of the crystallization must be but a very slight modification of the crystallization of the corresponding dry melt and the general course is determined at that stage.

The slow crystallization of basaltic magma with, say, 0.5 per cent volatiles might be expected, therefore, to be sensibly the same as the crystallization of the corresponding dry melt through a considerable proportion of its course. If the crystallization be fractional and there comes a stage when hornblende separates, certain volatiles being presumably essential constituents of that mineral, there is to be expected a significant modification of the course of crystallization as compared with that of the dry melt. Yet this modification need not be drastic. If, for example, the general course has been toward an enrichment of the liquid in alkaline feldspar it is possible (indeed it is rather probable, since hornblende bears a reaction relation to pyroxene) that the subsequent course will trend towards the same general goal though it will reach that goal by a somewhat modified route. And if fractionation be adequate to produce a liquid very rich in alkaline feldspar, with perhaps excess silica, the crystallization of this fraction may be notably different from the crystallization of the corresponding dry melt, especially if the water has in the meantime mounted to some 4 or 5 per cent or even more, as it presumably does in some cases. Nevertheless the attainment of this fraction has already been determined by processes in which the volatiles have played a relatively minor rôle. In the very late stages of the crystallization of this salic fraction, the volatiles may readily be supposed to assume not merely an important but the dominant rôle, but by this time, as we noted elsewhere, the stage of vein forming solutions has been reached and igneous differentiation proper is complete.

The principal functions of the volatiles during the crystallization of a deep-seated mass would thus appear to be to lower the temperatures of crystallization, to increase the fluidity, to facilitate the separation and interaction of phases, and to modify the course of crystallization only in moderate degree except in very late stages. Nothing that has been said here is to be taken as suggesting anything but a very important rôle for the volatiles in extrusive igneous phenomena, but such phenomena are a small proportion of igneous action as a whole and of little importance in the problem of igneous differentiation. One must agree heartily with Vogt's remark, "I cannot endorse the statement that the volatile components have been the important factor in magmatic differentiation."[1]

1 J. H. L. Vogt, *Jour. Geol.*, 30, 1922, p. 672.

CHAPTER XVII

PETROGENESIS AND THE PHYSICS OF THE EARTH

I T IS in many ways desirable to establish the connection of igneous activity with ascertained facts regarding the nature of the earth as a whole and if possible with the early history and the ultimate origin of the earth. Any system of petrogeny must, of course, be reconcilable with geophysical facts, in so far as these are facts, but it is a different matter to suppose that petrology must be based upon some chosen system of cosmogony. From the very nature of its subject-matter cosmogony must ever be less capable than petrology of reaching demonstrable conclusions. This is, perhaps, true of geophysics also, but in less degree. A brief survey of the data and of some present-day conclusions in geophysical matters may be desirable, together with some suggestion as to their connection with the advocated system of petrogenesis.

THE BROADER DENSITY RELATIONS

The average density of the earth is 5.52, whereas the average density of the rocks of the accessible crust is well below 3. From these facts it must be concluded that the material of the deeper layers of the earth is of a density considerably greater than 5.52. It has been considered that this great density of the interior can not be due to compression alone, the basis of estimate being the known density and compressibility of surface rocks and the indications as to change of compressibility afforded by earthquake waves.[1] The great density has therefore been assigned to a change of chemical composition, the common assumption being that the core is of a nickel-iron alloy analogous to the metallic constituent of meteorites. It is the contention of some investigators that the iron of meteorites is due to a secondary reducing action,[2] and if this be true the analogy of the earth with this type of meteorite substance loses some of its force. Indeed, the opinion that the great density of deep layers in the earth can not be due to compression alone takes no count of possible changes of state even of the more familiar kind, much less of the

1 Williamson and Adams, *Jour. Wash. Acad. Sci.*, 13, 1923, p. 419.
2 G. P. Merrill, *Proc. U.S. Nat. Museum*, 73, Art. 21, 1928, pp. 1-7.

possibility that, under the extreme conditions that may obtain deep within the earth, matter may be in the electronic condition supposed to exist in stars, where the density may be very high and where the ordinary distinctions of chemistry fall away since the electrons are not arranged in the groupings characteristic of chemical atoms. Opinion as to the deep layers of the earth must therefore be held in abeyance.

The sources of igneous activity can have no immediate connection with the deeper layers so that lack of knowledge of these layers is no serious matter from the petrologic point of view. Of shallower layers we have more definite indications because the properties of materials under the conditions there prevailing may be but little removed from their properties under surface conditions or under conditions which can be produced in the laboratory.

OBSERVATIONS THROWING LIGHT ON THE PHYSICAL
CONDITION OF EARTH SHELLS

The principal observations which have a bearing on the question of the character and condition of the materials of the outer portion of the earth may be listed under the following heads:

> Earthquake-wave propagation
> The geothermal gradient
> The radioactive content of rocks
> Tidal deformation of the earth
> Isostasy

The list includes phenomena which throw light on the mechanical properties and also on the thermal condition of earth substance, but since these are not independent of each other they are given together.

EARTHQUAKE-WAVE PROPAGATION

Indications as to the nature of materials at various depths is obtained from a study of the velocity of earthquake waves. The velocity of one set of waves which pass into the earth is a function of the compressibility and the density. Mathematical analysis of the apparent surface velocity has led to conclusions as to their velocity in the various layers into which their calculated course has penetrated and thus permits deductions as to the compressibility and the density of these layers. One can not obtain both compressibility and density from these results alone. The one must be known in order to obtain the other, but from the measured values of these properties in familiar rocks one may ascertain what rocks would have the appropriate combination of these properties to give the calculated velocities at various depths. These velocities increase with depth. The actual velocity of one type of wave (compression wave) near the surface is 5.6 km/sec and corresponds approximately with that calculated from the compressibility and density of granitic rocks as measured at ordinary temperatures. In addition to the general increase of velocity with depth it is said that there are abrupt increases occurring in certain layers. At one such layer the velocity appears to increase abruptly to 7.8 km/sec but there is no general agreement as to the depth at which this layer occurs. According to Gutenberg and others the discontinuity is at a depth of 60 km, but Jeffreys would place it at 37 km.[1] The velocity there obtaining (7.8 km/sec) is only a little less than that calculated from the properties of dunite at ordinary temperatures (8.2 km/sec).[2] A rock approaching dunite in composition, i.e., an ordinary peridotite, might give adequate velocities but apparently no other familiar rock would. The effect of temperature upon compressibility is not known and in stating a conclusion that the observed velocity indicates the presence of peridotite it is necessary to suppose that the effect of temperature is negligible, but since high temperature would probably affect the properties of any rock in such a way as to reduce the velocity, the existence of any familiar rock other than peridotite below the noted discontinuity becomes even less probable at high temperatures. This is the best approach to a solution that is offered and it is in qualitative harmony with what one might expect on petrologic grounds. The conclusion most commonly reached is that there is a granitic layer variously estimated at from 6 to 25 km in thickness which gradually changes through intermediate to basic rocks and these give place rather abruptly to an ultrabasic layer at a depth anywhere from 37 to 60 km. The geologic evidence as obtained in the deepest eroded sections indicates a preference for the larger of these figures.

A suggestion that has been made by some petrologists is that there is a zone of an eclogitic nature within the earth and some would place its upper limit at this shallow major discontinuity indicated by seismology. The properties of eclogite have not been determined so it is not known how this assumption may fit the earthquake data or where the zone may be more rationally placed, though the suggestion made is that it occurs above any peridotite zone that may exist.[3] In the supposed eclogite zone material of basaltic composition is regarded as occurring in the crystalline compounds characteristic of eclogite, i.e., principally as pyrope garnet and chloromelanite pyroxene. The eclogite is thus considered to be a high-pressure facies of basalt and in the

1 H. Jeffreys, *Nature*, 118, 1926, p. 443.
2 Adams and Gibson, *Proc. Nat. Acad. Sci.*, 12, 1926, pp. 275-83.
3 L. L. Fermor, *Geol. Mag. Decade VI*, 1, 1914, pp. 65-7; P. Eskola, *Norsk Geologisk Tidskrift*, 6, 1920, p. 182; V. M. Goldschmidt, *Vidensk. Selsk. Skr. I. Mat.-Naturv. Kl.*, 1922, No. 11, p. 4.

present state of knowledge the possibility of such a layer must be reckoned with.

The temperatures that may exist at various depths within the earth present a very difficult problem. Temperatures at very great depths, like physical and chemical conditions at those depths, need not concern us here. We have evidence from man-made sections of a continuous rise of temperature which varies considerably from place to place. The actual depth which these sections attain is utterly insignificant and while extrapolation of the temperature curve would rapidly carry us to very high temperatures, the real evidence of the existence of high temperatures anywhere in the earth comes from the fact that, at frequent intervals in the earth's history, rocks which are molten only above 1000° C have come from the depths into the outer crust in the liquid state. They have not only been injected into the crust but have also been poured out upon the earth's surface and the process is still going on, so that from time to time man has had the opportunity of seeing these molten rocks and even of measuring their temperature.

The earliest views as to the origin of this high temperature were that the earth had once been molten and was, indeed, still molten with the exception of a thin outer crust. With increasing knowledge of the physics of the earth it has been necessary to abandon this simple form of hypothesis. A modification was proposed by Kelvin in which blocks of the outer crust sank into the earth, at first remelting, but in doing so, gradually reducing the general temperature to a value where remelting was no longer possible and thus bringing about a honeycombed-solid condition of the whole earth. With the attainment of this relatively stable mechanical condition, cooling took place almost solely by conduction and it was supposed that, with the aid of the present thermal gradient and the mathematical theory of heat conduction, temperatures at a depth could be calculated, as well as the lapse of time since this stable condition was established. The actual time so calculated turned out to be so short that geologists were unable to accept it as adequate for the formation of the known thickness of sedimentary beds and for the development of life as recorded in them. With the discovery of radioactivity and the determination of the quantities of radioactive substances in rocks, the source of heat thus supplied made it possible to abandon this embarrassingly short period of time. However, it led to a difficulty of another kind, for the actual amount of radioactive substances in surface rocks is such that, if these rocks extended to great depths, the heat being lost by the earth, as indicated by the thermal gradient, would be far less than that produced. The earth would thus be heating up.

The plain facts are that there is nothing that can be called definite knowledge to preclude the possibility that the earth is heating up. Some geologists had on other grounds already abandoned the concept of a once-molten earth and had turned to an earth built up by the accumulation of cold planetesimals formed by the tidal disruption of the sun upon the approach of another star.[1] They found no great embarrassment in the discovery of the presence of radioactive substances and in the possibility that the earth might be heating up from that cause.

The present tendency is to return to the concept of an earth formerly molten. It is based on a reconsideration of the problem of tidal disruption[2] and an analogy of the history of the earth with the supposed history of other planets. Attempts have been renewed to deduce the story of the cooling of a molten earth.[3] The same assumption as that of Kelvin (or another which leads to the same result) has been made as to the early stages of cooling and the establishment of a stable condition which permits cooling only by conduction. In present attacks on the problem, the age of the earth is taken as known from the uranium-lead ratio in the oldest intrusive rocks. From this and from the observed thermal gradient the thickness of the radioactive shell is calculated, that is, the depth to which rock having the same quantity of radioactive matter as surface rock may be assumed to extend. The conclusion reached is that a thickness of 10-20 km of average granite, which type of rock has more radioactive substance than any other common type, would account for all the radioactive substance of the earth. These newer attacks upon the thermal problem of the earth thus lead to some conclusions as to the distribution of various kinds of rocks in the crust, imposing narrow limits to the possible downward extension of granite, and it is towards this aspect that attention is now drawn. Below the level noted there must be no radioactive substance or it must fall off so rapidly that its quantity soon becomes negligible. This result is not acceptable. Below the granite layer there must, in any rational scheme of things, be intermediate and basic rocks whose radioactive content is far from negligible. Below these again the ultrabasic rocks must be regarded as having the lower but still important quantity of radioactive substances usually found in ultrabasic rocks. A falling-off in radioactive content is to be expected but the drastic rate of fall indicated by the calculation is out of the question. Examination of the data used in the calculation shows that they should be modified in a direction

1 T. C. Chamberlin, in Chamberlin and Salisbury, *Geology*, II, 1905, pp. 1-81.
2 Jeans, *Problems of Cosmogony and Stellar Dynamics*, 1919.
3 A. Holmes, *Geol. Mag.*, 1915, pp. 60, 102; 1916, p. 265; H. Jeffreys, *The Earth*, Cambridge, 1924, p. 89; L. H. Adams, *Jour. Wash. Acad. Sci.*, 14, 1924, p. 459.

which mitigates the difficulty somewhat but probably without eliminating it altogether.

The calculation of the thickness of the radioactive shell depends almost solely on a knowledge of the present rate of heat loss from the earth and very little on the past history. With the aid of the thermal gradient and the conductivity of rocks the heat loss is calculated and from this the total radioactive content of the earth,[1] or better, the thickness of an outer shell with a content of radioactive substance equal to that found in acid rocks such as occur at the surface.[2] It is to be noted that a knowledge of the kind of rocks occurring in depth and the gradient there obtaining is unnecessary. It is sufficient to know the gradient in any layer and the conductivity of that layer. If a lower layer have a lower conductivity it will have a correspondingly higher gradient. The conductivity used should therefore be that of the kind of rocks in which the bore holes have been sunk and not an average conductivity of all rocks including those of supposed basic and ultrabasic shells. Usually the rocks concerned are sedimentary rocks more or less porous and impregnated with water.

The thermal gradient has been measured in continental areas and found to be variable. Where it is found unusually high this value is ordinarily rejected in estimating an average, on the assumption that the high value is due to exceptional causes. However, the high values are of equal significance with the low values in the problem of the earth's heat loss. Some two-thirds of the earth's surface is occupied by the oceans and there the thermal gradient is altogether unknown. The earth's "crust" is presumably thinner there and the gradient greater. It may be several times that in continental areas.

The assumption involved in the calculation is that all heat loss is by conduction, yet we know that there is loss from other causes. A single warm spring in Montana has a temperature only a little above the mean annual temperature, yet its flow is such that heat is transferred to the surface at a rate equal to the loss by conduction over 200 sq. kilometers.[3] Springs of such magnitude are few in number but the aggregate heat transfer of small warm springs, individually insignificant, may be very great. Hot springs transfer great quantities of heat.[4] Zies[5] estimates as a minimum that in the Valley of Ten Thousand Smokes 2.6 x 10^7 litres of steam were evolved per second in 1919. This steam, taken at 100° though much of it is far above that temperature, condensed to water and cooled to 10° C accounts for the bringing to the surface of about 10^9 cal per sec.

1 Rutherford, *Radioactivity*, Cambridge, 1905, p. 493.
2 Strutt, *Proc. Roy. Soc.*, A, Vol. 771, 1906, p. 475.
3 "Warm Spring, Montana," *U.S. Geol. Surv. .Water Supply Paper 557*, 1927, p. 82.
4 Allen and Day, "Steam Wells and Other Thermal Activity at 'The Geysers,' California," *Carnegie Inst. Wash. Pub. No.* 378, pp. 100-2.
5 E. G. Zies, *Jour. Geol.*, 32, 1924, pp. 309-10.

This small area thus accounts in steam alone, without the additional hot water circulation, for a heat loss of at least 1/3000 of all that lost by conduction over the land surface of the earth. The minimum estimate of Zies might not impossibly be increased several fold. What may be the total heat loss of all the "smoking" volcanoes on the earth there is no means of knowing but it may be very high. Every outflow of lava transfers heat to the surface by a method which does not depend upon its slow flow along a low temperature gradient. The total loss of heat from continental areas may easily be more than double that calculated from the thermal gradient. There is reason to believe that submarine flows may be more common than continental flows. On the whole the total loss of heat from the earth may be many times that calculated from the thermal gradient. The restriction to be placed upon the downward extension of radioactive matter need not therefore be as drastic as that indicated by the simple assumption made in the calculation.

The calculation involves, of course, the fundamental assumption that the earth is not now heating up. Of this we have no assurance. Even if the earth's history is one of general cooling there is nothing in this to preclude the possibility that at the present time it might be heating up or perhaps that sub-continental areas may be heating up but the whole earth cooling down.

Joly has attempted to solve the earth's thermal problem by abandoning altogether the concept of gradual cooling and avoiding the embarrassment of radioactive heating by assuming that there has been periodical discharge of heat in great quantity by outpourings of magma upon the ocean floor. These periods are separated by long periods of slow reheating of the earth by the heat of radioactive disintegration. The theory involves the periodic remelting of a basaltic substratum, which at the present time is crystalline and thus capable of transmitting distortional seismic waves. Joly attempts to explain isostasy, the rhythm of mountain-building and many other geologic observations with the aid of his theory.[1] The objection has been raised that there is no means of obtaining re-solidification of the basaltic substratum once it is melted. There are many other objections to details such as the supposed persistence of solid granite in contact with liquid basalt, Joly thus subscribing to the age-old fallacy that granite melts at a higher temperature than basalt. In spite of the various objections it is not improbable that an hypothesis involving periodic discharge of heat in great quantity will ultimately be preferred. Holmes now offers adherence to some such concept.[2] It is, on the whole, perhaps less unreasonable than the assump-

1 J. Joly, *The Surface History of the Earth*, Oxford, 1925.
2 *Geol. Mag.*, 62, 1925, p. 504.

tion of drastic rate of decrease of radioactive substance in depth that is involved in alternative hypotheses.

The significance of these general considerations, from the petrologic point of view, is that no definite knowledge is inconsistent with the view that there is a granitic shell as much as 25 km thick and a passage through intermediate to basic rocks of normal radioactive content without attaining the ultrabasic rocks until a depth of some 60 km is reached.

The newer methods of treating the earth's history as a problem in heat conductivity attempt to determine not only the distribution of radioactive matter but also the probable internal temperatures. Geological evidence brings to light many objections to regarding the heat-loss as due entirely to a flow along a low temperature gradient. There have always been periods of rise of molten magma into the outer crust of the earth, often of colossal magnitude. In these intrusive magmas heat was transferred by a method independent of the slow flow of conduction and its escape was facilitated by greatly steepened gradients. We can not therefore arrive at an adequate idea of temperatures within the earth by a method which assumes that at present and for hundreds of millions of years in the past, the sole heat-loss has been by conduction along a gradient which changed gradually and regularly according to the requirements of mathematical theory. There is no means of estimating the effects of intrusive and extrusive masses so there is no reliable means of estimating temperatures at depth.

Upon crustal temperatures definite knowledge is thus reduced to the fact of a rising temperature which somewhere within striking distance of the surface is such as to render possible the production of the most refractory magma we know—basaltic magma. What interpretation is to be put upon the expression "striking distance" we can not ascertain from any consideration of temperature distribution alone. There are, however, some indications which depend upon wholly different factors.

TIDAL DEFORMATION AND DISTORTIONAL SEISMIC WAVES

Two lines of evidence prove that, not only the earth as a whole, but also the individual shells of which it is composed have a very high degree of rigidity. The information comes from the manner of propagation of earthquake waves and from the magnitude of the tidal deformation of the earth's body. Some consideration has already been given to earthquake waves. Besides the compressional wave, whose velocity has been discussed and whose transmission would be accomplished even in a liquid medium, there is a distortional wave which can be transmitted only by material having elasticity of form or rigidity. This wave is freely transmitted in all layers (with the possible exception of a core) and indicates for all layers a high degree of rigidity. Tidal deformation of the earth is so small as to indicate an average rigidity greater than that of steel. The high rigidity is usually regarded as indicative of crystallinity and probably this is true for a considerable depth, but we can not be sure that excessively high pressure would not endow non-crystalline matter with high rigidity; indeed, highly rigid liquid (glass) has been assumed to exist at quite moderate depths, as we shall see.

Another type of evidence as to the properties of rock masses in depth is afforded by the condition known as isostasy. It is found by measurements of the force of gravity that the major elevations of the earth's surface are almost completely compensated by a deficiency of mass beneath them. The fact seems well established but the nature of the compensation and the depth at which it is complete are matters of interpretation upon which there is no general agreement. Hayford obtains about 120 km as the best depth of complete compensation on the assumption of uniform distribution of compensation but shows that values of from 60-300 km fit the data about as well. He also points out that the data give equal support to the assumption that all compensation takes place in a layer 10 miles thick whose top is at 25 miles depth. The net result is such as would be produced if the crust of the earth were floating in a fluid medium, the lighter segments floating with more free-board than the heavier, though, in the face of other evidence of rigidity, few go the length of advocating the intervention of a liquid. This flow under load may be reconciled with the small tidal deformation when it is realized that the forces induced by tidal pull are of short duration, whereas isostatic compensation results from forces applied for long periods.

THE SOURCE OF MAGMAS

It is not primarily the province of petrology to bring about reconciliation of such apparent discrepancies as may be found in data of this sort. Petrology as such is interested merely in the light these data may throw on the problem of the source of magmas.

There is, at present, a very distinct trend of thought towards identifying the problem of the source of magmas with the problem of the source of basaltic magmas. To be sure there is no general agreement upon the possibility of the derivation of all magmas and rocks from basaltic magma, nor among those who do accept the possibility is there agreement as to the method of derivation. Having regard for the probable constitution of the whole earth, as indicated in the foregoing outline of data, we can see little chance of escape from the necessity of deriving all magmas *ultimately* from matter much more basic than basalt and probably from peridotitic substance as represented in stony meteorites. However this may be, it can scarcely be denied that basaltic magma is of great importance and that a solution of the problem of the source of basaltic magma would be at least an important step. The present time itself seems to lie within a period of production of basaltic magma. It is also a period of rigid crustal shells. It seems necessary therefore to suppose that basaltic magma is now produced and perhaps has been in the past under conditions of general earth rigidity.

Those who are convinced of the fundamental character of basaltic magma have usually sought its source in a layer of basaltic composition. As to the state of this layer there is the widest divergence of views. According to certain views, as advocated principally by Daly, it is above its melting temperature at the pressure there prevailing, but this pressure is adequate to induce the proved high rigidity.[1] It is thus a glass but ready to become thin-fluid on relief of pressure. According to others it is crystalline and therefore below its melting temperature at the existing pressure but above its melting temperature at lower pressure and thus, again, ready to become fluid on relief of pressure. This latter view is still further modified by some few who hold that it is not only crystalline but exists in the high pressure facies, eclogite, liquid becoming available in a similar manner.[2] As already noted, a different suggestion as to the manner of liquefaction has been made by Joly who concludes that the basaltic layer is periodically remelted as a result of the secular accumulation of the heat of radioactive disintegration.[3]

Of these various methods of obtaining fluid basalt from a layer of basaltic composition Daly's is probably the best, for, as we shall see in the sequel, it is questionable whether basaltic liquid would be the usual product of remelting of a crystalline basaltic layer by whatever method. Nevertheless it is difficult to believe that the pressure existing at moderate depths, say 50-60 km, would be adequate to endow a basaltic liquid, above the melting temperature, with the rigidity that earthquake waves seem to demand. The maximum effect obtained by Bridgman on the viscosity of 43 pure liquids was a 10^7-fold increase produced by 12,000 atmospheres. This amount was not actually determined but is extrapolated from a 10^3-fold increase at 6000 atmospheres.[4] Bridgman used pure liquids, no doubt largely that his results might be definable and reproducible but presumably, also, in the belief that when freezing ensued it would be sharp and complete at a point and its occurrence, therefore, definitely ascertainable. But even a pure compound may melt or freeze incongruently and therefore through a range of temperatures (or pressures at a given temperature) so that it is by no means certain that Bridgman was not measuring the viscosity of a mush in the case of

1 R. A. Daly, *Proc. Amer. Phil. Soc.*, 64, 1925, p. 283.
2 Fermor, *op. cit.*, and A. Holmes, *Geol. Mag.*, 64, 1927, p. 266.
3 J. Joly, *Surface History of the Earth*, Oxford, 1925.
4 P. W. Bridgman, *Proc. Amer. Acad. Arts Sci.*, 61, 1926, pp. 57-99.

those substances which showed very great increase of viscosity with pressure. Even if the maximum indicated effect of 10^7-fold is a real increase of liquid viscosity, it is doubtful whether a corresponding increase in the viscosity of basaltic liquid over that existing at ordinary pressure would be adequate to endow it with rigidity to earthquake waves, and it is even more doubtful in the case of the distortional wave of longer period caused by tidal forces.

The matter may be viewed in another light. At ordinary temperature and pressure basaltic glass is, mechanically, a solid possessing a definite rigidity. The temperature coefficient of this rigidity is enormous and is such that a rise of about 1000° C wipes out all semblance of rigidity and converts it into a thin liquid. At high pressures and ordinary temperature basalt glass is again a rigid substance, but if it is to be such a substance at high temperatures and high pressures it is necessary to suppose that this enormous temperature coefficient of rigidity at low pressures does not exist at high pressures.

Adams and Gibson have measured the compressibility of basaltic glass at ordinary temperature and find that the compressibility and density are such that the calculated speed of longitudinal earthquake waves in it would be approximately 6.4 km/sec.[1] There is apparently a layer in the crust where such a velocity obtains but the temperature is undoubtedly rather high there, and if we assume the layer to be composed of basaltic glass, the implication is that the effect of temperature on rigidity is negligible. A crystalline rock of intermediate composition, a diorite, has such properties at ordinary temperature that it also would give a velocity of 6.4 km/sec. If we assume that the layer in question is crystalline diorite it is again necessary to believe that the effect of temperature on rigidity is negligible. But the making of this assumption in connection with crystalline diorite is a different matter from making it in connection with glassy basalt. It is exceedingly improbable that the temperature effect on glassy basalt would be negligible whereas it is within the bounds of probability that it would be small with a crystalline material not too close to the melting temperature. The preference is decidedly for crystalline diorite below its melting temperature in the layer where compressional waves have the 6.4 km/sec velocity rather than for glassy basalt above its melting temperature in that layer. Granting that the layer in question is crystalline diorite there seems to be no place either above it or below it for a layer of basaltic glass. Above crystalline diorite there could hardly be molten basalt in any persistent arrangement of temperatures, and below it earthquake wave velocities are apparently too high for basaltic glass.

The advantages of the supposed basaltic glass as a producer of basal-

1 Adams and Gibson, *Proc. Nat. Acad. Sci.*, 12, 1926, pp. 275-83.

tic magmas, as we know them, are not as great as they may seem to be at first glance. It requires relief of pressure to convert such glass into a thin liquid and it is equally possible for relief of pressure to convert crystalline material into liquid. Moreover, the liquid so produced from unsaturated basalt glass would be enormously superheated under the new pressure whereas the liquid so produced from crystalline rock would be just saturated under the new pressure.

The elastic properties of the earth appear to reduce us to the necessity of considering the remelting of crystalline material. As already stated it is not likely that crystalline basaltic substance would usually give basaltic liquid in any process of remelting. Secular reheating of the mass would give much liquid at a temperature well below that requisite for complete melting. The liquid portion would not have a basaltic composition, but would be more salic and would be capable of intrusion into the upper crust long before a temperature was attained adequate to remelt the whole basalt. The magma commonly injected into the crust, if it is to be produced by slow (radiothermal) reheating of crystalline basalt, would thus not ordinarily be of basaltic composition.

A fact not realized by any of those dealing with the subject is that remelting of crystalline basalt, consequent upon relief of pressure, would inevitably lead to the same result. The relief of pressure must be a decidedly transitory matter, for as soon as the liquid flows upward to the region of lower pressure, the superincumbent pressure at the zone of production of liquid must resume nearly its original value. In view of the brief interval of time during which the lowered pressure can be regarded as existing, the remelting must be approximately adiabatic. The mass undergoing remelting must find practically all the heat of melting within itself and its temperature must therefore be lowered. In the case of a pure silicate compound, already at its melting point under the higher pressure (but not melted), the temperature would fall to the melting point under the lower pressure and a fraction of the substance would melt which is approximately determined by the latent heat of melting and the specific heats of the liquid and solid. A crystalline silicate at its melting point, having a latent heat of melting of 100 cal/gm and a specific heat of both solid and liquid of 0.25 cal/gm would be melted to the extent of about 40 per cent of its mass by a release of pressure sufficient to lower its melting point 100° C. If it was originally at a temperature lower than the melting point at the higher pressure, a lesser amount would melt, and of course if it was at a temperature below the melting point at the lower pressure, no melting would occur. For basalt, or any material which melts through a temperature interval, the amount of melting (produced by the same change of conditions pictured above) would be somewhat less than that for a pure substance. Now the constants given above are roughly those of basalt, so it would require a relief of pressure corresponding with the unloading of 25 km of rock to bring about the melting of about 40 per cent of a basalt mass and the liquid so formed would not be basaltic. According to the recent trend of thought regarding internal temperatures of a continuously cooling earth, the temperature is far below the surface melting temperature of basalt at a depth of 25 km, in fact, the melting temperature of basalt is attained only at a depth of some 50-60 km.[1] Even through-striking fissures establishing a connection with the surface and extending to a depth of 75-90 km would reach a zone of temperature (according to such estimates) not more than about 100° above the surface melting temperature of basalt and could induce the melting of only about 40 per cent of a basaltic mass, were such a mass situated at that level, and the liquid produced would not be basaltic. But at 37-60 km we encounter material of a peridotitic character, according to the indications of earthquake phenomena, and if this be true the only source of basaltic magma that appears available on the doctrine of a continuously cooling earth lies in the peridotite zone, from which it must be produced by selective fusion. Neither assumption as to earth history can represent the full truth. Cooling has certainly not been by conduction alone but the simple rhythmic reheating advocated by Joly may be as far from the truth. There is therefore no assurance as to where the zone of possible production of liquid may be, but on the doctrine of periodic reheating by radiothermal energy we have seen that basaltic substance would give rise by selective fusion to non-basaltic liquid and that, at a time when there was enough liquid to render its intrusion possible, the liquid would be considerably more salic than basalt. Whichever assumption is made as to the thermal history of the earth, we find the most probable source of basaltic magma in the selective fusion of a portion of the peridotite layer. It is, perhaps, worth while to examine somewhat more closely the selective fusion of a peridotite layer.

PRODUCTION OF BASALTIC MAGMA BY SELECTIVE FUSION
OF PERIDOTITE

Washington has sought to obtain an idea of the composition of the earth's peridotite shell on the assumption that it approaches that of achondritic meteorites. The assumption appears to be reasonable and his figures, based on the average composition of 20 such meteorites, will here be accepted as a working basis.[2] The peridotite shell recalculated to simple minerals has, according to Washington, the following composition, omitting minor constituents.

---

[1] L. H. Adams, *Jour. Wash. Acad. Sci.*, 14, 1924, p. 468.
[2] H. S. Washington, *Amer. Jour. Sci.*, 9, 1925, pp. 357-63.

Olivine	12.82 per cent		
Hypersthene	43.92 " "		
Diopside	18.33 " "	⎫	
Anorthite	13.23 " "	⎬ 39 per cent	
Albite	5.83 " "		
Orthoclase	1.69 " "	⎭	

If this material were subject to refusion there would be a stage at which all the orthoclase, albite, anorthite and diopsidic pyroxene, and little else, would be liquefied and the liquid would constitute about 40 per cent of the mass.[1]

The composition of the liquid expressed as mineral molecules would be

Orthoclase	4.5
Albite	15.0
Anorthite	33.5
Ferriferous diopside	47.0

This liquid could therefore crystallize as 53 per cent plagioclase approaching Ab_1An_2 and 47 per cent pyroxene. In the actual content of the orthoclase molecule and the relative amounts of plagioclase and pyroxene such material is essentially basaltic, but it is perhaps not as close to the great lava floods of fissure eruptions, the plateau basalts, as material in which the plagioclase is somewhat less basic. The rock substance suffering selective fusion may, therefore, be assumed to differ somewhat from the peridotite shell as calculated by Washington, and this difference would be towards the probable average of all meteorites, for some meteorites have their plagioclase dominantly in the form of oligoclase. A source of plateau basalts in the selective fusion of peridotite substance analogous to meteoritic material is, therefore, quite within the bounds of possibility and even of probability. Even definite proof that the plateau basalts are exactly the material that would be produced by selective fusion of meteorite substance would not, of course, prove that they are so produced, because the fractional crystallization of meteoric matter would give the same result, so that the basaltic layer of a differentiated earth would be of essentially the same composition as the basalt formed by selective fusion. Having regard, however, for the fact that re-fusion would be selective, there would appear to be some preference for the assumption that basalt is derived by selective fusion of peridotite, if from crystalline matter of any kind.

It seems necessary to leave open the question whether selective fusion takes place as a result of release of pressure or as a result of reheating.

---

[1] More intensive action would result in the formation of a greater proportion of liquid. This would be formed by solution of hypersthene and olivine in the above.

It is often stated that the strength of rocks is such that no arch of any notable areal extent could be sustained and that release of pressure over any considerable mass of rock is therefore impossible. There is no assurance that the actual problem is so simple. In a shrinking earth there would presumably always be a tendency towards an arching of an outer layer of the crust and this tendency might be realizable if liquid were available to flow in beneath the arch, especially a heavy liquid which could float the arch. If in any way a connection were established between the zone of potential arching and a deeper zone of potential formation of liquid, the two actions might be realized simultaneously, the occurrence of the one being dependent upon the occurrence of the other. An outer shell of compression and of potential arching must be succeeded downward by a zone of tangential tension and it is possible that connection with a deeper zone of potential formation of liquid may be established by development of radial fissures in the zone of tangential tension. The level of no strain, lying between the zones of compression and of tension, would then constitute a level of important intrusive masses originating in the postulated manner. Suboceanic areas would have somewhat different mechanics and in them there would be no level of important intrusive masses. The action pictured would be sensibly the same as Daly's conception of abyssal injection[1] but the local and transient release of pressure, obtained when connection is established with the level of no strain, would be the controlling factor in the formation of basaltic liquid selectively from crystalline peridotite. Daly is under a similar necessity of finding a mechanism of release of pressure in order to convert pressure-rigid basaltic glass into fluid basalt.

On the other hand it may be that the postulated means of obtaining release of pressure is not available or is inadequate and that selective fusion of peridotite is periodically accomplished by reheating. Such fusion would be very different from the re-fusion of a basaltic mass. Having once suffered this action and the draining off of the liquid a mass of peridotite would thenceforth be barren. The zone of possible production of basaltic magma would thus migrate, presumably downward, and there would be some rationale to the assumption of the lapse of a considerable period of time before a new period of important magma development. Time would be required for accumulation of enough (radioactive) heat to melt an adequate proportion of a new layer of peridotite.

There is, however, an avenue of escape from the selective fusion of peridotite which depends on the possibility, suggested by Holmes,[2] that eclogite would have sensibly the same elastic properties as peridotite.

1 R. A. Daly, *Our Mobile Earth*, New York, 1926, pp. 134 *ff*.
2 A. Holmes, *Geol. Mag.*, 64, 1927, p. 266.

We might then assume that the 37-60 km discontinuity is the upper limit of an eclogite layer and that peridotite comes in only at greater depth. Holmes' location of the eclogite zone at 15 km depth under the continents and only 4 to 8 km depth under the ocean basins is hardly acceptable. It would necessitate that gabbroid magma would normally crystallize to eclogite at a depth of only 5 to 9 km beneath continental areas, and this we may feel reasonably sure is not true.

If there is eclogite at a level of from 37-60 km and if we may accept Holmes' further assumption that eclogite, eutectic-like, melts as a unit, then, by re-fusion of eclogite in any manner, basaltic magma might be expected. The whole question of eclogite as a possible high-pressure facies of gabbro is, however, still open. Moreover, the complex nature of the mineral components of eclogite renders it unlikely that the rock would melt as a unit; indeed, the field associations of igneous eclogites suggest that they are a special form of ultrabasic differentiate from gabbroid magma. In the present state of knowledge it would appear to be necessary to leave open the possibility of the production of basaltic magma by the non-selective fusion of eclogite whether by reheating or otherwise.

It is to be noted that the production of basaltic liquid by the remelting of eclogite and its production by selective fusion from peridotite are not mutually exclusive. The gabbroid constituents of the peridotite in the zone where liquid can be produced by release of pressure may exist in the eclogite facies. As such they would still constitute the low-melting portion and, upon production of liquid and its rise to higher levels, it could crystallize there as gabbro or give rise to a differentiation series.

The possibility that gabbroid substance may exist in the eclogite facies at no very great depth presents interesting features in connection with the problem of general earth-differentiation, if there be such. What the possible differentiates of eclogite may be we do not know but we may be very sure that they would not be what we know as normal, sub-alkaline rock series containing diorite and granite. The production of a granitic differentiate from basaltic magma depends upon the physico-chemical properties of the feldspar series and the hypersthene-augite series which crystallize from basaltic magma under lower pressures. Both of these are absent in eclogite. If there is a general level below which gabbroid substance must exist in the eclogite facies then there could be no tendency for that portion of the earth to give a granitic differentiate in any general earth-differentiation that may have occurred. Only when gabbroid liquid rises above that level is it possible to produce a granitic differentiate from it. If some of the recent, very low estimates of the thickness of granite in the crust should be substantiated, the explanation may lie in the very thin shell of the earth which can participate in the common type of differentiation. On the other hand if there is really a 25 km layer of granite it would seem necessary to place the minimum possible depth of a zone of eclogite at some 200 km.

Another feature of the selective fusion of peridotite material to give basaltic liquid may be noted. Even if the material fused is of a gabbroid and not an eclogitic character, it will have no tendency to differentiate towards granite or indeed to differentiate in any way so long as it remains as an interstitial liquid in contact with the peridotite minerals. It must there recrystallize (upon increase of pressure) to gabbroid constituents. Only when the liquid is separated from the peridotite minerals by rising to a higher level can it differentiate in any way.

On the whole the production of basaltic magma by selective fusion of peridotite at a depth probably as great as 75-100 km seems to be the preferable method in spite of the great difficulties involved, but it will be plain from the foregoing discussion that the problem of the origin of magmas is far from approaching definite solution. Geophysical data help very little. All that can be said is that they are apparently consistent with the derivation of basaltic magma in the manner described. The question of the parental nature of basaltic magma can not be given a definite answer. It is becoming increasingly probable that all magmas could be derived from the basaltic by crystallization-differentiation, but even were this a proven fact it would not follow that magmas are so derived. The decision on this point would still depend principally upon geologic considerations; indeed it was largely such considerations, rather than petrology, that led to the suggestion of the parental nature of basalt.

Geologic evidence shows that from Keewatin time to the present, a source of basaltic magma has been available, perhaps not continuously but certainly at very frequent intervals. Whether it is currently produced or whether it was produced long ago by earth-differentiation, we have no means of deciding. And so we must leave open the question whether the various differentiates of basaltic magma were currently produced or were in some cases produced by a primordial differentiation. Many granitic magmas may have their immediate origin in the remelting, say by deep burial, of a granite derived in more remote times from basic material. Yet in most igneous cycles that contain basaltic rocks occurring as regional dikes or as copious surface flows, and most cycles do, it seems necessary to appeal to basaltic parental magma and to the derivation from it of other magmas of the cycle. Crystallization-differentiation could not give approximate uniformity of composition in basalts if they were the basic differentiate of an intermediate magma. The compositions of all rocks more basic than the parental magma would be sporadic. The ultrabasic rocks, those more basic than basalt, are the rocks which show this character, whereas all rocks less basic tend to cluster around certain liquid lines of descent.[1] This is found to be true upon a general survey of rock compositions. If that were the whole story it might be due to a remote derivation from basaltic magma. But the same facts are true of most, if not all, individual igneous cycles so that there are rather strong reasons for regarding the magmas of these cycles as immediately derived from basaltic magma, itself currently produced.

The production of parental basaltic magma by selective fusion of more basic material, consequent upon release of pressure, would account satisfactorily for the general lack of liquid magmas produced by refusion of early crystals which sink into lower layers of a magma chamber, in other words for the lack of such liquids as the peridotitic. The liquid produced by release of pressure is just saturated, not for the conditions normally prevailing in the zone where it is produced, but for the conditions prevailing in the zone to which it may rise as a result of the release of pressure which has caused its production and beyond which it can not rise because the (back) pressure so created would cause recongelation at the source. If the magma rose as a result of any such action and spread out at a certain level in the crust as a cake-like mass (a magma chamber) it would be just saturated in its new surroundings and incapable of redissolving its own early crystals in any part of its mass. The mechanism which contributes to the production and rise of magmas may thus be one which insures that they shall be, ordinarily, saturated magmas. When the relief of pressure is accomplished by actually establishing a connection with the surface of the earth the magma which flows out is ordinarily found to be saturated, that is, to contain at least a small proportion of crystals.

If magmas came from a deep zone where they are unsaturated even under the conditions there prevailing they would have an abundance of superheat when they rose to higher levels. The evidence points to the general lack of such superheat even in the plateau basalts, which we may regard as parental magma, and suggests a preference, therefore, for the production of these magmas by selective fusion of peridotite material caused by release of pressure.

The main purpose of this chapter is not, however, to express advocacy of any particular mode of derivation of magma but rather to indicate how inadequate are the data now available to permit a definite decision on this point. The early stages of organic evolution—the early stages of the development of the human race—are shrouded in mystery. So it is with the early stages of the development of a magmatic cycle.

1 In this connection Fig. 37 may be re-examined with profit.

CHAPTER XVIII

THE CLASSIFICATION OF IGNEOUS ROCKS

FOR a long time petrologists have sought what has been termed a natural or genetic classification of igneous rocks. The search has not been rewarded with success; at least, so it is said. The hope seems to have been entertained that some new principle would be discovered upon which the genesis of rocks depends and that with its discovery rocks would be found capable of division into genetic classes which would transcend the supposedly artificial classifications based on the mineral make-up of the rocks. In the organic kingdom a classification had been built up—a purely artificial one based on the morphological characters—long before the evolutionary principle that governed their development was more than dimly appreciated. Upon full appreciation of this principle it was not found necessary to discard existing classification. For the most part the establishment of evolution gave merely a fresh impetus and a new meaning to existing classification. This lack of need of a new classification was due to the fact that classification had been based on morphologic characters and these were the fundamental units which responded to the factors controlling the process of development.

The situation will not be found very different in the case of rocks when a general understanding of the process of their development has been reached. The minerals of which rocks are composed are the expression of a response to the conditions of their genesis. A genetic classification can not transcend a classification based on mineral composition and by this is meant, of course, modal mineral composition.

Upon the actual kind of mineralogic classification a word may be said. There has long been a great clamor for emphasis upon the quantitative element in classification. This has been in some measure due to the use of the term, rock species, to designate a rock type, the implication being that there is something very definite and precise about the so-called rock species. If any analogy is to be drawn between the animal and the mineral kingdom it should consist rather in a comparison of mineral species with animal species. And if it is to be carried further this can be done by comparing rock types with faunas. Just as a fauna is a collection of animal species resulting from a response to external conditions and to each other, so a rock type is a collection of mineral species resulting from a response to external conditions and to each other. With this analogy in mind we get a rational view of the importance of the quantitative element in rock classification. How artificial a classification of faunas would be which was based on the ratio of foxes to hares, of hares to moles and so on! To be sure it is by no means accidental that the ratio of hares to foxes is 10 in a certain area and only 2 in another, but as compared with the broad factors controlling life in the two areas it is relatively accidental. A classification of the two faunas on the basis of the ratios noted might serve a utilitarian purpose from the viewpoint of a trapper, but it would have little scientific value. So it is not accidental that a rock is nearly pure olivine here and only 75 per cent olivine a few feet distant, but it is relatively accidental and should not be made a fundamental factor in classification.

We have at present a classification which is characterized by a looseness in its quantitative aspects. It takes the nature of the plagioclase, if present, as of fundamental importance. Since the nature of the plagioclase is an indication of the stage of development from an original magma, the use of that character is amply justified from the genetic point of view. It takes the nature of the colored constituent, whether olivine, pyroxene, hornblende or mica as a fundamental character also and, since this is again dependent upon the stage of development from an original magma, this feature of the classification is desirable. The same statement may be made of the use of the presence or absence of quartz and of other features of the classification. It is the ordinary, familiar, modal classification of rocks, often known as the Rosenbusch classification. The quantitative element is not lacking but it is relatively unimportant. The looseness of the classification is not due to our lack of knowledge of rocks but to our very knowledge of them. No implication is intended that our knowledge is all that could be desired. That can never be. But with increase of our knowledge of rocks as members of series and of these series as members of super-series there will come a greater appreciation of the desirability of loose classification. A mineralogical classification is a genetic classification since minerals are the fundamental genetic units and there is little prospect that the present loosely quantitative, modal mineral classification will be superseded. Other systems may serve a temporary or a collateral purpose, but the system now most commonly used is the natural system of classification of igneous rocks.

INDICES

GENERAL INDEX

abyssal injection, 317
accumulation of crystals, effects of, 26-27; *see also* settling of crystals
ADAMS, F. D., 216
ADAMS, L. H., 68, 183, 303, 305, 307, 313, 315
addition-subtraction diagrams, 76, 88
adiabatic reaction, 186
age of the earth, 306-307
albitization, 132
alkaline feldspars, enrichment of liquids in, 33, 46, 79-85
alkaline lines of descent, 129
alkaline rocks, 5; origin of, 207, 215, 216, 234-257
ALLAN, J. A., 257
ALLEN, E. T., 117, 308
allivalite, 164, 165
alloys, equilibrium in, 25
alnoite, origin of, 259-268
alumina, form of curve for, 97-117
aluminous sediments, reaction with magmas, 207-214
analyses of peridotites, 154
ANDERSEN, OLAF, 29, 41, 194
andesites, glassy groundmass of, 118, 120
anorthosite, 123, 124, 126, 166-167, 170-174; dikes of, 170
antipathies of minerals, 20
aphanites, importance of, 94, 117-122, 125, 126, 294; limits of composition of, 276
aphyric basalts, limitation of composition of, 140-143
apophysis of peridotite dike, 154
aqueo-igneous fusion, 282-302
ASKLUND, B., 13, 15-18, 80-83, 85-90, 110
assimilation, 75-78, 134; effects of, 175-223; of limestone, 267; summary of effects of, 220-223
associations of minerals, 20
augite. *See* pyroxene
AUROUSSEAU, M., 117

auto-intrusion, 168-170
autolith, 198
average of rock types (Daly), 12, 123

BÄCKSTROM, H., 15
BAILEY, E. B., 134; addendum by, 173; *see also* MULL AUTHORS
BAIN, G. W., 14
banded gabbro, 168-170, 173
BARLOW, A. E., 216
BARTH, TOM, 131, 278
BARTRUM, J. A., 95
basalt, melting temperature, 298, 309; selective fusion of, 314
basaltic glass, compressibility of, 312; layer of, in earth, 312-313; velocity of seismic waves in, 313
basaltic liquid, rigidity of, 312, 313
basaltic magma, action upon inclusions, 199-201; early separation of augite from, 64-69, 91, 105; fractional crystallization of, 63-91; limitations of composition of, 140-143; non-fractional crystallization of, 79; parental nature of, 21, 63, 67, 75, 215, 319, 320; quaternary analogue of, 65-66; saturated condition of, 320
basaltic rocks, aphyric, limitation of composition of, 140-143
basaltic substratum, re-fusion of, 309, 314
basalts, normative feldspar in, 137, 139-141, 144; olivine of, 159-164
BAYLEY, W. S., 214
Beemerville, New Jersey, rocks of, 257
BEGER, P., 258
biotite, importance of formation of, 79-85; reactions leading to, 83-85
border facies, 21, 22; of peridotites, 150
BOWEN, N. L., 14, 22, 26, 27, 28, 29, 33, 36, 44, 45, 50, 54, 142, 167, 168, 173, 174, 177, 211, 213, 220, 234, 241, 259
BRIDGMAN, P. W., 312
BRÖGGER, W. C., 4, 131, 220

INDEX OF SYSTEMS

Each system is given once only. It appears under the component whose initial letter is highest in the alphabet. Thus the system, CaO-Al₂O₃-SiO₂, should be sought under Al₂O₃, and the system, diopside-anorthite-albite, under albite.

INDEX OF COMPONENTS AND COMPOUNDS
IN THE SYSTEMS

A CATALOGUE OF SELECTED
DOVER SCIENCE BOOKS

A CATALOGUE OF SELECTED
DOVER SCIENCE BOOKS

Physics: The Pioneer Science, Lloyd W. Taylor. Very thorough non-mathematical survey of physics in a historical framework which shows development of ideas. Easily followed by laymen; used in dozens of schools and colleges for survey courses. Richly illustrated. Volume 1: Heat, sound, mechanics. Volume 2: Light, electricity. Total of 763 illustrations. Total of cvi + 847pp.
60565-5, 60566-3 Two volumes, Paperbound 5.50

THE RISE OF THE NEW PHYSICS, A. d'Abro. Most thorough explanation in print of central core of mathematical physics, both classical and modern, from Newton to Dirac and Heisenberg. Both history and exposition: philosophy of science, causality, explanations of higher mathematics, analytical mechanics, electromagnetism, thermodynamics, phase rule, special and general relativity, matrices. No higher mathematics needed to follow exposition, though treatment is elementary to intermediate in level. Recommended to serious student who wishes verbal understanding. 97 illustrations. Total of ix + 982pp.
20003-5, 20004-3 Two volumes, Paperbound $6.00

INTRODUCTION TO CHEMICAL PHYSICS, John C. Slater. A work intended to bridge the gap between chemistry and physics. Text divided into three parts: Thermodynamics, Statistical Mechanics, and Kinetic Theory; Gases, Liquids and Solids; and Atoms, Molecules and the Structure of Matter, which form the basis of the approach. Level is advanced undergraduate to graduate, but theoretical physics held to minimum. 40 tables, 118 figures. xiv + 522pp.
62562-1 Paperbound $4.00

BASIC THEORIES OF PHYSICS, Peter C. Bergmann. Critical examination of important topics in classical and modern physics. Exceptionally useful in examining conceptual framework and methodology used in construction of theory. Excellent supplement to any course, textbook. Relatively advanced.
Volume 1. Heat and Quanta. Kinetic hypothesis, physics and statistics, stationary ensembles, thermodynamics, early quantum theories, atomic spectra, probability waves, quantization in wave mechanics, approximation methods, abstract quantum theory. 8 figures. x + 300pp.   60968-5 Paperbound $2.50
Volume 2. Mechanics and Electrodynamics. Classical mechanics, electro- and magnetostatics, electromagnetic induction, field waves, special relativity, waves, etc. 16 figures, viii + 260pp.   60969-3 Paperbound $2.75

FOUNDATIONS OF PHYSICS, Robert Bruce Lindsay and Henry Margenau. Methods and concepts at the heart of physics (space and time, mechanics, probability, statistics, relativity, quantum theory) explained in a text that bridges gap between semi-popular and rigorous introductions. Elementary calculus assumed. "Thorough and yet not over-detailed," *Nature*. 35 figures. xviii + 537 pp.
60377-6 Paperbound $3.50

FUNDAMENTAL FORMULAS OF PHYSICS, edited by Donald H. Menzel. Most useful reference and study work, ranges from simplest to most highly sophisticated operations. Individual chapters, with full texts explaining formulae, prepared by leading authorities cover basic mathematical formulas, statistics, nomograms, physical constants, classical mechanics, special theory of relativity, general theory of relativity, hydrodynamics and aerodynamics, boundary value problems in mathematical physics, heat and thermodynamics, statistical mechanics, kinetic theory of gases, viscosity, thermal conduction, electromagnetism, electronics, acoustics, geometrical optics, physical optics, electron optics, molecular spectra, atomic spectra, quantum mechanics, nuclear theory, cosmic rays and high energy phenomena, particle accelerators, solid state, magnetism, etc. Special chapters also cover physical chemistry, astrophysics, celestian mechanics, meteorology, and biophysics. Indispensable part of library of every scientist. Total of xli + 787pp.
60595-7, 60596-5 Two volumes, Paperbound $6.00

INTRODUCTION TO EXPERIMENTAL PHYSICS, William B. Fretter. Detailed coverage of techniques and equipment: measurements, vacuum tubes, pulse circuits, rectifiers, oscillators, magnet design, particle counters, nuclear emulsions, cloud chambers, accelerators, spectroscopy, magnetic resonance, x-ray diffraction, low temperature, etc. One of few books to cover laboratory hazards, design of exploratory experiments, measurements. 298 figures. xii + 349pp.
(EBE) 61890-0 Paperbound $3.00

CONCEPTS AND METHODS OF THEORETICAL PHYSICS, Robert Bruce Lindsay. Introduction to methods of theoretical physics, emphasizing development of physical concepts and analysis of methods. Part I proceeds from single particle to collections of particles to statistical method. Part II covers application of field concept to material and non-material media. Numerous exercises and examples. 76 illustrations. x + 515pp.
62354-8 Paperbound $4.00

AN ELEMENTARY TREATISE ON THEORETICAL MECHANICS, Sir James Jeans. Great scientific expositor in remarkably clear presentation of basic classical material: rest, motion, forces acting on particle, statics, motion of particle under variable force, motion of rigid bodies, coordinates, etc. Emphasizes explanation of fundamental physical principles rather than mathematics or applications. Hundreds of problems worked in text. 156 figures. x + 364pp. 61839-0 Paperbound $2.75

THEORETICAL MECHANICS: AN INTRODUCTION TO MATHEMATICAL PHYSICS, Joseph S. Ames and Francis D. Murnaghan. Mathematically rigorous introduction to vector and tensor methods, dynamics, harmonic vibrations, gyroscopic theory, principle of least constraint, Lorentz-Einstein transformation. 159 problems; many fully-worked examples. 39 figures. ix + 462pp. 60461-6 Paperbound $3.50

THE PRINCIPLE OF RELATIVITY, Albert Einstein, Hendrick A. Lorentz, Hermann Minkowski and Hermann Weyl. Eleven original papers on the special and general theory of relativity, all unabridged. Seven papers by Einstein, two by Lorentz, one each by Minkowski and Weyl. "A thrill to read again the original papers by these giants," *School Science and Mathematics.* Translated by W. Perret and G. B. Jeffery. Notes by A. Sommerfeld. 7 diagrams. viii + 216pp.
60081-5 Paperbound $2.25

EINSTEIN'S THEORY OF RELATIVITY, Max Born. Relativity theory analyzed, explained for intelligent layman or student with some physical, mathematical background. Includes Lorentz, Minkowski, and others. Excellent verbal account for teachers. Generally considered the finest non-technical account. vii + 376pp.
60769-0 Paperbound $2.75

PHYSICAL PRINCIPLES OF THE QUANTUM THEORY, Werner Heisenberg. Nobel Laureate discusses quantum theory, uncertainty principle, wave mechanics, work of Dirac, Schroedinger, Compton, Wilson, Einstein, etc. Middle, non-mathematical level for physicist, chemist not specializing in quantum; mathematical appendix for specialists. Translated by C. Eckart and F. Hoyt. 19 figures. viii + 184pp.
60113-7 Paperbound $2.00

PRINCIPLES OF QUANTUM MECHANICS, William V. Houston. For student with working knowledge of elementary mathematical physics; uses Schroedinger's wave mechanics. Evidence for quantum theory, postulates of quantum mechanics, applications in spectroscopy, collision problems, electrons, similar topics. 21 figures. 288pp.
60524-8 Paperbound $3.00

ATOMIC SPECTRA AND ATOMIC STRUCTURE, Gerhard Herzberg. One of the best introductions to atomic spectra and their relationship to structure; especially suited to specialists in other fields who require a comprehensive basic knowledge. Treatment is physical rather than mathematical. 2nd edition. Translated by J. W. T. Spinks. 80 illustrations. xiv + 257pp.
60115-3 Paperbound $2.00

ATOMIC PHYSICS: AN ATOMIC DESCRIPTION OF PHYSICAL PHENOMENA, Gaylord P. Harnwell and William E. Stephens. One of the best introductions to modern quantum ideas. Emphasis on the extension of classical physics into the realms of atomic phenomena and the evolution of quantum concepts. 156 problems. 173 figures and tables. xi + 401pp.
61584-7 Paperbound $3.00

ATOMS, MOLECULES AND QUANTA, Arthur E. Ruark and Harold C. Urey. 1964 edition of work that has been a favorite of students and teachers for 30 years. Origins and major experimental data of quantum theory, development of concepts of atomic and molecular structure prior to new mechanics, laws and basic ideas of quantum mechanics, wave mechanics, matrix mechanics, general theory of quantum dynamics. Very thorough, lucid presentation for advanced students. 230 figures. Total of xxiii + 810pp.
61106-X, 61107-8 Two volumes, Paperbound $6.00

INVESTIGATIONS ON THE THEORY OF THE BROWNIAN MOVEMENT, Albert Einstein. Five papers (1905-1908) investigating the dynamics of Brownian motion and evolving an elementary theory of interest to mathematicians, chemists and physical scientists. Notes by R. Fürth, the editor, discuss the history of study of Brownian movement, elucidate the text and analyze the significance of the papers. Translated by A. D. Cowper. 3 figures. iv + 122pp.
60304-0 Paperbound $1.50

MATHEMATICAL FOUNDATIONS OF STATISTICAL MECHANICS, A. I. Khinchin. Introduction to modern statistical mechanics: phase space, ergodic problems, theory of probability, central limit theorem, ideal monatomic gas, foundation of thermodynamics, dispersion and distribution of sum functions. Provides mathematically rigorous treatment and excellent analytical tools. Translated by George Gamow. viii + 179pp. 60147-1 Paperbound $2.50

INTRODUCTION TO PHYSICAL STATISTICS, Robert B. Lindsay. Elementary probability theory, laws of thermodynamics, classical Maxwell-Boltzmann statistics, classical statistical mechanics, quantum mechanics, other areas of physics that can be studied statistically. Full coverage of methods; basic background theory. ix + 306pp. 61882-X Paperbound $2.75

DIALOGUES CONCERNING TWO NEW SCIENCES, Galileo Galilei. Written near the end of Galileo's life and encompassing 30 years of experiment and thought, these dialogues deal with geometric demonstrations of fracture of solid bodies, cohesion, leverage, speed of light and sound, pendulums, falling bodies, accelerated motion, etc. Translated by Henry Crew and Alfonso de Salvio. Introduction by Antonio Favaro. xxiii + 300pp. 60099-8 Paperbound $2.25

FOUNDATIONS OF SCIENCE: THE PHILOSOPHY OF THEORY AND EXPERIMENT, Norman R. Campbell. Fundamental concepts of science examined on middle level: acceptance of propositions and axioms, presuppositions of scientific thought, scientific law, multiplication of probabilities, nature of experiment, application of mathematics, measurement, numerical laws and theories, error, etc. Stress on physics, but holds for other sciences. "Unreservedly recommended," *Nature* (England). Formerly *Physics: The Elements*. ix + 565pp. 60372-5 Paperbound $4.00

THE PHASE RULE AND ITS APPLICATIONS, Alexander Findlay, A. N. Campbell and N. O. Smith. Findlay's well-known classic, updated (1951). Full standard text and thorough reference, particularly useful for graduate students. Covers chemical phenomena of one, two, three, four and multiple component systems. "Should rank as the standard work in English on the subject," *Nature*. 236 figures. xii + 494pp. 60091-2 Paperbound $3.50

THERMODYNAMICS, Enrico Fermi. A classic of modern science. Clear, organized treatment of systems, first and second laws, entropy, thermodynamic potentials, gaseous reactions, dilute solutions, entropy constant. No math beyond calculus is needed, but readers are assumed to be familiar with fundamentals of thermometry, calorimetry. 22 illustrations. 25 problems. x + 160pp. 60361-X Paperbound $2.00

TREATISE ON THERMODYNAMICS, Max Planck. Classic, still recognized as one of the best introductions to thermodynamics. Based on Planck's original papers, it presents a concise and logical view of the entire field, building physical and chemical laws from basic empirical facts. Planck considers fundamental definitions, first and second principles of thermodynamics, and applications to special states of equilibrium. Numerous worked examples. Translated by Alexander Ogg. 5 figures. xiv + 297pp. 60219-2 Paperbound $2.50

MICROSCOPY FOR CHEMISTS, Harold F. Schaeffer. Thorough text; operation of microscope, optics, photomicrographs, hot stage, polarized light, chemical procedures for organic and inorganic reactions. 32 specific experiments cover specific analyses: industrial, metals, other important subjects. 136 figures. 264pp.
61682-7 Paperbound $2.50

OPTICKS, Sir Isaac Newton. A survey of 18th-century knowledge on all aspects of light as well as a description of Newton's experiments with spectroscopy, colors, lenses, reflection, refraction, theory of waves, etc. in language the layman can follow. Foreword by Albert Einstein. Introduction by Sir Edmund Whittaker. Preface by I. Bernard Cohen. cxxvi + 406pp.
60205-2 Paperbound $4.00

LIGHT: PRINCIPLES AND EXPERIMENTS, George S. Monk. Thorough coverage, for student with background in physics and math, of physical and geometric optics. Also includes 23 experiments on optical systems, instruments, etc. "Probably the best intermediate text on optics in the English language," *Physics Forum*. 275 figures. xi + 489pp.
60341-5 Paperbound $3.50

PHYSICAL OPTICS, Robert W. Wood. A classic in the field, this is a valuable source for students of physical optics and excellent background material for a study of electromagnetic theory. Partial contents: nature and rectilinear propagation of light, reflection from plane and curved surfaces, refraction, absorption and dispersion, origin of spectra, interference, diffraction, polarization, Raman effect, optical properties of metals, resonance radiation and fluorescence of atoms, magneto-optics, electro-optics, thermal radiation. 462 diagrams, 17 plates. xvi + 846pp.
61808-0 Paperbound $4.50

MIRRORS, PRISMS AND LENSES: A TEXTBOOK OF GEOMETRICAL OPTICS, James P. C. Southall. Introductory-level account of modern optical instrument theory, covering unusually wide range: lights and shadows, reflection of light and plane mirrors, refraction, astigmatic lenses, compound systems, aperture and field of optical system, the eye, dispersion and achromatism, rays of finite slope, the microscope, much more. Strong emphasis on earlier, elementary portions of field, utilizing simplest mathematics wherever possible. Problems. 329 figures. xxiv + 806pp.
61234-1 Paperbound $5.00

THE PSYCHOLOGY OF INVENTION IN THE MATHEMATICAL FIELD, Jacques Hadamard. Important French mathematician examines psychological origin of ideas, role of the unconscious, importance of visualization, etc. Based on own experiences and reports by Dalton, Pascal, Descartes, Einstein, Poincaré, Helmholtz, etc. xiii + 145pp.
20107-4 Paperbound $1.50

INTRODUCTION TO CHEMICAL PHYSICS, John C. Slater. A work intended to bridge the gap between chemistry and physics. Text divided into three parts: Thermodynamics, Statistical Mechanics, and Kinetic Theory; Gases, Liquids and Solids; and Atoms, Molecules and the Structure of Matter, which form the basis of the approach. Level is advanced undergraduate to graduate, but theoretical physics held to minimum. 40 tables, 118 figures. xiv + 522pp.
62562-1 Paperbound $4.00

CONTRIBUTIONS TO THE FOUNDING OF THE THEORY OF TRANSFINITE NUMBERS, Georg Cantor. The famous articles of 1895-1897 which founded a new branch of mathematics, translated with 82-page introduction by P. Jourdain. Not only a great classic but still one of the best introductions for the student. ix + 211pp.

60045-9 Paperbound $2.50

ESSAYS ON THE THEORY OF NUMBERS, Richard Dedekind. Two classic essays, on the theory of irrationals, giving an arithmetic and rigorous foundation; and on transfinite numbers and properties of natural numbers. Translated by W. W. Beman. iii + 115pp.   21010-3 Paperbound $1.75

GEOMETRY OF FOUR DIMENSIONS, H. P. Manning. Part verbal, part mathematical development of fourth dimensional geometry. Historical introduction. Detailed treatment is by synthetic method, approaching subject through Euclidean geometry. No knowledge of higher mathematics necessary. 76 figures. ix + 348pp.

60182-X Paperbound $3.00

AN INTRODUCTION TO THE GEOMETRY OF N DIMENSIONS, Duncan M. Y. Sommerville. The only work in English devoted to higher-dimensional geometry. Both metric and projectiv properties of n-dimensional geometry are covered. Covers fundamental ideas of incidence, parallelism, perpendicularity, angles between linear space, enumerative geometry, analytical geometry, polytopes, analysis situs, hyperspacial figures. 60 diagrams. xvii + 196pp.   60494-2 Paperbound $2.00

THE THEORY OF SOUND, J. W. S. Rayleigh. Still valuable classic by the great Nobel Laureate. Standard compendium summing up previous research and Rayleigh's original contributions. Covers harmonic vibrations, vibrating systems, vibrations of strings, membranes, plates, curved shells, tubes, solid bodies, refraction of plane waves, general equations. New historical introduction and bibliography by R. B. Lindsay, Brown University. 97 figures. lviii + 984pp.

60292-3, 60293-1 Two volumes, Paperbound $6.00

ELECTROMAGNETIC THEORY: A CRITICAL EXAMINATION OF FUNDAMENTALS, Alfred O'Rahilly. Critical analysis and restructuring of the basic theories and ideas of classical electromagnetics. Analysis is carried out through study of the primary treatises of Maxwell, Lorentz, Einstein, Weyl, etc., which established the theory. Expansive reference to and direct quotation from these treatises. Formerly *Electromagnetics*. Total of xvii + 884pp.

60126-9, 60127-7 Two volumes, Paperbound $6.00

ELEMENTARY CONCEPTS OF TOPOLOGY, Paul Alexandroff. Elegent, intuitive approach to topology, from the basic concepts of set-theoretic topology to the concept of Betti groups. Stresses concepts of complex, cycle and homology. Shows how concepts of topology are useful in math and physics. Introduction by David Hilbert. Translated by Alan E. Farley. 25 figures. iv + 57pp.

60747-X Paperbound $1.25

THE ELEMENTS OF NON-EUCLIDEAN GEOMETRY, Duncan M. Y. Sommerville. Presentation of the development of non-Euclidean geometry in logical order, from a fundamental analysis of the concept of parallelism to such advanced topics as inversion, transformations, pseudosphere, geodesic representation, relation between parataxy and parallelism, etc. Knowledge of only high-school algebra and geometry is presupposed. 126 problems, 129 figures. xvi + 274pp.

60460-8 Paperbound $2.50

NON-EUCLIDEAN GEOMETRY: A CRITICAL AND HISTORICAL STUDY OF ITS DEVELOPMENT, Roberto Bonola. Standard survey, clear, penetrating, discussing many systems not usually represented in general studies. Easily followed by non-specialist. Translated by H. Carslaw. Bound in are two most important texts: Bolyai's "The Science of Absolute Space" and Lobachevski's "The Theory of Parallels," translated by G. B. Halsted. Introduction by F. Enriques. 181 diagrams. Total of 431pp.

60027-0 Paperbound $3.00

ELEMENTS OF NUMBER THEORY, Ivan M. Vinogradov. By stressing demonstrations and problems, this modern text can be understood by students without advanced math backgrounds. "A very welcome addition," *Bulletin, American Mathematical Society.* Translated by Saul Kravetz. Over 200 fully-worked problems. 100 numerical exercises. viii + 227pp.

60259-1 Paperbound $2.50

THEORY OF SETS, E. Kamke. Lucid introduction to theory of sets, surveying discoveries of Cantor, Russell, Weierstrass, Zermelo, Bernstein, Dedekind, etc. Knowledge of college algebra is sufficient background. "Exceptionally well written," *School Science and Mathematics.* Translated by Frederick Bagemihl. vii + 144pp.

60141-2 Paperbound $1.75

A TREATISE ON THE DIFFERENTIAL GEOMETRY OF CURVES AND SURFACES, Luther P. Eisenhart. Detailed, concrete introductory treatise on differential geometry, developed from author's graduate courses at Princeton University. Thorough explanation of the geometry of curves and surfaces, concentrating on problems most helpful to students. 683 problems, 30 diagrams. xiv + 474pp.

60667-8 Paperbound $3.50

AN ESSAY ON THE FOUNDATIONS OF GEOMETRY, Bertrand Russell. A mathematical and physical analysis of the place of the a priori in geometric knowledge. Includes critical review of 19th-century work in non-Euclidean geometry as well as illuminating insights of one of the great minds of our time. New foreword by Morris Kline. xx + 201pp.

60233-8 Paperbound $2.50

INTRODUCTION TO THE THEORY OF NUMBERS, Leonard E. Dickson. Thorough, comprehensive approach with adequate coverage of classical literature, yet simple enough for beginners. Divisibility, congruences, quadratic residues, binary quadratic forms, primes, least residues, Fermat's theorem, Gauss's lemma, and other important topics. 249 problems, 1 figure. viii + 183pp.

60342-3 Paperbound $2.00

AN ELEMENTARY INTRODUCTION TO THE THEORY OF PROBABILITY, B. V. Gnedenko and A. Ya. Khinchin. Introduction to facts and principles of probability theory. Extremely thorough within its range. Mathematics employed held to elementary level. Excellent, highly accurate layman's introduction. Translated from the fifth Russian edition by Leo Y. Boron. xii + 130pp.

60155-2 Paperbound $2.00

SELECTED PAPERS ON NOISE AND STOCHASTIC PROCESSES, edited by Nelson Wax. Six papers which serve as an introduction to advanced noise theory and fluctuation phenomena, or as a reference tool for electrical engineers whose work involves noise characteristics, Brownian motion, statistical mechanics. Papers are by Chandrasekhar, Doob, Kac, Ming, Ornstein, Rice, and Uhlenbeck. Exact facsimile of the papers as they appeared in scientific journals. 19 figures. v + 337pp. 6⅛ x 9¼.

60262-1 Paperbound $3.50

STATISTICS MANUAL, Edwin L. Crow, Frances A. Davis and Margaret W. Maxfield. Comprehensive, practical collection of classical and modern methods of making statistical inferences, prepared by U. S. Naval Ordnance Test Station. Formulae, explanations, methods of application are given, with stress on use. Basic knowledge of statistics is assumed. 21 tables, 11 charts, 95 illustrations. xvii + 288pp.

60599-X Paperbound $2.50

MATHEMATICAL FOUNDATIONS OF INFORMATION THEORY, A. I. Khinchin. Comprehensive introduction to work of Shannon, McMillan, Feinstein and Khinchin, placing these investigations on a rigorous mathematical basis. Covers entropy concept in probability theory, uniqueness theorem, Shannon's inequality, ergodic sources, the E property, martingale concept, noise, Feinstein's fundamental lemma, Shanon's first and second theorems. Translated by R. A. Silverman and M. D. Friedman. iii + 120pp.

60434-9 Paperbound $1.75

INTRODUCTION TO SYMBOLIC LOGIC AND ITS APPLICATION, Rudolf Carnap. Clear, comprehensive, rigorous introduction. Analysis of several logical languages. Investigation of applications to physics, mathematics, similar areas. Translated by Wiliam H. Meyer and John Wilkinson. xiv + 214pp.

60453-5 Paperbound $2.50

SYMBOLIC LOGIC, Clarence I. Lewis and Cooper H. Langford. Probably the most cited book in the literature, with much material not otherwise obtainable. Paradoxes, logic of extensions and intensions, converse substitution, matrix system, strict limitations, existence of terms, truth value systems, similar material. vii + 518pp.

60170-6 Paperbound $4.50

VECTOR AND TENSOR ANALYSIS, George E. Hay. Clear introduction; starts with simple definitions, finishes with mastery of oriented Cartesian vectors, Christoffel symbols, solenoidal tensors, and applications. Many worked problems show applications. 66 figures. viii + 193pp.

60109-9 Paperbound $2.50

GUIDE TO THE LITERATURE OF MATHEMATICS AND PHYSICS, INCLUDING RELATED WORKS ON ENGINEERING SCIENCE, Nathan Grier Parke III. This up-to-date guide puts a library catalog at your fingertips. Over 5000 entries in many languages under 120 subject headings, including many recently available Russian works. Citations are as full as possible, and cross-references and suggestions for further investigation are provided. Extensive listing of bibliographical aids. 2nd revised edition. Complete indices. xviii + 436pp.

60447-0 Paperbound $3.00

INTRODUCTION TO ELLIPTIC FUNCTIONS WITH APPLICATIONS, Frank Bowman. Concise, practical introduction, from familiar trigonometric function to Jacobian elliptic functions to applications in electricity and hydrodynamics. Legendre's standard forms for elliptic integrals, conformal representation, etc., fully covered. Requires knowledge of basic principles of differentiation and integration only. 157 problems and examples, 56 figures. 115pp.   60922-7 Paperbound $1.50

THEORY OF FUNCTIONS OF A COMPLEX VARIABLE, A. R. Forsyth. Standard, classic presentation of theory of functions, stressing multiple-valued functions and related topics: theory of multiform and uniform periodic functions, Weierstrass's results with additiontheorem functions. Riemann functions and surfaces, algebraic functions, Schwarz's proof of the existence-theorem, theory of conformal mapping, etc. 125 figures, 1 plate. Total of xxviii + 855pp. 6⅛ x 9¼.

61378-X, 61379-8 Two volumes, Paperbound $6.00

THEORY OF THE INTEGRAL, Stanislaw Saks. Excellent introduction, covering all standard topics: set theory, theory of measure, functions with general properties, and theory of integration emphasizing the Lebesgue integral. Only a minimal background in elementary analysis needed. Translated by L. C. Young. 2nd revised edition. xv + 343pp.   61151-5 Paperbound $3.00

THE THEORY OF FUNCTIONS, *Konrad Knopp. Characterized as "an excellent introduction . . . remarkably readable, concise, clear, rigorous" by the* Journal of the American Statistical Association *college text.*

A COURSE IN MATHEMATICAL ANALYSIS, Edouard Goursat. *The entire "Cours d'analyse" for students with one year of calculus, offering an exceptionally wide range of subject matter on analysis and applied mathematics. Available for the first time in English. Definitive treatment.*

VOLUME I: Applications to geometry, expansion in series, definite integrals, derivatives and differentials. Translated by Earle R. Hedrick. 52 figures. viii + 548pp.   60554-X Paperbound $5.00

VOLUME II, PART I: Functions of a complex variable, conformal representations, doubly periodic functions, natural boundaries, etc. Translated by Earle R. Hedrick and Otto Dunkel. 38 figures. x + 259pp.   60555-8 Paperbound $3.00

VOLUME II, PART II: Differential equations, Cauchy-Lipschitz method, non-linear differential equations, simultaneous equations, etc. Translated by Earle R. Hedrick and Otto Dunkel. 1 figure. viii + 300pp.   60556-6 Paperbound $3.00

VOLUME III, PART I: Variation of solutions, partial differential equations of the second order. Poincaré's theorem, periodic solutions, asymptotic series, wave propagation, Dirichlet's problem in space, Newtonian potential, etc. Translated by Howard G. Bergmann. 15 figures. x + 329pp.   61176-0 Paperbound $3.50

VOLUME III, PART II: Integral equations and calculus of variations: Fredholm's equation, Hilbert-Schmidt theorem, symmetric kernels, Euler's equation, transversals, extreme fields, Weierstrass's theory, etc. Translated by Howard G. Bergmann. Note on Conformal Representation by Paul Montel. 13 figures. xi + 389pp.
61177-9 Paperbound $3.00

ELEMENTARY STATISTICS: WITH APPLICATIONS IN MEDICINE AND THE BIOLOGICAL SCIENCES, Frederick E. Croxton. Presentation of all fundamental techniques and methods of elementary statistics assuming average knowledge of mathematics only. Useful to readers in all fields, but many examples drawn from characteristic data in medicine and biological sciences. vii + 376pp.
60506-X Paperbound $2.50

ELEMENTS OF THE THEORY OF FUNCTIONS. A general background text that explores complex numbers, linear functions, sets and sequences, conformal mapping. Detailed proofs. Translated by Frederick Bagemihl. 140pp.
60154-4 Paperbound $1.50

THEORY OF FUNCTIONS, PART I. Provides full demonstrations, rigorously set forth, of the general foundations of the theory: integral theorems, series, the expansion of analytic functions. Translated by Federick Bagemihl. vii + 146pp.
60156-0 Paperbound $1.50

INTRODUCTION TO THE THEORY OF FOURIER'S SERIES AND INTEGRALS, Horatio S. Carslaw. A basic introduction to the theory of infinite series and integrals, with special reference to Fourier's series and integrals. Based on the classic Riemann integral and dealing with only ordinary functions, this is an important class text. 84 examples. xiii + 368pp.   60048-3 Paperbound $3.00

AN INTRODUCTION TO FOURIER METHODS AND THE LAPLACE TRANSFORMATION, Philip Franklin. Introductory study of theory and applications of Fourier series and Laplace transforms, for engineers, physicists, applied mathematicians, physical science teachers and students. Only a previous knowledge of elementary calculus is assumed. Methods are related to physical problems in heat flow, vibrations, eletcrical transmission, electromagnetic radiation, etc. 828 problems with answers. Formerly *Fourier Methods*. x + 289pp.   60452-7 Paperbound $2.75

INFINITE SEQUENCES AND SERIES, Konrad Knopp. Careful presentation of fundamentals of the theory by one of the finest modern expositors of higher mathematics. Covers functions of real and complex variables, arbitrary and null sequences, convergence and divergence. Cauchy's limit theorem, tests for infinite series, power series, numerical and closed evaluation of series. Translated by Frederick Bagemihl. v + 186pp.   60153-6 Paperbound $2.00

INTRODUCTION TO THE DIFFERENTIAL EQUATIONS OF PHYSICS, Ludwig Hopf. No math background beyond elementary calculus is needed to follow this classroom or self-study introduction to ordinary and partial differential equations. Approach is through classical physics. Translated by Walter Nef. 48 figures. v + 154pp.
60120-X Paperbound $1.75

DIFFERENTIAL EQUATIONS FOR ENGINEERS, Philip Franklin. For engineers, physicists, applied mathematicians. Theory and application: solution of ordinary differential equations and partial derivatives, analytic functions. Fourier series, Abel's theorem, Cauchy Riemann differential equations, etc. Over 400 problems deal with electricity, vibratory systems, heat, radio; solutions. Formerly *Differential Equations for Electrical Engineers*. 41 illustrations. vii + 299pp.
60601-5 Paperbound $2.50

THEORY OF FUNCTIONS, PART II. Single- and multiple-valued functions; full presentation of the most characteristic and important types. Proofs fully worked out. Translated by Frederick Bagemihl. x + 150pp.   60157-9 Paperbound $1.50

PROBLEM BOOK IN THE THEORY OF FUNCTIONS, I. More than 300 elementary problems for independent use or for use with "Theory of Functions, I." 85pp. of detailed solutions. Translated by Lipman Bers. viii + 126pp.
60158-7 Paperbound $1.50

PROBLEM BOOK IN THE THEORY OF FUNCTIONS, II. More than 230 problems in the advanced theory. Designed to be used with "Theory of Functions, II" or with any comparable text. Full solutions. Translated by Frederick Bagemihl. 138pp.
60159-5 Paperbound $1.75

INTRODUCTION TO THE THEORY OF EQUATIONS, Florian Cajori. Classic introduction by leading historian of science covers the fundamental theories as reached by Gauss, Abel, Galois and Kronecker. Basics of equation study are followed by symmetric functions of roots, elimination, homographic and Tschirnhausen transformations, resolvents of Lagrange, cyclic equations, Abelian equations, the work of Galois, the algebraic solution of general equations, and much more. Numerous exercises include answers. ix + 239pp.   62184-7 Paperbound $2.75

LAPLACE TRANSFORMS AND THEIR APPLICATIONS TO DIFFERENTIAL EQUATIONS, N. W. McLachlan. Introduction to modern operational calculus, applying it to ordinary and partial differential equations. Laplace transform, theorems of operational calculus, solution of equations with constant coefficients, evaluation of integrals, derivation of transforms, of various functions, etc. For physics, engineering students. Formerly *Modern Operational Calculus*. xiv + 218pp.
60192-7 Paperbound $2.50

PARTIAL DIFFERENTIAL EQUATIONS OF MATHEMATICAL PHYSICS, Arthur G. Webster. Introduction to basic method and theory of partial differential equations, with full treatment of their applications to virtually every field. Full, clear chapters on Fourier series, integral and elliptic equations, spherical, cylindrical and ellipsoidal harmonics, Cauchy's method, boundary problems, method of Riemann-Volterra, many other basic topics. Edited by Samuel J. Plimpton. 97 figures. vii + 446pp.   60263-X Paperbound $3.00

PRINCIPLES OF STELLAR DYNAMICS, Subrahmanyan Chandrasekhar. Theory of stellar dynamics as a branch of classical dynamics; stellar encounter in terms of 2-body problem, Liouville's theorem and equations of continuity. Also two additional papers. 50 illustrations. x + 313pp. 5⅝ x 8⅜.
60659-7 Paperbound $3.00

CELESTIAL OBJECTS FOR COMMON TELESCOPES, T. W. Webb. The most used book in amateur astronomy: inestimable aid for locating and identifying hundreds of celestial objects. Volume 1 covers operation of telescope, telescope photography, precise information on sun, moon, planets, asteroids, meteor swarms, etc.; Volume 2, stars, constellations, double stars, clusters, variables, nebulae, etc. Nearly 4,000 objects noted. New edition edited, updated by Margaret W. Mayall. 77 illustrations. Total of xxxix + 606pp.
20917-2, 20918-0 Two volumes, Paperbound $5.50

A SHORT HISTORY OF ASTRONOMY, Arthur Berry. Earliest times through the 19th century. Individual chapters on Copernicus, Tycho Brahe, Galileo, Kepler, Newton, etc. Non-technical, but precise, thorough, and as useful to specialist as layman. 104 illustrations, 9 portraits, xxxi + 440 pp.
20210-0 Paperbound $3.00

ORDINARY DIFFERENTIAL EQUATIONS, Edward L. Ince. Explains and analyzes theory of ordinary differential equations in real and complex domains: elementary methods of integration, existence and nature of solutions, continuous transformation groups, linear differential equations, equations of first order, non-linear equations of higher order, oscillation theorems, etc. "Highly recommended," *Electronics Industries*. 18 figures. viii + 558pp.
60349-0 Paperbound $4.00

DICTIONARY OF CONFORMAL REPRESENTATIONS, H. Kober. Laplace's equation in two dimensions for many boundary conditions; scores of geometric forms and transformations for electrical engineers, Joukowski aerofoil for aerodynamists, Schwarz-Christoffel transformations, transcendental functions, etc. Twin diagrams for most transformations. 447 diagrams. xvi + 208pp. 6⅛ x 9¼.
60160-9 Paperbound $2.50

ALMOST PERIODIC FUNCTIONS, A. S. Besicovitch. Thorough summary of Bohr's theory of almost periodic functions citing new shorter proofs, extending the theory, and describing contributions of Wiener, Weyl, de la Vallée, Poussin, Stepanoff, Bochner and the author. xiii + 180pp.
60018-1 Paperbound $2.50

AN INTRODUCTION TO THE STUDY OF STELLAR STRUCTURE, S. Chandrasekhar. A rigorous examination, using both classical and modern mathematical methods, of the relationship between loss of energy, the mass, and the radius of stars in a steady state. 38 figures. 509pp.
60413-6 Paperbound $3.75

INTRODUCTION TO THE THEORY OF GROUP'S OF FINITE ORDER, Robert D. Carmichael. Progresses in easy steps from sets, groups, permutations, isomorphism through the important types of groups. No higher mathematics is necessary. 783 exercises and problems. xiv + 447pp.
60300-8 Paperbound $4.00

ELEMENTARY MATHEMATICS FROM AN ADVANCED STANDPOINT: VOLUME II—
GEOMETRY, Feliex Klein. Using analytical formulas, Klein clarifies the precise
formulation of geometric facts in chapters on manifolds, geometric and higher
point transformations, foundations. "Nothing comparable," *Mathematics Teacher*.
Translated by E. R. Hedrick and C. A. Noble. 141 figures. ix + 214pp.

(USO) 60151-X Paperbound $2.25

ENGINEERING MATHEMATICS, Kenneth S. Miller. Most useful mathematical techniques for graduate students in engineering, physics, covering linear differential
equations, series, random functions, integrals, Fourier series, Laplace transform,
network theory, etc. "Sound and teachable," Science. 89 figures. xii + 417pp.
6 x 8½. 61121-3 Paperbound $3.00

INTRODUCTION TO ASTROPHYSICS: THE STARS, Jean Dufay. Best guide to observational astrophysics in English. Bridges the gap between elementary populariza-
tions and advanced technical monographs. Covers stellar photometry, stellar spectra
and classification, Hertzsprung-Russell diagrams, Yerkes 2-dimensional classification, temperatures, diameters, masses and densities, evolution of the stars. Translated by Owen Gingerich. 51 figures, 11 tables. xii + 164pp.

60771-2 Paperbound $2.50

INTRODUCTION TO BESSEL FUNCTIONS, Frank Bowman. Full, clear introduction to
properties and applications of Bessel functions. Covers Bessel functions of zero
order, of any order; definite integrals; asymptotic expansions; Bessel's solution to
Kepler's problem; circular membranes; etc. Math above calculus and fundamentals
of differential equations developed within text. 636 problems. 28 figures. x +
135pp. 60462-4 Paperbound $1.75

DIFFERENTIAL AND INTEGRAL CALCULUS, Philip Franklin. A full and basic introduction, textbook for a two- or three-semester course, or self-study. Covers para-
metric functions, force components in polar coordinates, Duhamel's theorem,
methods and applications of integration, infinite series, Taylor's series, vectors and
surfaces in space, etc. Exercises follow each chapter with full solutions at back
of the book. Index. xi + 679pp. 62520-6 Paperbound $4.00

THE EXACT SCIENCES IN ANTIQUITY, O. Neugebauer. Modern overview chiefly
of mathematics and astronomy as developed by the Egyptians and Babylonians.
Reveals startling advancement of Babylonian mathematics (tables for numerical
computations, quadratic equations with two unknowns, implications that Pytha-
gorean theorem was known 1000 years before Pythagoras), and sophisticated
astronomy based on competent mathematics. Also covers transmission of this
knowledge to Hellenistic world. 14 plates, 52 figures. xvii + 240pp.

22332-9 Paperbound $2.50

THE THIRTEEN BOOKS OF EUCLID'S ELEMENTS, translated with introduction and
commentary by Sir Thomas Heath. Unabridged republication of definitive edition
based on the text of Heiberg. Translator's notes discuss textual and linguistic
matters, mathematical analysis, 2500 years of critical commentary on the Elements.
Do not confuse with abridged school editions. Total of xvii + 1414pp.

60088-2, 60089-0, 60090-4 Three volumes, Paperbound $9.50

ASTRONOMY AND COSMOGONY, Sir James Jeans. Modern classic of exposition, Jean's latest work. Descriptive astronomy, atrophysics, stellar dynamics, cosmology, presented on intermediate level. 16 illustrations. Preface by Lloyd Motz. xv + 428pp.   60923-5 Paperbound $3.50

EXPERIMENTAL SPECTROSCOPY, Ralph A. Sawyer. Discussion of techniques and principles of prism and grating spectrographs used in research. Full treatment of apparatus, construction, mounting, photographic process, spectrochemical analysis, theory. Mathematics kept to a minimum. Revised (1961) edition. 110 illustrations. x + 358pp.   61045-4 Paperbound $3.50

THEORY OF FLIGHT, Richard von Mises. Introduction to fluid dynamics, explaining fully the physical phenomena and mathematical concepts of aeronautical engineering, general theory of stability, dynamics of incompressible fluids and wing theory. Still widely recommended for clarity, though limited to situations in which air compressibility effects are unimportant. New introduction by K. H. Hohenemser. 408 figures. xvi + 629pp.   60541-8 Paperbound $5.00

AIRPLANE STRUCTURAL ANALYSIS AND DESIGN, Ernest E. Sechler and Louis G. Dunn. Valuable source work to the aircraft and missile designer: applied and design loads, stress-strain, frame analysis, plates under normal pressure, engine mounts, landing gears, etc. 47 problems. 256 figures. xi + 420pp.
                                                    61043-8 Paperbound $3.50

PHOTOELASTICITY: PRINCIPLES AND METHODS, H. T. Jessop and F. C. Harris. An introduction to general and modern developments in 2- and 3-dimensional stress analysis techniques. More advanced mathematical treatment given in appendices. 164 figures. viii + 184pp. 6⅛ x 9¼.   (USO) 60720-8 Paperbound $2.50

THE MEASUREMENT OF POWER SPECTRA FROM THE POINT OF VIEW OF COMMUNICATIONS ENGINEERING, Ralph B. Blackman and John W. Tukey. Techniques for measuring the power spectrum using elementary transmission theory and theory of statistical estimation. Methods of acquiring sound data, procedures for reducing data to meaningful estimates, ways of interpreting estimates. 36 figures and tables. Index. x + 190pp.   60507-8 Paperbound $2.50

GASEOUS CONDUCTORS: THEORY AND ENGINEERING APPLICATIONS, James D. Cobine. An indispensable reference for radio engineers, physicists and lighting engineers. Physical backgrounds, theory of space charges, applications in circuit interrupters, rectifiers, oscillographs, etc. 83 problems. Over 600 figures. xx + 606pp.   60442-X Paperbound $3.75

Prices subject to change without notice.

Available at your book dealer or write for free catalogue to Dept. Sci, Dover Publications, Inc., 180 Varick St., N.Y., N.Y. 10014. Dover publishes more than 150 books each year on science, elementary and advanced mathematics, biology, music, art, literary history, social sciences and other areas.